329
Topics in Current Chemistry

Editorial Board:

K.N. Houk, Los Angeles, CA, USA
C.A. Hunter, Sheffield, UK
M.J. Krische, Austin, TX, USA
J.-M. Lehn, Strasbourg, France
S.V. Ley, Cambridge, UK
M. Olivucci, Siena, Italy
J. Thiem, Hamburg, Germany
M. Venturi, Bologna, Italy
P. Vogel, Lausanne, Switzerland
C.-H. Wong, Taipei, Taiwan
H.N.C. Wong, Shatin, Hong Kong
H. Yamamoto, Chicago, IL, USA

For further volumes:
http://www.springer.com/series/128

Aims and Scope

The series Topics in Current Chemistry presents critical reviews of the present and future trends in modern chemical research. The scope of coverage includes all areas of chemical science including the interfaces with related disciplines such as biology, medicine and materials science.

The goal of each thematic volume is to give the non-specialist reader, whether at the university or in industry, a comprehensive overview of an area where new insights are emerging that are of interest to larger scientific audience.

Thus each review within the volume critically surveys one aspect of that topic and places it within the context of the volume as a whole. The most significant developments of the last 5 to 10 years should be presented. A description of the laboratory procedures involved is often useful to the reader. The coverage should not be exhaustive in data, but should rather be conceptual, concentrating on the methodological thinking that will allow the non-specialist reader to understand the information presented.

Discussion of possible future research directions in the area is welcome.

Review articles for the individual volumes are invited by the volume editors.

Readership: research chemists at universities or in industry, graduate students.

John M. Pezzuto · Nanjoo Suh
Editors

Natural Products in Cancer Prevention and Therapy

With contributions by

V.M. Adhami · A.S. Agyeman · I. Almazari · J.J. Bernard ·
J.-G. Chen · T.-Y. Chen · A.H. Conney · A.T. Dinkova-Kostova ·
P.A. Egner · J.W. Fahey · F. Fuentes · C. Gerhauser · K.A. Gold ·
J.D. Groopman · W.K. Hong · T.W. Kensler · T.O. Khor · E.S. Kim ·
A.-N.T. Kong · J.H. Lee · S.M. Lippman · Y.-R. Lou · Y.-P. Lu ·
H. Mukhtar · P. Nghiem · L. Shu · G.D. Stoner · Z.-Y. Su · N. Suh ·
Y.-J. Surh · P. Talalay · K. Visvanathan · G.C. Wagner · J.L. Wang ·
L.-S. Wang · I.I. Wistuba · C.S. Yang

Editors
John Pezzuto
College of Pharmacy
University of Hawaii
Hilo, Hawaii
USA

Nanjoo Suh
Ernest Mario School of Pharmacy
Department of Chemical Biology
Rutgers University
Piscataway
New Jersey
USA

ISSN 0340-1022   ISSN 1436-5049 (electronic)
ISBN 978-3-642-34574-6   ISBN 978-3-642-34575-3 (eBook)
DOI 10.1007/978-3-642-34575-3
Springer Heidelberg New York Dordrecht London

Library of Congress Control Number: 2012952677

© Springer-Verlag Berlin Heidelberg 2013

This work is subject to copyright. All rights are reserved by the Publisher, whether the whole or part of the material is concerned, specifically the rights of translation, reprinting, reuse of illustrations, recitation, broadcasting, reproduction on microfilms or in any other physical way, and transmission or information storage and retrieval, electronic adaptation, computer software, or by similar or dissimilar methodology now known or hereafter developed. Exempted from this legal reservation are brief excerpts in connection with reviews or scholarly analysis or material supplied specifically for the purpose of being entered and executed on a computer system, for exclusive use by the purchaser of the work. Duplication of this publication or parts thereof is permitted only under the provisions of the Copyright Law of the Publisher's location, in its current version, and permission for use must always be obtained from Springer. Permissions for use may be obtained through RightsLink at the Copyright Clearance Center. Violations are liable to prosecution under the respective Copyright Law.

The use of general descriptive names, registered names, trademarks, service marks, etc. in this publication does not imply, even in the absence of a specific statement, that such names are exempt from the relevant protective laws and regulations and therefore free for general use.

While the advice and information in this book are believed to be true and accurate at the date of publication, neither the authors nor the editors nor the publisher can accept any legal responsibility for any errors or omissions that may be made. The publisher makes no warranty, express or implied, with respect to the material contained herein.

Printed on acid-free paper

Springer is part of Springer Science+Business Media (www.springer.com)

Preface

Although cancer may be experienced by any age group, incidence increases with time, suggesting in many cases there is a prolonged period from the time of initiation to the time of invasive and metastatic cancer. Accordingly, numerous opportunities for intervention are apparent, either through primary prevention at early stages or through therapeutic interventions during later stages of carcinogenesis. In general, cancer chemoprevention is considered as the use of drugs, vitamins, or other agents to reduce the risk of (or delay the development or recurrence of) cancer. The concept of implementing cancer chemoprevention through the use of nontoxic agents, dietary and natural sources, has emerged as an appropriate strategy for controlling disease progression.

Research in the area of cancer chemoprevention has grown over the past few decades, and this has become a rather specialized field of study. Phytochemicals from natural products are recognized as promising agents that play a role in cancer prevention as well as in cancer therapy. Dietary constituents as well as natural products have been demonstrated to modulate common signaling pathways in cancer development. These naturally occurring compounds could become important agents for the prevention of various types of cancer.

In the field of natural products in cancer prevention and therapy, there is the need to review the progress that has been made during the last 50 years and to identify the challenges ahead.

This volume of *Topics in Current Chemistry* addresses the hurdles and challenges in the practice of human cancer prevention in the general population. The process of slowing the progression of cancer is applicable to many cancers with long latency. Although cancer chemoprevention has proven to be a successful strategy in animals, its application to humans has met with limited success.

In this volume, Hasan Mukhtar and his colleague discuss various challenges associated with chemoprevention of cancer with focus on studies with green tea. Allan Conney and his colleague discuss the inhibition of ultraviolet B radiation (UVB)-induced nonmelanoma skin cancer by discovery of a path from tea to caffeine to exercise to decreased tissue fat. From the inhibitory effects of tea and caffeine in UVB-induced skin carcinogenesis, his group further demonstrated the role of increased locomotor activity and decreased tissue fat in skin cancer.

Gary Stoner summarizes the beneficial effects of berries in the prevention of esophageal squamous cell carcinoma in rodents as well as recent data from a human clinical trial in China. He concludes that the use of berry preparations might be a practical approach for the prevention of esophageal squamous cell carcinoma in China and, potentially, other high-risk regions for this disease. Young-Joon Surh demonstrates that cancer chemopreventive and therapeutic potential of guggulsterone, a phytosterol derived from the gum resin of guggul plants. With anti-inflammatory, antioxidative properties, and cancer chemopreventive and therapeutic potential, the underlying molecular mechanisms and chemopreventive/therapeutic targets of guggulsterone were discussed.

In the context of cancer prevention approaches, the importance of chemoprotection against cancer by isothiocyanates is discussed by several investigators. Albena Dinkova-Kostova notes that the isothiocyanates are among the most extensively studied chemoprotective agents, and the Cruciferae family represents a rich source of glucosinolates. There have been numerous examples of the chemoprotective effects of isothiocyanates in a number of animal models of experimental carcinogenesis at various organ sites and against carcinogens of several different types. She indicates that the efficient protection in tumorigenesis and metastasis might be due to multiple mechanisms, involving the Keap1/Nrf2/ARE and NF-κB pathways. The Keap1-Nrf2 signaling pathway is further discussed as a key target for cancer prevention by Thomas Kensler. He reports the ongoing clinical evaluation of broccoli or broccoli sprouts rich in either sulforaphane or its precursor form in plants for cancer prevention in Qidong, China. He indicates that interventions with well-characterized preparations of broccoli sprouts may enhance the detoxication of aflatoxins and air-borne toxins, which may in turn attenuate cancer in targeted populations.

Tony Kong also discusses dietary phytochemicals and cancer chemoprevention focusing on the oxidative stress, Nrf2 and epigenomics. His recent studies show that dietary phytochemicals possess cancer chemopreventive potential through the induction of Nrf2-mediated antioxidant/detoxification enzymes and anti-inflammatory signaling pathways to protect organisms against cellular damage caused by oxidative stress. He concludes that the advancement and development of dietary phytochemicals in cancer chemoprevention research requires the better understanding of the Nrf2-mediated antioxidant, detoxification, and anti-inflammatory systems and corresponding in vitro and in vivo epigenetic mechanisms. Clarissa Gerhäuser summarizes important epigenetic approaches in her extensive review of in vitro and in vivo data on natural products and cancer prevention. A role of epigenetic regulation in cancer chemoprevention and new challenges in future nutri-epigenetic research are also discussed.

CS Yang and his colleagues argue the importance of understanding of differential effects of specific forms of tocopherols in cancer prevention. Many epidemiological studies have suggested that a low vitamin E nutritional status is associated with increased cancer risk. However, several recent large-scale human trials have produced negative results in cancer prevention and therapy with α-tocopherol. He notes that a better understanding of the biological activities of different forms of tocopherols is needed. For safe and inexpensive cancer prevention with tocopherols, use of a naturally occurring tocopherol mixture is suggested for broad anticancer activity of various types of cancer.

Scott Lippman discusses the evolution of chemoprevention research in exciting new directions. Since large chemoprevention trials in unselected patients have often been negative, this trend promises to be reversed by more-focused and novel trial designs emphasizing the identification of molecular targets and predictive biomarkers. He points out the importance of clinical study designs, relevant biomarkers, and surrogate endpoints in new prevention trials. His review in this issue highlights several promising natural agents and how early clinical development may elucidate their role in personalized cancer chemoprevention. Kathryn Gold emphasizes the need for personalizing cancer prevention through a reverse migration strategy. She proposes a new approach to drug development, drawing on the experience in the treatment of advanced cancer to bring agents, biomarkers, and study designs into the prevention setting. She concludes that personalized therapy may develop more effective, tolerable chemoprevention by identifying molecular drivers of cancer and using matched targeted agents.

We would like to thank the authors of this volume for their excellent contributions. This special issue was supported in part by NIH Grant R13 CA159733 "Natural products in the prevention of cancer," awarded by the National Cancer Institute, Congressionally directed funding P116Z100211 awarded by the US Department of Education, the College of Pharmacy, University of Hawaii at Hilo.

Piscataway, NJ Nanjoo Suh
Hilo, HI John M. Pezzuto

Contents

Chemoprevention of Esophageal Squamous Cell Carcinoma with Berries 1
Gary D. Stoner and Li-Shu Wang

Cancer Prevention by Different Forms of Tocopherols 21
Chung S. Yang and Nanjoo Suh

Cancer Chemopreventive and Therapeutic Potential of Guggulsterone 35
Inas Almazari and Young-Joon Surh

Inhibition of UVB-Induced Nonmelanoma Skin Cancer: A Path from Tea to Caffeine to Exercise to Decreased Tissue Fat 61
Allan H. Conney, You-Rong Lou, Paul Nghiem, Jamie J. Bernard,
George C. Wagner, and Yao-Ping Lu

Cancer Chemoprevention and Nutri-Epigenetics: State of the Art and Future Challenges 73
Clarissa Gerhauser

A Perspective on Dietary Phytochemicals and Cancer Chemoprevention: Oxidative Stress, Nrf2, and Epigenomics 133
Zheng-Yuan Su, Limin Shu, Tin Oo Khor, Jong Hun Lee, Francisco Fuentes,
and Ah-Ng Tony Kong

Keap1–Nrf2 Signaling: A Target for Cancer Prevention by Sulforaphane 163
Thomas W. Kensler, Patricia A. Egner, Abena S. Agyeman, Kala Visvanathan,
John D. Groopman, Jian-Guo Chen, Tao-Yang Chen, Jed W. Fahey,
and Paul Talalay

ix

Chemoprotection Against Cancer by Isothiocyanates: A Focus on the Animal Models and the Protective Mechanisms 179
Albena T. Dinkova-Kostova

Human Cancer Chemoprevention: Hurdles and Challenges 203
Vaqar Mustafa Adhami and Hasan Mukhtar

Personalizing Lung Cancer Prevention Through a Reverse Migration Strategy 221
Kathryn A. Gold, Edward S. Kim, Ignacio I. Wistuba, and Waun K. Hong

Natural-Agent Mechanisms and Early-Phase Clinical Development 241
Janet L. Wang, Kathryn A. Gold, and Scott M. Lippman

Index 253

Chemoprevention of Esophageal Squamous Cell Carcinoma with Berries

Gary D. Stoner and Li-Shu Wang

Abstract Esophageal squamous cell carcinoma (SCC) is responsible for about one-seventh of all cancer-related mortality worldwide. This disease has a multifactorial etiology involving numerous environmental, genetic, and dietary factors. The 5-year survival from esophageal SCC is poor because the disease has usually metastasized at the time of diagnosis. Clinical investigations have shown that primary chemoprevention of this disease is feasible; however, only a few agents have shown efficacy. The Fischer 344 (F-344) rat model of esophageal SCC has been used extensively to investigate the pathophysiology of the disease and to identify chemopreventive agents of potential use in human trials. Multiple compounds that inhibit tumor initiation and/or tumor progression in the rat model have been identified. These include the isothiocyanates which inhibit the metabolic activation of esophageal carcinogens and agents that inhibit the progression of dysplastic lesions to cancer including inhibitors of inducible nitric oxide synthase (iNOS), cyclooxygenase-2 (COX-2), vascular endothelial growth factor (VEGF), and c-Jun (a component of activator protein-1 [AP-1]). The present review deals principally with the use of berry preparations for the prevention of esophageal SCC in rodents, and summarizes recent data from a human clinical trial in China. Our results suggest that the use of berry preparations might be a practical approach to the prevention of esophageal SCC in China and, potentially, other high risk regions for this disease.

Keywords Berries · Chemoprevention · Esophagus · Squamous Cell Carcinoma

G.D. Stoner (✉) and L.-S. Wang
Department of Medicine, Division of Hematology and Oncology, Medical College of Wisconsin, Milwaukee, WI 53226, USA
e-mail: gstoner@mcw.edu

Contents

1 Introduction 2
2 Epidemiology of Esophageal SCC 2
3 Etiology of Esophageal SCC 3
4 Strategies to Prevent Esophageal SCC 4
5 Rat Esophagus Tumor Model 4
   5.1 Tumor Induction with *N*-Nitrosomethylbenzylamine 4
   5.2 Genetic Alterations in NMBA-Treated Rat Esophagus 6
6 Chemoprevention of Rat Esophageal Tumors with Berries and Berry Components 8
7 Chemoprevention of Human Esophageal SCC 13
8 Effects of Freeze-Dried Strawberries on Dysplastic Lesions of the Esophagus 13
9 Effects of Strawberries on Protein Expression Levels of iNOS, COX-2, pNFκB-p65, and Phospho-S6 in Esophageal Mucosa 15
10 Summary 15
11 Conclusions 16
References 17

Abbreviations

Bax   BCL-2-associated X protein
Bcl-2 B-Cell lymphoma 2

1 Introduction

Esophageal cancer in humans occurs worldwide with a variable geographic distribution and ranks seventh as a cause of cancer mortality [1]. The two main types of esophageal cancer are squamous cell carcinoma (SCC) and adenocarcinoma. SCCs represent about 90% of esophageal malignancy worldwide, although adenocarcinomas are more prevalent in the USA [2]. Epithelial dysplasia, characterized by an accumulation of atypical cells with nuclear abnormalities and loss of polarity, is the principal precursor lesion of esophageal SCC [3, 4]. Like many other epithelial cancers, esophageal SCC develops through a progressive sequence from mild, moderate to severe dysplasia, carcinoma in situ, and invasive carcinoma [5–7]. Most esophageal cancer patients present with advanced metastatic disease at the time of diagnosis [8] which results in a poor prognosis; only one in five esophageal SCC patients survive more than 3 years after initial diagnosis [9, 10].

2 Epidemiology of Esophageal SCC

The incidence of esophageal SCC shows a marked variation in geographical distribution. The highest risk areas include western and northern China, Japan, Iran, Iraq, southeastern Africa, Uruguay, France, and parts of South America

[3, 11–13]. Half of all esophageal SCC in the world occurs in China and principally in high-risk areas of Henan and Shandong provinces. In Linxian county (Henan province) the age-adjusted annual mortality rates from esophageal SCC have been as high as 151/100,000 for males and 115/100,000 for females [14]. Studies in these high-risk areas have identified specific environmental factors as etiological agents of the disease. Males have a threefold to fourfold greater risk for developing esophageal SCC than females and, in the USA, the incidence of esophageal SCC is five times higher in African Americans than in Caucasians [15].

3 Etiology of Esophageal SCC

Risk factors involved in the etiology of esophageal SCC have been summarized in detail in previous reviews [12, 16, 17] and will be discussed briefly here. Two major risk factors are tobacco smoking and alcohol consumption. Several tobacco carcinogens, including certain nitrosamines and polycyclic hydrocarbons, may be causally related to the disease [18, 19]. Alcohol consumption has been shown to increase the risk for esophageal SCC amongst tobacco users [20]. Consumption of salt-cured, salt-pickled, and moldy food is also implicated in the development of this disease because of the injurious effects of salt on the epithelium of the esophagus and the frequent presence of N-nitrosamine carcinogens and/or fungal toxins in the food [21]. Studies in China and South Africa suggest that N-nitroso compounds and their precursors are etiological agents for esophageal SCC [22, 23]. N-Nitrosamine compounds have been identified in the diets and gastric juice collected from subjects in Henan province, China [24]. O^6-Methylguanine adducts have been detected in the DNA of normal esophageal tissue obtained from esophageal cancer patients in China, further substantiating the role of methylating nitrosamines in the development of esophageal SCC [25].

Other factors associated with the etiology of esophageal SCC include vitamin and trace mineral deficiencies [23, 26]. Plasma levels of vitamins A, C, and E are frequently reduced in patients with the disease. There is an inverse relationship between esophageal cancer mortality and levels of zinc, selenium, and other trace elements in foods [12]. Diets high in starch but low in fruits and vegetables have also been linked to an increased risk for esophageal SCC [8, 27]. Consumption of temperature hot beverages, such as tea, and fungal invasion in esophageal tissues leading to localized inflammation and irritation may be additional promoting factors for the disease [26]. Finally, a role for human papilloma virus (HPV) has been suggested in the etiology of SCC of the esophagus [28], although a recent study in Australia of 222 esophageal SCC patients indicated that only eight tested positive for HPV (six cases of HPV-16 and two cases of HPV-35) [29].

4 Strategies to Prevent Esophageal SCC

An important approach to the prevention of esophageal SCC is through changes in lifestyle, especially the avoidance of tobacco and alcohol use. For populations living in high-risk regions, additional benefits could be realized by (1) the elimination of high-salt foods and foods that may be contaminated with microbial toxins and nitrosamine compounds, (2) the increased consumption of vegetables and fruit, (3) educational efforts to inform populations of the major risk factors for the disease and steps they might take to reduce their risk, and (4) perhaps most importantly, the continued and expanded use of endoscopic surveillance of the esophagus of individuals in high-risk regions to identify premalignant lesions and malignant tumors and take appropriate clinical measures to deal with them.

Chemoprevention may be another feasible approach to the prevention of esophageal SCC, especially in the high incidence areas of the world where carcinogen exposure is high. Animal models are important for the identification of putative chemopreventive agents for specific organ sites as well as for determining their mechanisms of action. Section 5 of this chapter focuses principally on a description of a rat model of esophageal SCC and its use for the evaluation of berries and berry preparations for chemoprevention of esophageal cancer. Previous review articles have summarized investigations on the ability of individual synthetic and naturally-occurring compounds to prevent the development of esophageal tumors in rats [12, 16, 17]. The reader is referred to these reviews for discussions of individual compounds.

5 Rat Esophagus Tumor Model

5.1 Tumor Induction with N-Nitrosomethylbenzylamine

The Fischer-344 rat has proven to be a valuable animal model for studies of the molecular biology and chemoprevention of esophageal SCC [12, 18, 30]. The most potent inducer of esophageal tumors in the F344 rat is the nitrosamine compound, N-nitrosomethylbenzylamine (NMBA), a procarcinogen that must be metabolically activated to induce tumors in the esophagus (Fig. 1). The metabolism of NMBA leads ultimately to the formation of a methylcarbonium ion that methylates guanine residues at the N^7 and O^6 positions [12]. The O^6-methylguanine adduct is particularly important for carcinogenesis since it is poorly repaired and leads to single base mispairing in DNA. Repeat NMBA dosing results in esophageal tumor formation in rats within 20–25 weeks (Fig. 2). Several preneoplastic lesions produced in NMBA-treated rat esophagus closely mimic lesions observed in the human disease. These lesions include simple hyperplasia, leukoplakia, and epithelial dysplasia (Fig. 3). Squamous papilloma is the predominant tumor type seen in the rat esophagus whereas papillomas are rarely observed in the human esophagus. The incidence

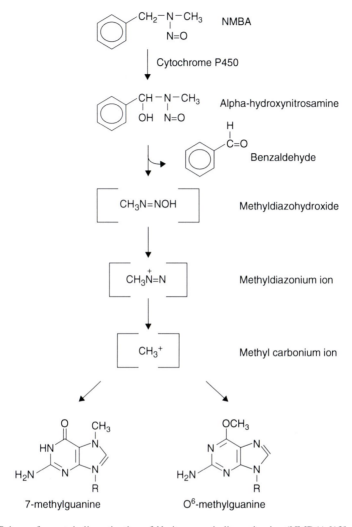

Fig. 1 Schema for metabolic activation of *N*-nitrosomethylbenzylamine (NMBA) [12]

Fig. 2 Appearance of rat esophageal lesions at the termination of a 25 week bioassay. There are several papillomas on the surface of the esophagus (*black arrows*). The lesion on the *lower left* was found to be a carcinoma upon histopathological analysis (*white arrow*) [31]

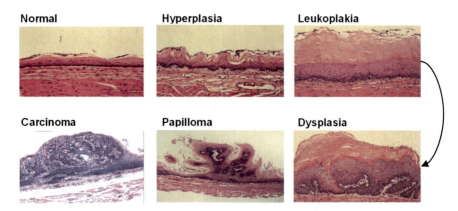

Fig. 3 Histopathology of normal rat esophagus and NMBA-induced lesions in rat esophagus [31]

of SCC in the rat esophagus is low because the animals often succumb to the occlusive effects of papillomas in their esophagi before carcinomas can develop. In a typical bioassay, subcutaneous administration of NMBA at either 0.3 or 0.5 mg/kg body weight three times per week for 5 weeks, or once per week for 15 weeks, results in a 100% tumor incidence at 20–25 weeks. On average, these two doses of NMBA will produce from two to four or four to ten tumors per esophagus, respectively, at 25 weeks. Our laboratory and others have used this model to develop surrogate end-point biomarkers, identify novel targets for intervention, and evaluate putative chemoprevention agents against esophageal SCC.

5.2 Genetic Alterations in NMBA-Treated Rat Esophagus

Genetic analyses of NMBA-induced rat esophageal tumors have identified multiple molecular alterations in the conversion of normal esophagus to cancer, and these events have been discussed in detail [16, 17]. In contrast to human esophageal SCCs, >90% of NMBA-induced rat esophageal papillomas have a G:C → A:T transition mutation in codon 12 of the H-*ras* gene [32, 33]. This mutation is consistent with the formation of O^6-methylguanine adducts in DNA. Interestingly, mutational activation of the Harvey-*ras* (H-*ras*) gene is present infrequently (~5%) in premalignant dysplastic lesions; however, the high frequency of the mutation in papillomas suggests that it is important for the progression of some premalignant lesions to papillomas [34]. The high occurrence of H-*ras* gene mutations in esophageal papillomas is consistent with the observed increases in protein expression levels of p44/42 mitogen-activated protein kinase (Erk 1/2) [35] and both mRNA and protein expression levels of c-Jun, a component of the transcription activator, activator protein-1 (AP-1) [36, 37]. Other studies have demonstrated elevations in cyclin D1 and cyclin E mRNA levels in rat esophageal papillomas, and immunohistochemical staining revealed extensive nuclear staining for both G_1 cyclins

[38–41]. These observations suggest that cell cycle regulation is altered during rat esophageal tumorigenesis. Deregulated expression of transforming growth factor β1 (TGF-β1) and increased expression of epidermal growth factor receptor (EGFR) has also been documented in these tumors [39]. As in human esophageal tumors, G:C → A:T transition mutations have been observed in the *p*53 tumor suppressor gene in ~30% of rat esophageal papillomas [32, 38]. Undoubtedly, all of the above molecular changes are responsible, at least in part, for the increased epithelial cell proliferation rates observed in NMBA-induced preneoplastic lesions and papillomas compared to normal rat esophagus as indicated by immunohistochemical staining for proliferating cell nuclear antigen (PCNA) [42] or Ki-67 [35].

Immunohistochemical staining for apoptotic cells by terminal deoxynucleotidyl transferase dUTP nick end labeling (TUNEL), and protein expression levels of B-cell lymphoma 2 (Bcl-2) and BCL-2-associated X protein (Bax) by Western blot, revealed no significant differences in these biomarkers between NMBA-induced preneoplastic tissues and untreated rat esophagus [35]. However, TUNEL staining for apoptotic cells and the protein expression levels of Bcl-2 and Bax were increased in papillomas relative to preneoplastic tissues.

Elevated levels of cyclooxygenase-2 (COX-2), prostaglandin E_2 (PGE_2), inducible nitric oxide synthase (iNOS), nitrate/nitrite, and the nuclear translocation of nuclear factor kappa B (NFκB)-p50 in preneoplastic tissues and papillomas of NMBA-treated rat esophagus have also been reported [35–37, 43–45]. The mRNA and protein levels of COX-2 and iNOS and the nuclear translocation of NFκB p50 increased with progression of dysplastic lesions to papillomas indicating that they play a functional role in esophageal tumorigenesis. Recently, both cyclooxygenase and lipoxygenase pathways of arachidonic acid metabolism were found to be upregulated in rat esophagus following treatment with NMBA [46]. Alterations in the expression of these genes are undoubtedly associated with the inflammatory changes observed in preneoplastic lesions and papillomas in NMBA-treated rat esophagus.

Immunohistochemical staining for CD34 revealed a marked increased in microvessel density (MVD) or angiogenesis in preneoplastic tissues and papillomas of NMBA-treated rat esophagus when compared to normal esophagus [36, 37]. Western blot analysis indicated that NMBA-induced angiogenesis is associated with increased protein expression levels of vascular endothelial growth factor-C (VEGF-C) [36, 37] and HIF-1α [35].

The identification of alterations in genes associated with cell proliferation, cell cycle, apoptosis, inflammation, and angiogenesis during the progression of normal esophagus > preneoplastic lesions (dysplasia) > papilloma in NMBA-treated F344 rats provides a significant number of potential biomarkers to evaluate the efficacy of chemopreventive agents. We have attempted to take advantage of this information in recent years when evaluating the efficacy of berry formulations for the prevention of esophageal SCC.

Table 1 Freeze-dried berries as effective agents against NMBA-induced esophageal tumorigenesis in F344 rat when administered in the diet before, during and after NMBA treatment [31]

Berry types	Experimental protocols	Tumor incidence (% inhibition)	Tumor multiplicity (% inhibition)
BRBs[a]	Complete	8–22%	40–50%
STRWs[b]	Complete	20%	24–56%

[a]BRBs = freeze-dried black raspberries
[b]STRWs = freeze-dried strawberries

6 Chemoprevention of Rat Esophageal Tumors with Berries and Berry Components

In 1990 our laboratory reported on the ability of the naturally-occurring polyphenol, ellagic acid (EA), to inhibit NMBA-induced tumorigenesis in the F-344 rat esophagus when given in the diet before, during, and after treatment of the rats with NMBA [47]. At that time, it was known that EA was present in strawberries, blackberries [48], grapes, walnuts, and Brazil nuts [49], but its quantity in these and other fruits and nuts had not been determined. We therefore extracted EA from a series of freeze-dried (lyophilized) fruits and nuts using either acetone/water or methanol and found the highest concentrations (520–1,800 µg/g dry weight) in blackberries (BB), red raspberries, strawberries (STRW), walnuts, and pecans [50]. EA was present in the pulp and seed of the berries, but none was detected in the juice. Based upon these observations, we decided to freeze-dry berries to increase the concentration of EA and other potential inhibitory agents in them because berries are about 80–90% water. The freeze-dried berries were ground into a fine powder and the berry powder mixed into synthetic American Institute of Nutrition-76A (AIN-76A) diet for subsequent administration to the animals. In an initial toxicity study we observed that the administration of either BRB or STRW to F-344 rats at 10% of the diet for 9 months resulted in no observable clinical toxicity. Histopathological analysis of all major organs revealed no obvious toxic effects in any of the organs. In addition, there was a 10–15% reduction in blood cholesterol in rats killed after 9 months of berry treatment [51].

In a series of experiments, both freeze-dried STRW and BRB, at 5% and 10% of the diet, produced a 50–60% inhibition of NMBA-induced tumor development in the rat esophagus when administered in the diet before, during, and after NMBA treatment (Table 1) [42, 52, 53]. This inhibition was similar to that seen in earlier experiments with pure EA [47], suggesting that the inhibitory effects of the berries were due to their content of EA. However, analysis of both STRW and BRB diets indicated that the EA content in the berry diets was much lower than the dietary concentrations of pure EA (0.4–4.0 mg/kg) used in the initial study of Mandal and Stoner. It became apparent, therefore, that other components in the berries were responsible, at least in part, for their cancer inhibitory effects. In this regard, subsequent biofractionation studies have shown that the anthocyanins in BRB

Table 2 Some chemopreventive agents in black raspberry powder [56]

Components	Berry samples analyzed[a] 1997	2001	2006
Minerals			
Calcium	215.00	175.00	188.00
Selenium	<5.00	<5.00	<5.00
Zinc	2.69	2.34	2.16
Vitamins			
A from carotene	n.d.[b]	915.00	132.00
Ascorbic acid	4.40	1.10	6.60
α-Carotene	<0.02	<0.02	<0.03
β-Carotene	<0.02	0.06	<0.07
α-Tocopherol	n.d.	n.d.	10.40
β-Tocopherol	n.d.	n.d.	3.51
γ-Tocopherol	n.d.	n.d.	11.20
Folate	0.06	0.08	0.14
Sterols			
β-Sitosterol	80.10	88.80	110.00
Campesterol	3.40	5.90	5.50
Simple phenols			
Ellagic acid	166.30	185.00	225.00
Ferulic acid	17.60	<5.00	47.10
p-Coumaric acid	9.23	6.82	6.92
Chlorogenic acid	n.d.	n.d.	0.14
Quercetin	n.d.	43.60	36.50
Anthocyanins (complex phenols)			
Cyanidin-3-O-glucoside	n.d.	250.00	278.50
Cyanidin-3-O-sambubioside	n.d.	220.00	56.00
Cyanidin-3-O-rutinoside	n.d.	2,002.00	1,790.00
Cyanidin-3-O-xylosylrutinoside	n.d.	510.00	853.50

[a]Components reported in mg/100 g dry weight, except selenium in μg/100 g, and vitamin A in IU. Data from crop years 1997, 2001, and 2006
[b]n.d. = not determined

are more crucial for their chemopreventive effects in the rat esophagus than the ellagitannins (the natural forms of ellagic acid) [35, 54, 55]. Moreover, the alcohol-insoluble (fiber) fractions of several berry types (BRB, STRW, and blueberries [BB]) were nearly as effective as the anthocyanin fractions in inhibiting rat esophageal tumorigenesis [35]. To date, the components responsible for the inhibitory effects of berry fiber on esophageal carcinogenesis have not been identified. Finally, it is not known to what degree other potential chemopreventive agents in berries such as the simple phenols (e.g., quercitin, coumaric acid, chlorogenic acid, ferulic acid), vitamins (A, C, E, folic acid), minerals (calcium, selenium, magnesium, zinc), phytosterols (β-sitosterol, campesterol), and other compounds are responsible for their cancer inhibitory effects [31, 56]. Table 2 lists berry components that are routinely measured in yearly batches of BRB used in our studies. Because the BRB

Table 3 Freeze-dried berries as effective agents against NMBA-induced esophageal tumorigenesis in F344 rat in a post-initiation scheme [31]

Berry types	Experimental protocols	Tumor incidence (% inhibition)	Tumor multiplicity (% inhibition)
BRBs[a]	Post-initiation	40–47%	40–60%
STRWs[b]	Post-initiation	0%	31–38%

[a]BRBs = Freeze-dried black raspberries
[b]STRWs = Freeze-dried strawberries

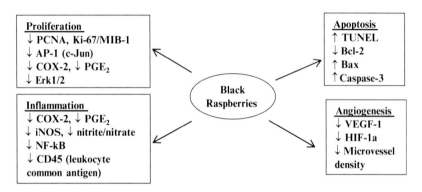

Fig. 4 Effects of BRB on cellular events and associated genes in NMBA-treated rat esophagus [56]

have been obtained from the same farm for the past 20 years, are of the same cultivar (Jewel variety), and are grown in the same field, the variation in the content of most of these components does not exceed 25–30% [56].

At the same dietary concentrations (5% and 10%), STRW and BRB significantly reduced esophageal tumor multiplicity by more than 30% when administered in a post-initiation scheme, indicating the ability of the berries to inhibit tumor progression in the esophagus (Table 3) [42, 52, 53]. By reducing the dose of NMBA, subsequent studies with BRB at the same dietary concentrations have demonstrated up to 50–60% inhibition of NMBA-induced tumorigenesis in the rat esophagus [35, 55]. In an early study, BRB reduced the PCNA labeling index in NMBA-treated esophagus, indicating their ability to reduce the growth rate of preneoplastic cells [42]. Further mechanistic studies indicated that BRB diets down-regulated NMBA-induced COX-2, i-NOS, c-Jun, and VEGF-C mRNA and protein expression levels in the esophagus, and this correlated with reduced levels of PGE_2, nitrate/nitrite, and MVD, respectively, and with tumor multiplicity [36, 37]. The extent of COX-2 inhibition by BRB was similar to that observed with a specific COX-2 inhibitor, L-748706, under development by Merck, Inc., indicating the potency of the berries for down-regulating COX-2 [57]. Recent studies indicate that BRB protectively modulate genes associated with apoptosis (Bcl-2, Bax) in NMBA-treated rat esophagus [35]. Figure 4 summarizes the effects of a 5% BRB diet on genes associated with cell proliferation, inflammation, apoptosis, and angiogenesis in NMBA-treated rat esophagus as determined by Real-Time Polymerase Chain Reaction (RT-PCR) and Western blot analysis.

cDNA microarray analysis has been used to identify additional genes in NMBA-treated rat esophagus whose expression is affected by a BRB diet. In an initial study, BRB were evaluated for their effects on the expression of esophageal genes during the initiation stage of cancer development [58]. Male F344 rats 4–5 weeks old were fed either control AIN-76A diet or control diet supplemented with 5% BRB for 3 weeks. During the 3rd week, one-half of all rats fed either the control diet or the berry supplemented diet received three subcutaneous injections of NMBA. Esophagi from control and berry fed rats were harvested 24 h after the third injection of NMBA and the epithelium was stripped of the submucosal and muscularis layers. RNA microarrays for more than 41,000 transcripts revealed that treatment with NMBA only for 1 week led to the dysregulation of 2,261 genes in the esophageal epithelium, and the berry diet restored 462 of these genes to near-normal levels of expression, regardless of whether they were upregulated or downregulated. These 462 genes included genes associated with signal transduction, cell proliferation/cell cycle, inflammation, differentiation, adhesion and motility, apoptosis, and angiogenesis. Relative to the control diet, treatment with the 5% BRB diet alone altered the expression levels of only 36 genes, suggesting that the berries produce only modest effects on the rat esophagus [59]. In a follow-up study, the effects of a 5% BRB diet on gene expression in NMBA-induced preoplastic esophagus and papillomas was determined [46]. Esophagi from control, NMBA-treated, and NMBA + 5% BRB-treated rats were collected at the end of a 35-week bioassay. Treatment with the 5% BRB diet reduced the number of dysplastic lesions and the number and size of esophageal papillomas in NMBA-treated rats. When compared to esophagi from control rats, NMBA treatment led to the differential expression of 4,807 genes in preoplastic esophagus and 17,846 genes in esophageal papillomas. Dietary BRB modulated 626 of the 4,807 differentially expressed genes in preoplastic esophagus and 625 of the 17,846 differentially expressed genes in esophageal papillomas towards normal levels of expression. In both preoplastic esophagus and in papillomas, BRB modulated the mRNA expression of genes associated with carbohydrate and lipid metabolism, cell proliferation and death, inflammation and many other cellular functions (Table 4). In these same tissues, Western blot analysis revealed that the BRB positively modulated the expression of proteins associated with cell proliferation, apoptosis, inflammation, angiogenesis, and both cyclooxygenase and lipoxygenase pathways of arachidonic acid metabolism. Interestingly, matrix metalloproteinases involved in tissue invasion and metastasis and proteins associated with cell–cell adhesion, were also positively modulated by BRB. Genes commonly modulated by BRB at both time points (1 week and 35 weeks) were keratin 6 alpha, keratin 11, keratin 14, keratin 17, cadmium inducible gene 1L, amphiregulin protein kinase, and cGMP-dependent, type II. Four of the seven genes are keratin genes associated with squamous cell differentiation, and the other three genes are involved in cellular communication. These results suggest that the modulation of genes involved in cell differentiation

Table 4 Cellular functions in preneoplastic esophagus (PE) and in esophageal papillomas from NMBA-treated rats whose genes are restored to near normal levels of expression by BRBs [46]

Cellular functions	PE[a] No. of genes[b]	p Value	Papilloma No. of genes	p Value
Carbohydrate metabolism	27	4.32E-06	3	2.10E-02
Lipid metabolism	28	1.18E-06	9	2.10E-02
Inflammatory response	16	6.71E-04	29	1.19E-02
Cell cycle	16	4.33E-04	18	3.66E-04
Cell death	20	4.31E-04	10	1.19E-02
Cellular movement	23	3.54E-04	9	2.10E-02
Cell morphology	39	1.88E-04	9	2.10E-02
Cellular growth and proliferation	29	1.71E-04	4	2.10E-02
Cell-to-cell signaling and interaction	27	1.69E-04	13	2.10E-02
Post-translational modification	3	3.38E-03	5	6.25E-03
Cell signaling	29	2.53E-03	7	3.24E-02
Vitamin and mineral metabolism	17	2.51E-03	8	3.24E-02
DNA repair	9	1.71E-03	14	3.58E-04
Energy production	5	1.54E-03	1	4.16E-02
Nucleic acid metabolism	11	1.51E-03	8	2.10E-02

[a]PE is the entire esophagus from NMBA-treated rats following removal of papillomas. It is composed of normal and hyperplastic epithelium, and foci of dysplastic change

[b]Genes were assigned to specific cellular functions using Ingenuity's pathway analysis software. The count column indicates the number of genes from the two sets of 626 and 625 genes in PE and in papillomas (see Results), respectively, that are associated with the specific cellular function

and communication may be important for the preventative effects of BRB in models of squamous cell carcinogenesis [46].

The relative ability of seven different berry types (acai, black raspberries, blueberries, goji, noni, red raspberries, and strawberries) to prevent NMBA-induced tumorigenesis in rat esophagus when added at 5% of AIN-76A diet was recently assessed [60]. These berry types were chosen for the study because they vary markedly in their type and content of chemopreventive constituents, including anthocyanins, carotenoids, ellagitannins, phytosterols, and stilbenes. Because black raspberries contain high levels of both anthocyanins and ellagitannins, we predicted that they would be more effective than the other berry types in preventing NMBA-induced esophageal carcinogenesis. Intriguingly, however, all seven berry types were about equally effective in inhibiting NMBA-induced tumorigenesis in the esophagus (Table 5). They also reduced levels of the serum cytokines, interleukin 5 (IL-5) and growth-related-oncogene/keratinocyte-associated chemokine (GRO/KC), the rat homolog for human interleukin-8 (IL-8), and these effects were associated with increased serum antioxidant capacity. Although this data is preliminary, it suggests that multiple berry types possess compounds with anticancer potential. In that regard, the residue (fiber) fractions of BRB, STRW, and BB were about equally effective in preventing NMBA-induced esophageal cancer [55] and, thus, the "equalizer" amongst the different berry types might reside in their content and specific types of fiber.

Table 5 Effect of different berry types on NMBA-induced esophageal tumors in F-344 rats when administered at 5% of the diet [60]

Group	Diet	NMBA (0.3 mg/kg/inj)	Tumor incidence (%)	Tumor multiplicity (mean ± SE)	Tumor size (mm^3) (mean ± SE)
1	AIN-76A control diet	–	0	0	0
2	AIN-76A	+	95	2.15 ± 0.41	11.69 ± 5.07
3	AIN-76A + 5% BRBs	+	60[a]	1.07 ± 0.28[b]	7.50 ± 2.46
4	AIN-76A + 5% BLBs	+	63[a]	1.00 ± 0.32[b]	9.21 ± 7.01
5	AIN-76A + 5% STRWs	+	75[a]	1.25 ± 0.32[b]	8.58 ± 3.46
6	AIN-76A + 5% RRBs (WGO2)	+	75[a]	1.19 ± 0.28[b]	6.72 ± 1.85
7	AIN-76A + 5% RRBs (Meeker)	+	63[a]	0.88 ± 0.27[b]	9.07 ± 3.86
8	AIN-76A + 5% noni	+	60[a]	1.10 ± 0.41[b]	7.93 ± 3.21
9	AIN-76A + 5% wolfberry	+	63[a]	0.94 ± 0.27[b]	5.73 ± 1.24
10	AIN-76A + 5% acai	+	75[a]	1.19 ± 0.25[b]	5.26 ± 2.15

BRBs black raspberries, *BLBs* blueberries, *STRWs* strawberries, *RRBs* red raspberries
[a]Significantly lower than Group 2 (NMBA only) as determined by χ2 test ($P < 0.05$)
[b]Significantly lower than Group 2 (NMBA only) as determined by analysis of variance ($P < 0.05$)

7 Chemoprevention of Human Esophageal SCC

Chemoprevention of human esophageal SCC is focused mainly on blocking the progression of premalignant lesions, such as epithelial dysplasia, to malignant SCC. With the availability of cytological and endoscopic screening techniques, it is possible to identify high-risk individuals for esophageal cancer and take appropriate measures to reduce their risk. The progressive use of endoscopy in high-risk areas of China has proven useful in identifying individuals with premalignant lesions and improving their survival by clinical intervention. For example, esophageal dysplasias classified as "severe" by routine histopathology are often removed by surgical intervention, a practice that has saved many lives in China. Individuals with premalignant lesions of the esophagus have also been accrued to chemoprevention trials and we have discussed the results of these trials in detail in a previous review [16]. This chapter, therefore, will be confined to a discussion of results from a randomized phase II trial of lyophilized strawberries for their effects on dysplastic lesions of the esophagus [61].

8 Effects of Freeze-Dried Strawberries on Dysplastic Lesions of the Esophagus

A randomized phase II chemoprevention trial in a high-risk population in China was conducted to determine the ability of lyophilized strawberries to influence the development of dysplastic lesions [61]. The primary endpoint of the trial was to

Table 6 Effect of 60 g ($N = 36$) or 30 g ($N = 36$) freeze-dried strawberries on histological grade of esophageal precancerous lesions [61]

Treatment (g)[a]	Normal Before/after (%)	Hyperplasia Before/after (%)	Mild dysplasia Before/after (%)	Moderate dysplasia Before/after (%)
30	0/2 (5.5)	0/3 (8.3)	30 (83.3)/24 (66.7)	6 (16.7)/7 (19.4)
60	0/19 (52.7)	0/9 (25)	31 (86.1)/5 (13.9)[b]	5 (13.9)/3 (8.3)

[a]Daily for 6 months
[b]Significantly lower than before strawberry treatment as determined by the McNemar test ($P < 0.0001$)

evaluate the effects of the berries on histologic grade of the lesions. Although most preclinical and clinical studies on the chemoprevention of esophageal cancer have been conducted with black raspberries, strawberries were chosen for this trial for the following reasons: (1) in the rat model of NMBA-induced esophageal SCC, strawberries were nearly as effective as BRBs in preventing the development of esophageal papillomas, (2) strawberries are the principal berry type grown in China, and therefore they are readily available for human consumption throughout the year at relatively low cost, and (3) the Chinese government was reluctant to permit the transit of black raspberry powder across the border into China because of the concern that the powder might contain some viable seed from which BRB could be grown randomly in China.

Freeze-dried strawberries (*Fragaria ananassa*) obtained from the California Strawberry Commission were shipped from the Ohio State University to Beijing, China where they were kept frozen at $-20\,°C$ until used in the trial. The trial was conducted in the Henan and Shandong provinces of China where the population is at high risk for the development of esophageal SCC. Seventy-five subjects identified by endoscopy to have dysplastic esophageal premalignant lesions were randomly assigned to receive freeze-dried strawberry powder at a total dose of either 30 g/day (15 g/2x/day; 37 subjects) or 60 g/day (30 g/2x/day; 38 subjects) for 6 months; the powder was mixed with water and consumed orally in the morning and in the evening. Subjects were encouraged to drink the berry slurry slowly over a period of about 30 min to increase the probability of localized absorption of berry compounds into the dysplastic lesions. After 6 months, potential changes in histological grade of the lesions were assessed in a blind fashion. For this purpose, the esophageal epithelium was characterized into three categories: normal, hyperplasia, and dysplasia (mild and moderate). Histopathological analysis indicated that the consumption of strawberry powder at 30 g/day did not significantly affect the overall histologic grade of the dysplastic lesions (Table 6). The consumption of 60 g/day, however, reduced the histologic grade of dysplastic premalignant lesions in 29 of the 36 patients. The strawberry powder was well tolerated at both dose levels, with no observed toxic effects or serious adverse events.

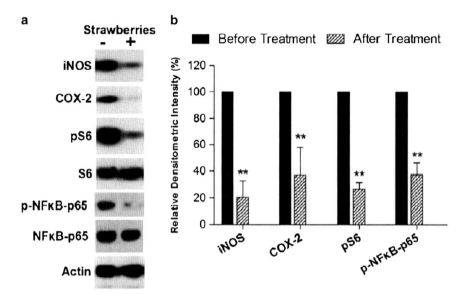

Fig. 5 Effect of freeze-dried strawberries on the protein expression of iNOS, COX-2, p-NFκB-p65, and pS6 as determined by Western blot analysis. Representative blots are shown in (**a**). The values are relative densitometric intensity expressed as mean ± SE. **$P < 0.001$ as determined by Student's t-test when compared with the expression level before strawberry treatment (**b**) [61]

9 Effects of Strawberries on Protein Expression Levels of iNOS, COX-2, pNFκB-p65, and Phospho-S6 in Esophageal Mucosa

Freeze-dried strawberries at 60 g/day also significantly reduced protein expression levels of iNOS by 79.5%, COX-2 by 62.9%, pNFκB-p65 by 62.6%, and phospho-S6 (pS6) by 73.2% in esophageal mucosa (Fig. 5). The effects of reducing the protein expression levels of iNOS, COX-2, and NF-κB on inflammation, cell proliferation and gene transcription are discussed above. Mammalian target of rapamycin (mTOR) is a key regulator of cell proliferation so downregulation of this gene could lead to reduced cell growth [62]. pS6 is one of the downstream targets of mTOR and measurement of pS6 by Western blot has been used to assess the activity of mTOR [63]. The protein expression levels of iNOS, COX-2, pNFκB-p65, and phospho-S6 were not affected in esophageal mucosa obtained from subjects who consumed 30 g/day. Freeze-dried strawberries (60 g/day) also significantly inhibited the Ki-67 labeling index by 37.9%.

10 Summary

These results indicate the potential of freeze-dried strawberry powder for preventing human esophageal SCC, supporting further clinical testing of this natural agent, especially in China and other developing countries. However, a companion article to [61], in *Cancer Prevention Research*, points out several relevant concerns about moving forward with additional clinical evaluation of freeze-dried strawberries as such, and recommends that, ultimately, additional trials be conducted with individual agents or combinations of agents in strawberries. These concerns are well taken and, ultimately, it should be possible to identify an agent or combination of agents that elicits little or no toxicity, is reasonable in cost, and equally or more efficacious than whole berries in regressing dysplastic lesions in the human esophagus.

11 Conclusions

The survival rate for esophageal SCC worldwide has not improved markedly in the past several decades in spite of advances in surgical techniques, radiotherapy, and chemotherapy. Prevention is clearly an important approach to reduce the incidence and mortality from this disease. Lifestyle changes, especially the avoidance of tobacco and alcohol, and the elimination of moldy and salty foods, would have a major effect in reducing the incidence and mortality from this disease. In addition, the increased consumption of vegetables and fruit, especially in high-risk areas, could well provide sources of preventative agents as well as reduce dietary deficiencies associated with the disease. Chemoprevention is another approach that deserves more attention in that the overall number of chemopreventive agents tested in humans against esophageal SCC is limited. Special emphasis should be placed on the identification of additional molecular targets in premalignant (dysplastic) lesions for chemopreventive modulation. Mechanistic studies using the F-344 rat model of esophageal carcinogenesis can provide important leads as to new targets for chemoprevention. In this regard, studies demonstrating the chemopreventive efficacy of agents that modulate the expression levels of iNOS, c-Jun (AP-1), COX-2, NFκB, mTOR, and VEGF in esophageal tissues could provide additional leads for agents that might be efficacious in humans. Freeze-dried berries modulate all of these markers as well as many others in esophageal carcinogenesis in both animals and humans. Additional preclinical studies, therefore, are needed to identify individual compounds in different berry types with chemopreventive potential for the esophagus, including the anthocyanins and their metabolites (e.g., protocatechuic acid [PCA]; [64, 65]), urolithins from ellagic acid [66, 67], pterostilbene from blueberries, which has been shown to inhibit colon cancer in rats by protectively modulating a wide range of relevant molecular biomarkers [68, 69], kaempferol or other phytosterols [70], carotenoids such as β-carotene, and polysaccharides in fiber. These compounds, alone or in combination, could be very effective for the prevention of esophageal SCC in humans.

References

1. Parkin DM, Bray FI, Devesa SS (2001) Cancer burden in the year 2000. The global picture. Eur J Cancer 37:S4–S66
2. Souza RF (2002) Molecular and biologic basis of upper gastrointestinal malignancy – esophageal carcinoma. Surg Oncol Clin N Am 11:257–272
3. Krasna MJ, Wolfer RS (1996) Esophageal carcinoma: diagnosis, evaluation and staging. In: Aisner J, Arriagada R, Green MR, Martini N, Perry MC (eds) Comprehensive textbook of thoracic oncology. Williams and Wilkins, Baltimore, pp 563–584
4. Wang GQ, Abnet CC, Shen Q et al (2005) Histological precursors of oesophageal squamous cell carcinoma: results from a 13 year prospective follow up study in a high risk population. Gut 54:187–192
5. Anani PA, Gardiol D, Savary M et al (1991) An extensive morphological and comparative study of clinically early and obvious squamous cell carcinoma of the esophagus. Pathol Res Pract 187:214–219
6. Kuwano H, Watanabe M, Sadanaga N et al (1993) Squamous epithelial dysplasia associated with squamous cell carcinoma of the esophagus. Cancer Lett 72:141–147
7. Shu YJ, Uan XQ, Jin SP (1981) Further investigation of the relationship between dysplasia and cancer of the esophagus. Chin Med 6:39–41
8. Layke JC, Lopez PP (2006) Esophageal cancer: a review and update. Am Fam Physician 73:2187–2194
9. Polednak AP (2003) Trends in survival for both histologic types of esophageal cancer in US surveillance, epidemiology and end results areas. Int J Cancer 105:98–100
10. Younes M, Henson DE, Ertan A et al (2002) Incidence and survival trends of esophageal carcinoma in the United States: racial and gender differences by histological type. Scand J Gastroenterol 37:1359–1366
11. Schottenfeld D (1984) Epidemiology of cancer of the esophagus. Semin Oncol 11:92–100
12. Stoner GD, Gupta A (2001) Etiology and chemoprevention of esophageal squamous cell carcinoma. Carcinogenesis 22:1737–1746
13. Yang CS (1980) Research on esophageal cancer in China: a review. Cancer Res 40:2633–2640
14. Munoz N, Buiatti E (1996) Chemoprevention of oesophageal cancer. IARC Sci Publ 136:27–33
15. Pickens A, Orringer MB (2003) Geographical distribution and racial disparity in esophageal cancer. Ann Thorac Surg 76:S1367–S1369
16. Stoner GD, Wang LS, Chen T (2007) Chemoprevention of esophageal squamous cell carcinoma. Toxicol Appl Pharm 224:337–349
17. Stoner GD, Wang LS, Zikri N et al (2007) Cancer prevention with freeze-dried berries and berry components. Semin Cancer Biol 17:403–410
18. Hecht SS, Stoner GD (1996) Lung and esophageal carcinogenesis. In: Aisner J, Arriagada R, Green MR, Martini N, Perry MC (eds) Comprehensive textbook of thoracic oncology. Williams and Wilkins, Baltimore, pp 25–50
19. Tuyns AJ (1982) Epidemiology of esophageal cancer in France. In: Pfeifer CJ (ed) Cancer of the esophagus. CRC Press, Boca Raton, pp 3–22
20. Tuyns AJ (1980) Recherches concernant les facteurs etiologiques du cancer de l'oesophage dans l'ouest de la France. Bull Cancer 67:15–28
21. Ribeiro U Jr, Posner MC, Safatle-Ribeiro AV et al (1996) Risk factors for squamous cell carcinoma of the esophagus. Br J Surg 83:1174–1185
22. Li MH, Ji C, Cheng SJ (1986) Occurrence of nitroso compounds in fungi-contaminated foods: a review. Nutr Cancer 8:63–69
23. Lu SH, Chui SX, Yang WX (1991) Relevance of N-nitrosamines to esophageal cancer in China. In: O'Neill IK, Chen J, Bartsch H (eds) Relevance to human cancer of N-nitroso compounds, tobacco smoke and mycotoxin. IARC Scientific Publications no. 105. IARC, Lyon, pp 11–17

24. Yang WX, Pu J, Lu SH et al (1992) Studies on the exposure level of nitrosamines in the gastric juice and its inhibition in high risk areas of esophageal cancer. Chin J Oncol 14:407–410
25. Umbenhauer D, Wild CP, Montesano R et al (1985) O^6-Methyldeoxyguanosine in oesophageal DNA among individuals at high risk of oesophageal cancer. Int J Cancer 36:661–665
26. Li MN, Cheng SJ (1984) Etiology of carcinoma of the esophagus. In: Huang GJ, Kai WY (eds) Carcinoma of the esophagus and gastric cardia. Springer, Berlin, pp 26–51
27. Freeman HJ (2004) Risk of gastrointestinal malignancies and mechanisms of cancer development with obesity and its treatment. Best Pract Res Clin Gastroenterol 18:1167–1175
28. Togawa K, Jaskiewicz K, Takahashi H et al (1994) Human papillomavirus DNA sequences in esophageal squamous cell carcinoma. Gastroenterology 107:128–136
29. Antonsson A, Nancarrow DJ, Brown IS et al (2010) High-risk human papillomavirus in esophageal squamous cell carcinoma. Cancer Epidemiol Biomarkers Prev 19:2080–2087
30. Beer DG, Stoner GD (1998) Clinical models of chemoprevention for the esophagus. Hematol Oncol Clin North Am 12:1055–1077
31. Stoner GD, Chen T, Kresty LA et al (2006) Protection against esophageal cancer in rodents with lyophilized berries: potential mechanisms. Nutr Cancer 54:33–46
32. Lozano JC, Nakazawa H, Cros MP et al (1994) G to A mutations in p53 and Ha-ras genes in esophageal papillomas induced by N-nitrosomethylbenzylamine in two strains of rats. Mol Carcinog 9:33–39
33. Wang Y, You M, Reynolds SH et al (1990) Mutational activation of the cellular Harvey ras oncogene in rat esophageal papillomas induced by methylbenzylnitrosamine. Cancer Res 50:1591–1595
34. Liston BW, Gupta A, Nines R et al (2001) Incidence and effects of Ha-*ras* codon 12 G → A transition mutations in preneoplastic lesions induced by N-nitrosomethylbenzylamine (NMBA) in the rat esophagus. Mol Carcinog 32:1–8
35. Wang LS, Hecht SS, Carmella S et al (2009) Anthocyanins in black raspberries prevent esophageal tumors in rats. Cancer Prev Res 2:84–93
36. Chen T, Hwang H, Rose ME et al (2006) Chemopreventive properties of black raspberries in N-nitrosomethylbenzylamine-induced rat esophageal tumorigenesis: down-regulation of cyclooxygenase-2, inducible nitric oxide synthase, and c-Jun. Cancer Res 66:2853–2859
37. Chen T, Rose ME, Hwang H et al (2006) Black raspberries inhibit N-nitrosomethylbenzylamine (NMBA)-induced angiogenesis in rat esophagus parallel to the suppression of COX-2 and iNOS. Carcinogenesis 27:2301–2307
38. Wang D, Weghorst CM, Calvert RJ et al (1996) Mutations in the p53 tumor suppressor gene in rat esophageal papillomas induced by N-nitrosomethylbenzylamine. Carcinogenesis 17:625–630
39. Wang QS, Sabourin CLK, Kresty LA et al (1996) Dysregulation of transforming growth factor β1 expression in N-nitrosomethylbenzylamine-induced rat esophageal tumorigenesis. Int J Oncol 9:473–479
40. Wang QS, Sabourin CLK, Wang H et al (1996) Overexpression of cyclin D1 and cyclin E in N-nitrosomethylbenzylamine-induced rat esophageal tumorigenesis. Carcinogenesis 17:1583–1588
41. Youssef EM, Hasuma T, Morishima Y et al (1997) Overexpression of cyclin D1 in rat esophageal carcinogenesis model. Jpn J Cancer Res 88:18–25
42. Kresty LA, Morse MA, Morgan C et al (2001) Chemoprevention of esophageal tumorigenesis by dietary administration of lyophilized black raspberries. Cancer Res 61:6112–6119
43. Carlton PS, Gopalakrishnan R, Gupta A et al (2002) Piroxicam is an ineffective inhibitor of N-nitrosomethylbenzylamine-induced tumorigenesis in the rat esophagus. Cancer Res 62:4376–4382
44. Chen T, Stoner GD (2004) Inducible nitric oxide synthase expression in *N*-nitrosomethylbenzylamine (NMBA)-induced rat esophageal tumorigenesis. Mol Carcinog 40:232–240
45. Yu HP, Liu L, Shi LY, Lu WH, Xu SQ (2004) Expression of cyclooxygenase-2 in esophageal squamous cell carcinogenesis. Zhonghua Yu Fang Yi Xue Za Zhi 38:22–25

46. Wang LS, Dombkowski AA, Rocha C et al (2011) Mechanistic basis for the chemopreventive effects of black raspberries at a late stage of rat esophageal carcinogenesis. Mol Carcinog 50:291–300
47. Mandal S, Stoner GD (1990) Inhibition of N-nitrosomethylbenzylamine-induced esophageal tumorigenesis in rats by ellagic acid. Carcinogenesis 11:55–61
48. Haslam E (1977) Symmetry and promiscuity in procyanidin biochemistry. Phytochemistry 16:1625–1645
49. Bate-Smith EC (1972) Detection and determination of ellagitannins. Phytochemistry 11:1153–1156
50. Daniel EM, Krupnick AS, Heur YH et al (1989) Extraction, stability and quantitation of ellagic acid in various fruits and nuts. J Food Comp Anal 2:338–349
51. Harris GK, Gupta A, Nines RG et al (2002) Effects of lyophilized black raspberries on azoxymethane-induced colon cancer and 8-hydroxy-2-deoxyguanosine levels in Fischer-344 rats. Nutr Cancer 40:125–133
52. Carlton PS, Kresty LA, Siglin JC et al (2001) Inhibition of N-nitrosomethylbenzylamine-induced tumorigenesis in the rat esophagus by dietary freeze-dried strawberries. Carcinogenesis 22:441–446
53. Stoner GD, Kresty LA, Carlton PS et al (1999) Isothiocyanates and freeze-dried strawberries as inhibitors of esophageal cancer. Toxicol Sci 52S:95–100
54. Hecht SS, Huang C, Stoner GD et al (2006) Identification of cyanidin glycosides as constituents of freeze-dried black raspberries which inhibit anti-benzo(a)pyrene-7,8-diol-9,10-epoxide induced NFκB and AP-1 activity. Carcinogenesis 27:1617–1626
55. Wang LS, Hecht SS, Carmella S et al (2010) Berry ellagitannins may not be sufficient for prevention of tumors in the rodent esophagus. J Agric Food Chem 58:3992–3995
56. Stoner GD (2009) Foodstuffs for preventing cancer: the preclinical and clinical development of berries. Cancer Prev Res 2:187–194
57. Stoner GD, Qin H, Chen T et al (2005) The effects of L-748706, a selective cyclooxygenase-2 inhibitor on N-nitrosomethylbenzylamine-induced rat esophageal tumorigenesis. Carcinogenesis 26:1590–1595
58. Stoner GD, Dombkowski AA, Reen RK et al (2008) Carcinogen-altered genes in rat esophagus positively modulated to normal levels of expression by both black raspberries and phenethyl isothiocyanate. Cancer Res 68:6460–6467
59. Lechner JF, Reen RK, Dombkowski A et al (2008) Effects of a black raspberry diet on gene expression in the rat esophagus. Nutr Cancer 60:61–69
60. Stoner GD, Wang LS, Seguin C et al (2010) Multiple berry types prevent N-nitrosomethylbenzylamine-induced esophageal cancer in rats. Pharm Res 27:1138–1145
61. Chen T, Yan F, Qian J et al (2012) Randomized phase II trial of lyophilized strawberries in patients with dysplastic precancerous lesions of the esophagus. Cancer Prev Res 5:41–50
62. Wu P, Hu YZ (2010) P13K/Akt/mTOR pathway inhibitors in cancer: a perspective on clinical progress. Curr Med Chem 17:4326–4341
63. Sun Q, Chen X, Ma J et al (2011) Mammalian target of rapamycin up-regulation of pyruvate kinase isoenzyme type M2 is critical for aerobic glycolysis and tumor growth. Proc Natl Acad Sci USA 108:4129–4134
64. Hirose Y, Tanaka T, Kawamori T et al (1995) Chemoprevention of urinary bladder cancer by the natural phenolic compound protocatechuic acid in rats. Carcinogenesis 16:2337–2342
65. Tanaka T, Kojima T, Suzui M et al (1993) Chemoprevention of colon carcinogenesis by the natural product of a simple phenolic compound protocatechuic acid: suppressing effects on tumor development and biomarkers expression of colon tumorigenesis. Cancer Res 53:3908–3913
66. Gonzales-Sarrias A, Gimenez-Bastida JA, Garcia-Conesa MT et al (2010) Occurrence of urolithins, gut microbiota ellagic acid metabolites and proliferation markers expression response in the human prostate gland upon consumption of walnuts and pomegranate juice. Mol Nutr Food Res 54:311–322

67. Sharma M, Li L, Celver J et al (2010) Effects of fruit ellagitannin extracts, ellagic acid, and their colonic metabolite, urolithin A, on Wnt signaling. J Agric Food Chem 58:3965–3969
68. Paul S, DeCastro AJ, Lee HJ et al (2010) Dietary intake of pterostilbene, a constituent of blueberries, inhibits the β-catenin/p65 downstream signaling pathway and colon carcinogenesis in rats. Carcinogenesis 31:1272–1278
69. Soo N, Paul S, Hao X et al (2007) Pterostilbene, an active constituent of blueberries, suppresses aberrant crypt foci formation in the azoxymethane-induced colon carcinogenesis model in rats. Clin Cancer Res 13:350–355
70. Luo H, Rankin GO, Liu L et al (2009) Kaempferol inhibits angiogenesis and VEGF expression through both HIF dependent and independent pathways in human ovarian cancer cells. Nutr Cancer 61:554–563

Cancer Prevention by Different Forms of Tocopherols

Chung S. Yang and Nanjoo Suh

Abstract Many epidemiological studies have suggested that a low vitamin E nutritional status is associated with increased cancer risk. However, several recent large-scale human trials with high doses of α-tocopherol (α-T) have produced disappointing results. This points out the need for a better understanding of the biological activities of the different forms of tocopherols. Using a naturally occurring tocopherol mixture (γ-TmT) that is rich in γ-T, we demonstrated the inhibition of chemically induced lung, colon, and mammary cancer formation as well as the growth of xenograft tumors derived from human lung and prostate cancer cells. This broad anticancer activity of γ-TmT has been attributed mainly to the trapping of reactive oxygen and nitrogen species and inhibition of arachidonic acid metabolism. Activation of peroxisome proliferator-activated receptor γ (PPARγ) and the inhibition of estrogen signaling have also been observed in the inhibition of mammary cancer development. δ-T has been shown to be more active than γ-T in inhibiting the growth of human lung cancer cells in a xenograft tumor model and the development of aberrant crypt foci in azoxymethane-treated rats, whereas α-T is not effective in these models. The higher inhibitory activities of δ-T and γ-T (than α-T) are proposed to be due to their trapping of reactive nitrogen species and their capacity to generate side-chain degradation products, which retain the intact chromanol ring structure and could have cancer preventive activities.

Keywords Breast · Colon · Inhibition · Lung cancer · Prostate · Tocopherols · Vitamin E

C.S. Yang (✉) and N. Suh
Department of Chemical Biology, Center for Cancer Prevention Research, Ernest Mario School of Pharmacy, Rutgers, The State University of New Jersey, Piscataway, NJ, USA
e-mail: csyang@rci.rutgers.edu

Contents

1 Introduction 22
2 Studies on Tocopherols and Cancer in Humans 23
   2.1 Observational Epidemiological Studies 23
   2.2 Intervention Trials with α-Tocopherol 25
3 Inhibition of Tumorigenesis by Single Forms and Mixtures of Tocopherol
   in Animal Models 26
   3.1 Inhibition of Lung Carcinogenesis and Tumor Growth 26
   3.2 Inhibition of Colon Inflammation and Tumorigenesis 27
   3.3 Inhibition of Mammary Carcinogenesis 28
   3.4 Inhibition of Prostate Carcinogenesis and Tumor Growth 28
4 Possible Mechanisms of Action 29
5 Concluding Remarks 29
References 30

1 Introduction

Tocopherols, collectively known as vitamin E, are a family of fat-soluble phenolic compounds. Each tocopherol contains a chromanol ring system and a phytyl chain containing 16 carbons. Depending upon the number and position of methyl groups on the chromanol ring, they exist as α-, β-, γ-, or δ-tocopherols (α-T, β-T, γ-T, and δ-T) [1]. Their structures are shown in Fig. 1. α-T is trimethylated at the 5-, 7-, and 8-positions of the chromanol ring, whereas γ-T is dimethylated at the 7- and 8-positions and δ-T is methylated at the 8-position. The hydrocarbon tail and ring structure provide the lipophilicity for tocopherols to be incorporated into the lipid bilayers of biological membranes. The phenolic group in the chromanol moiety effectively quenches lipid free radicals by one electron reduction. The resulting tocopherol phenoxy radical can be reduced by ascorbic acid or glutathione to regenerate the phenolic group. This is probably the most important physiological antioxidant mechanism to protect the integrity of biological membranes. The unmethylated carbons at 5- and 7-positions at the chromonol ring are electrophilic centers that effectively react with oxygen and nitrogen species (RONS). All the tocopherols are antioxidants; however, γ-T and δ-T are more effective than α-T in trapping reactive nitrogen species [2–4]. The formation of 5-nitro-γ-T, 5-nitro-δ-T, 7-nitro-δ-T, and 5,7-dinitro-δ-T have been reported [5].

The major dietary sources of tocopherols are vegetable oils, such as oils from corn, soybean, sesame, cottonseeds, and nuts [6, 7]. In these oils, γ-T is three to five times more abundant than α-T, and δ-T is as abundant as α-T, whereas β-T exists in only minute amounts. Upon ingestions, these tocopherols are incorporated into the chylomicrons and transported to the liver via the lymphatic system. The transfer of tocopherols in the liver to very low-density lipoproteins is mediated by a specific α-T transfer protein, which preferentially transfers α-T over γ-T, and δ-T is even less effectively transferred [8]. As a consequence, α-T is efficiently secreted into the circulation and transported to nonhepatic tissues, and is the most abundant form of vitamin E in the blood and tissues. The blood and tissue levels of γ-T are much lower, and those of δ-T are even lower.

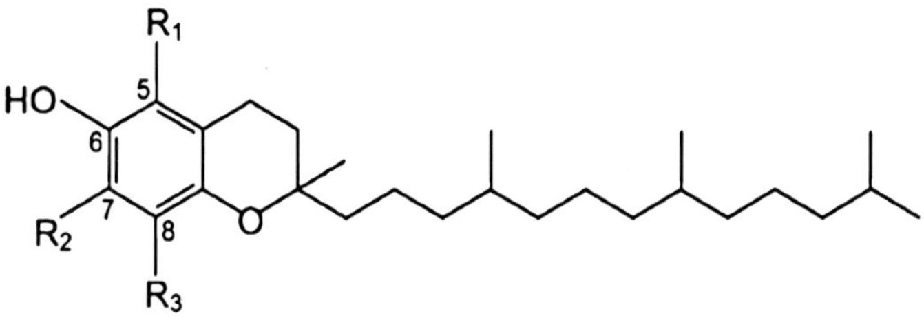

α-Tocopherol   R1 = R2 = R3 = CH3
β-Tocopherol   R1 = R3 = CH3, R2 = H
γ-Tocopherol   R2 = R3 = CH3, R1 = H
δ-Tocopherol   R1 = R2 = H, R3 = CH3

Fig. 1 Structures of tocopherols

Because α-T is the most abundant form of tocopherols in blood and tissues and has the highest activity in the classical fertility-restoration assay, α-T is generally considered to be "the vitamin E." Therefore, many studies on vitamin E have been conducted with α-tocopheryl acetate. However, the results of many of the animal studies are inconsistent, and the results of some of the human intervention studies are disappointing and at variance with those from observation epidemiological studies (reviewed in [9]). In recent years it has been recognized that γ-T and δ-T have beneficial health effects beyond α-T [9–12]. Our collaborative team at Rutgers has demonstrated the broad cancer preventive activities of a γ-T-rich mixture of tocopherols (γ-TmT) as well as pure δ-T and γ-T [13–20]. In this chapter we will discuss the cancer preventive activities of different forms of tocopherols, based on our recent results from animal studies, and their implications to human cancer prevention.

2 Studies on Tocopherols and Cancer in Humans

2.1 Observational Epidemiological Studies

Because of the involvement of RONS in carcinogenesis, the antioxidant nutrients tocopherols have been suggested to have cancer preventive functions. There are many studies that are in support of this concept, but some studies are not (reviewed in [9]). For example, of the three reported cohort studies on lung cancer, two studies found a significant inverse association between dietary intake of vitamin E and risk of lung cancer [21–23]. In both of these studies, the cancer preventive effects were found in current smokers, suggesting a protective effect of vitamin E against insults from cigarette smoking. In four case–control studies on lung cancer, three studies found lower serum α-T levels in lung cancer patients than in matched controls [9]. In a recent case–control study, Mahabir et al. observed that the odds ratios of lung cancer for increasing quartiles of dietary α-T intake were 1.0, 0.63, 0.58, and 0.39, respectively (P for trend <0.0001) [24]. The authors concluded that α-T accounts for 34–53% reduction in lung cancer risk [24]. Since the intake of γ-T was also increased in proportion to α-T in the diet, and at higher quantities, the beneficial effect could also be due to γ-T or the combined effects of all the forms of tocopherols. γ-T is three to four times more abundant than α-T and δ-T could also be more abundant than α-T in the American diet.

Of the six cohort studies on colorectal cancer reviewed, two studies showed an inverse association between vitamin E intake and colorectal cancer risk [25, 26]. For example, in the Iowa Women's Health Study [25], a high intake of vitamin E was associated with a low risk of colon cancer (P for trend <0.0001). This study also found that the protective effect was stronger in subjects under the age of 65 years than in subjects older than that. Of the two case–control studies, one found an inverse association between supplementary vitamin E intake and colorectal cancer risk [27], but the other did not find a protective effect of dietary or supplementary vitamin E against colorectal cancer [28].

Of the 14 case–control studies on prostate cancer reviewed, seven showed an inverse association between dietary or blood levels of tocopherols and risk of prostate cancer [9]. In two nested case–control studies (CLUE I and CLUE II), serum levels of γ-T, but not α-T, were inversely associated with prostate cancer risk [29, 30]. In CLUE I, serum levels of γ-T were significantly lower in subjects who developed prostate cancers than subjects who did not ($P = 0.02$), but no dose-response trend was observed. In CLUE II, a strong inverse association between γ-T and prostate cancer risk was observed ($P = 0.0001$) [29]. Out of the six cohort studies examining the association between dietary or supplementary vitamin E intake and prostate cancer risk, none found any significant association. In the National Institutes of Health-American Association of Retired Persons Diet and Health Study, dietary γ-T and δ-T were found to be significantly related to a reduced risk of advanced prostate cancer (RR: 0.68; 95% CI: 0.56–0.84 for γ-T and RR: 0.8; 95% CI: 0.67–0.96 for δ-T), but supplemental vitamin E (α-T) intake beyond dietary sources was not related to prostate cancer risk [31].

In 24 case–control studies on the relationship between the use of vitamin E supplementation and breast cancer; 11 studies found a risk reduction; however, 13 studies did not find an association [32]. In the Shanghai Breast Cancer Study, it was found that vitamin E supplement may reduce the risk of breast cancer among women who have low dietary intake [33]. In 12 cohort studies there was no association between vitamin E supplementation and breast cancer risk [32]. In one cohort study, the European Prospective Investigation into Cancer and Nutrition (EPIC) trial observed that vitamin E did not reduce breast cancer risk, but there was a weak risk reduction in post-menopausal women [34]. Previously, detailed assessments revealed that vitamin E (α-T) supplements did not protect against breast cancer [35, 36]. Recently, Fulan et al. performed a meta-analysis on 38 studies between vitamin E and breast cancer [37]. For case–control studies, dietary vitamin E and total vitamin E reduced breast cancer risk by 18% and 11%, respectively [37]. When the cohort studies were pooled with the case–control studies, dietary vitamin E and total vitamin E both became nonsignificant [37]. Thus, a conclusion remains elusive between breast cancer and vitamin E. The term "vitamin E" is used loosely, and a distinction in these case–control and cohort studies needs to clarify which variant of vitamin E is utilized. Thus, epidemiological evidence between different forms of vitamin E and breast cancer is limited.

2.2 Intervention Trials with α-Tocopherol

There have been many intervention trials to study the effects of vitamin E supplementation on cancer. However, the results from several large-scale intervention studies with α-T have been disappointing [38–41]. For example, in the Women's Health Study with 39,876 healthy US women aged 45 years or older, the administration of 600 IU of α-T on alternate days did not significantly affect the incidence of colon, lung, or total cancers [38]. In the Physicians' Health Study II Randomized Control Trial, supplementation with vitamin E (400 IU of α-T every other day) or vitamin C (500 mg synthetic ascorbic acid) to physicians for 8 years did not reduce the risk of prostate cancer or all other cancers [39].

The Alpha-Tocopherol, Beta-Carotene (ATBC) Cancer Prevention Study was initially designed to investigate the prevention of lung cancer in male smokers with a daily supplement of 50 IU of all-racemic-α-tocopheryl acetate and 20 mg of β-carotene in a two-by-two design [42]. Supplementation with α-T or β-carotene, or both, for 5–8 years did not produce a significant effect on the incidence of lung cancer [42]. However, α-T supplementation was significantly associated with the reduced incidence of prostate cancer (as a secondary endpoint) and higher serum α-T was associated with a reduced risk of prostate cancer (RR: 0.80; 95% CI: 0.66–0.96 for highest vs lowest quintile; P for trend = 0.03) [43–45]. These results encouraged the launching of the selenium and vitamin E cancer prevention trial (SELECT), in which 35,533 men from 427 study sites in the United States, Canada, and Puerto Rico were randomized between August 2001 and June 2004 [40]. These healthy individuals (ages >55 years old, and for blacks >50 years old) were allocated into four groups and took 400 IU all-rac α-tocopheryl acetate or 200 μg selenium from L-selenomethionine daily in a two-by-two design for an average of 5.5 years. However, the result showed that the supplementations did not prevent prostate or other cancers [40]. It was noted that the α-T supplement caused a 50% decrease in median plasma γ-T levels [40]. In the recently published follow-up (for 7–12 years) results of this study, subjects receiving α-T had a hazard ratio of 1.17 for developing prostate cancer [41]. A possible interpretations of the result is that supplementation of a nutrient to a population that is already adequate in this nutrient may not produce any beneficial effects. It is also possible that supplementation of a large quantity of α-T decreases the blood and tissue levels of γ-T, which has been suggested to have stronger anti-inflammatory and cancer preventive activities [9–12, 46, 47]. Other possible mechanisms have also been discussed [48], but the exact reasons for these negative results are not known. Nevertheless, the disappointing outcome of these large-scale trials reflects our lack of understanding of the biological activities of tocopherols and points to the need for systematic studies of the disease preventive activities of the different forms of tocopherols.

3 Inhibition of Tumorigenesis by Single Forms and Mixtures of Tocopherol in Animal Models

Previous cancer prevention studies in different animal models with pure α-T have obtained inconsistent results [9]. On the other hand, recent studies from our research team at Rutgers University have demonstrated the inhibitory effect of γ-TmT against lung, colon, mammary gland, and prostate cancers [13–20]. γ-TmT is a by-product in the distillation of vegetable oil and usually contains (per gram) 130 mg α-T, 15 mg β-T, 568 mg γ-T, and 243 mg δ-T. Some of our studies are discussed in the following sections.

3.1 Inhibition of Lung Carcinogenesis and Tumor Growth

In studying the lung cancer preventive activity of γ-TmT, we treated A/J mice (6 weeks old) with a tobacco carcinogen, 4-(methylnitrosamino)-1-(3-pyridyl)-1-butanone (NNK), plus benzo[a]pyrene (B[a]P), a ubiquitous environmental pollutant, at doses of 2 μmol each, by oral gavage weekly from weeks 1 to 8. At week 19, the mice in the control group (on the AIN93M diet) developed an average of 21 tumors per mouse [13]. Treatment of the mice with 0.3% γ-TmT in the diet during the entire experimental period lowered the tumor multiplicity to 14.8 (30% inhibition, $p < 0.05$). γ-TmT treatment also significantly reduced the average tumor volume and tumor burden by 50% and 55%, respectively [13]. In a second study, lung tumorigenesis was induced by i.p. injection of two doses of NNK (100 mg/kg on week 1 and 75 mg/kg on week 2). The 0.3% γ-TmT diet was given during the carcinogen-treatment stage, the post-initiation stage, or the entire experimental period. γ-TmT treatment during these three time periods all reduced the tumor multiplicity (17.1, 16.7, and 14.7 tumors per mouse, respectively, as compared to 20.8 in the control group; $p < 0.05$). Moreover, the tumor burden was significantly reduced by γ-TmT treatment given during the tumor initiation stage or during the entire experimental period by 36% and 43% inhibition, respectively [13].

In the NNK plus B[a]P-treated model, dietary γ-TmT treatment significantly increased the apoptotic index (based on cleaved-caspase 3 positive cells) from 0.09% to 0.25% in the lung tumors, whereas the treatment did not affect apoptosis in nontumorous lung tissues. Dietary γ-TmT treatment also significantly decreased the percentage of cells with positive immunostaining for 8-hydroxydeoxyguanine (8-oxo-dG) (from 26% to 17%), a marker for oxidative DNA damage, as well as for γ-H2AX (from 0.51% to 0.23%), a reflection of double-strand break-induced DNA repair. The plasma levels of prostaglandin E2 (PGE2) and leukotriene B4 (LTB4) were markedly elevated in the tumor-bearing A/J mice at week 19 as compared to mice that received no carcinogen treatment. γ-TmT treatment resulted in lower plasma levels of PGE2 (by 61%, $p < 0.05$) and LTB4 (by 12.7%, $p < 0.1$). These results demonstrate the antioxidant and anti-inflammatory activities of γ-TmT.

The antiangiogenic activity of dietary γ-TmT was demonstrated with antiendothelial cell CD31 antibodies. CD31-labeled capillary clusters and blood vessels were observed mainly in the peripheral area of the lung adenomas, and dietary γ-TmT reduced the microvessel density (blood vessels/mm^2) from 375 to 208 ($p < 0.05$) [13].

When 0.3% γ-TmT was given to NCr nu/nu mice in the diet 1 day after implantation of human lung H1299 cells (1×10^6 cells injected subcutaneously per site to both flanks of the mouse), an inhibition of xenograft tumor growth was observed [13]. After 6 weeks, the tumor size and weight were significantly reduced by 56% and 47%, respectively, as compared to the control group. The γ-TmT treatment also caused a 3.3-fold increase in apoptotic index as well as a 52% decrease in 8-oxo-dG-positive cells and a 57% decrease in γ-H2AX-positive cells in the xenograft tumors. Strong cytoplasm staining of nitrotyrosine was observed in xenograft tumors, and the staining intensity was decreased by 44% in mice that received γ-TmT. The γ-TmT treatment also reduced the plasma LTB4 level by 36.5% ($p < 0.05$) [13].

In a similar experiment, the effectiveness of different forms of pure tocopherols in the inhibition of H1299 xenograft tumor growth was compared [14]. Pure δ-T was found to be most effective, showing dose-response inhibition when given at 0.17% and 0.3% in the diet, and pure γ-T and γ-TmT were less effective. Studies of H1299 cells in culture also showed that δ-T was more effective than γ-T and γ-TmT in inhibiting cell growth, whereas α-T was not effective [13]. In another transplanted tumor study, dietary 0.1% and 0.3% γ-TmT were found to inhibit the growth of subcutaneous tumors (formed by injection of murine lung cancer CL13 cells) in A/J mice by 54% and 80%, respectively, on day 50 [15].

3.2 Inhibition of Colon Inflammation and Tumorigenesis

Previous studies concerning the effect of α-T on colon carcinogenesis have yielded mostly negative results [9]. Recently, we studied the effect of γ-TmT in the colons of mice that had been treated with azoxymethane (AOM) and dextran sulfate sodium (DSS) [16]. Dietary γ-TmT treatment (0.3% in the diet) resulted in a significantly lowered colon inflammation index (52% of the control) on day 7, and reduced the number of colon adenomas (to 9% of the control) on week 7. γ-TmT treatment also resulted in higher apoptotic indexes in adenomas, lower PGE2, LTB4, and nitrotyrosine levels in the colon, and lower PGE2, LTB4, and 8-isoprostane levels in the plasma on week 7. In the second experiment, with AOM/DSS-treated mice sacrificed on week 21, dietary γ-TmT treatment significantly inhibited adenocarcinoma and adenoma formation in the colon (to 17–33% of the control). In the third experiment, mice received dietary treatment with 0%, 0.1%, and 0.3% γ-TmT in the AIN 93 M basal diet. One week later, 1% DSS was given to mice in drinking water for 1 week to induce inflammation, and a dose-dependent anti-inflammation by γ-TmT treatment was also observed [16]. These studies demonstrate the antiinflammatory and anticarcinogenic activities of γ-TmT in the colon.

3.3 Inhibition of Mammary Carcinogenesis

In previous studies on mammary carcinogenesis, four studies showed a protective effect of α-T [49–52], but one study showed no effect [53]. Recently, we demonstrated that dietary administration of γ-TmT significantly inhibited N-methyl-N-nitrosourea-induced mammary tumorigenesis in rats [17, 18]. We found that mammary tumor growth and tumor multiplicity, as well as a proliferation marker, proliferating cell nuclear antigen (PCNA), were markedly decreased by administration of γ-TmT. Administration of 0.1%, 0.3%, or 0.5% γ-TmT dose-dependently suppressed mammary tumor development and growth [17]. The inhibition of mammary tumorigenesis was associated with increased expression of p21, p27, cleaved caspase-3, and PPAR-γ, whereas Akt and the estrogen-dependent signaling pathways in mammary tumors were significantly decreased by γ-TmT treatment [17]. Furthermore, in N-methyl-N-nitrosourea-treated rats, dietary γ-TmT, γ-T, and δ-T decreased PCNA levels while increasing the levels of cleaved-caspase 3 in mammary tumors, but α-T was not active [32].

Our in vitro data showed that treatment with γ-TmT, γ-, and δ-T inhibited cell proliferation in MCF-7 breast cancer cells in a dose-dependent manner, while α-T did not [17]. In MCF-7 and T47D breast cancer cells, γ-TmT, γ-T, and, more strongly, δ-T enhance the transactivation of peroxisome proliferator-activated receptor (PPAR)-γ [17]. Since PPARγ transactivation can be suppressed by ERα binding to the PPAR response element [54], the inhibition of ERα expression by tocopherols may result in the activation of PPARγ. Thus, tocopherols may indirectly activate PPARγ, and possibly through this pathway may interfere with ER-α expression, inhibit cell cycle progression, and induce apoptosis to prevent breast cancer. The inhibitory activities of γ-T and δ-T, but not α-T, have also been demonstrated in breast cancer cell lines by other authors [41, 54–56]. In a xenograft model, γ-T treatment inhibited tumor growth and enhanced apoptosis of tumor cells [57].

3.4 Inhibition of Prostate Carcinogenesis and Tumor Growth

Barve et al. demonstrated the inhibition of prostate carcinogenesis in the TRAMP model by 0.1% γ-TmT in the diet [20]. During the development of prostate cancer in the TRAMP mouse, loss of expression of Nrf2 and related cell protective enzymes was observed, and γ-TmT treatment prevented the loss [20]. Takahashi et al. demonstrated that γ-T (0.005% or 0.01% in the diet), but not α-T, decreased the number of adenocarcinomas in the ventral lobe in the transgenic rat for adenocarcinoma of prostate (TRAP) model [58] and the inhibitory action was associated with enhanced apoptosis (activation of caspase-3 and caspase-7). In collaboration with Dr. Xi Zheng and others, we also demonstrated the dose-dependent inhibition of LNCaP prostate cancer growth by γ-TmT (0.1%, 0.3%, and 0.5% in the diet) in a xenograft tumor model in severe combined immunodeficient (SCID) mice [19]. The inhibition was associated with suppressed cell mitosis and stimulated apoptosis (activation of caspase-3).

4 Possible Mechanisms of Action

As reviewed previously [9], many mechanisms have been proposed for the actions of tocopherols. Since our recent results show that γ-T and δ-T effectively inhibit carcinogenesis and xenograft tumor growth, but α-T does not, an important mechanistic issue is why γ-T and δ-T are more active than α-T. All tocopherols are antioxidant. However, the unmethylated 5-position of the chromanol ring enables γ-T and δ-T to quench reactive nitrogen species. In addition, because γ-T and δ-T are less effectively transported to the blood, they are prone to side-chain degradation by the ω-oxidation/β-oxidation pathway. The resulting metabolites, retaining the intact chromanol ring structure, have been reported to have interesting biological activities [9, 12]. The long chain metabolites have been shown to inhibit cyclooxygenase-2 activity [59]. In mice and rats receiving δ-T or γ-T supplementation, short-chain metabolites, δ- or γ-carboxyethyl hydroxychroman, and carboxymethylbutyl hydroxychroman have been found in blood and tissues at micromolar concentrations [14]. These metabolites, without the hydrophobic phytyl chain, may effectively trap RONS in the cytosol.

The activation of PPARγ and the inhibition of ERα-dependent estrogen signaling may play a role in the inhibition of mammary carcinogenesis. It has been shown that PPARγ was more effectively activated by γ-T and δ-T in comparison to α-T [17]. γ-T and δ-T have also been shown to be more active than α-T in inhibiting the growth and inducing apoptosis of different cancer cell lines [9]. For the former action, cell cycle arrest at the S phase and related decrease in cyclin D1, cyclin E, p27, p21, and p16 have been reported [9, 17]. For the induction of apoptosis, activation of caspase-2 and caspase-9, the involvement of caspase-independent pathways, and interruption of de novo synthesis of sphingolipids, have been proposed [9]. Other mechanisms for cancer prevention that contribute to the high activity of δ-T over γ-T in contrast to the very low or null activity of α-T still remain to be discovered.

5 Concluding Remarks

Based on epidemiological and animal studies, we may suggest that at the nutritional level, α-T, being an antioxidant nutrient, contributes to the cancer preventive activity. At the supra-nutritional level, however, γ-T and δ-T are cancer preventive, but α-T is not. The lack of cancer preventive activity of α-T is consistent with many previous studies in animal models [9] and may explain why disappointing results were observed in some recent large scale human trials with α-T [38–41, 60]. The decrease of γ-T levels in the blood and nonhepatic tissues by high doses of α-T has been well demonstrated in animal models and humans [9, 40]. When a high dose of α-T is used, it may decrease the blood and tissue levels of γ-T and diminish its cancer preventive activity [40, 41]. α-T may also increase the cancer incidence if it competes with γ-T and δ-T for binding to molecular targets that are important for cancer prevention. In future intervention trials, high doses of γ-T may also not be suitable because this may decrease the blood and tissue levels of α-T, as has been shown in animals [14]. In the light of the broad cancer preventive activity of γ-TmT and its general availability, this or similar tocopherol mixtures may have a high potential for practical application. These mixtures, with different tocopherols, existing at ratios approximately equal to those in our diet, may have an advantage over pure tocopherols.

Acknowledgment This work was supported by US NIH grants CA122474, CA133021, CA141756, and John L. Colaizzi Chair endowment Fund. We acknowledge the contribution of Drs. Guangxun Li, Zhihong Yang, and Gang Lu for their contribution to research on this topic.

References

1. Traber MG (2006) Vitamin E. In: Bowman BA, Russell RM (eds) Present knowledge in nutrition, 9th edn. ILSI Press, Washington, DC, pp 211–219
2. Christen S, Woodall AA, Shigenaga MK, Southwell-Keely PT, Duncan MW, Ames BN (1997) Gamma-tocopherol traps mutagenic electrophiles such as NO(X) and complements alpha-tocopherol: physiological implications. Proc Natl Acad Sci USA 94:3217–3222
3. Cooney RV, Franke AA, Harwood PJ, Hatch-Pigott V, Custer LJ, Mordan LJ (1993) Gamma-tocopherol detoxification of nitrogen dioxide: superiority to alpha-tocopherol. Proc Natl Acad Sci USA 90:1771–1775
4. Jiang Q, Lykkesfeldt J, Shigenaga MK, Shigeno ET, Christen S, Ames BN (2002) Gamma-tocopherol supplementation inhibits protein nitration and ascorbate oxidation in rats with inflammation. Free Radic Biol Med 33:1534–1542
5. Patel A, Liebner F, Netscher T, Mereiter K, Rosenau T (2007) Vitamin E chemistry. Nitration of non-alpha-tocopherols: products and mechanistic considerations. J Org Chem 72:6504–6512
6. U.S. (2005) Department of Commerce and U.S. Bureau of the Census. Fat and oils: production, consumption, and stocks 2004
7. Eitenmiller RR, Lee J (2004) Vitamin E: food chemistry, composition, and analysis. Marcel Dekker, New York
8. Qian J, Morley S, Wilson K, Nava P, Atkinson J, Manor D (2005) Intracellular trafficking of vitamin E in hepatocytes: the role of tocopherol transfer protein. J Lipid Res 46:2072–2082
9. Ju J, Picinich SC, Yang Z, Zhao Y, Suh N, Kong AN, Yang CS (2009) Cancer preventive activities of tocopherols and tocotrienols. Carcinogenesis 31:533–542
10. Jiang Q, Christen S, Shigenaga MK, Ames BN (2001) Gamma-tocopherol, the major form of vitamin E in the US diet, deserves more attention. Am J Clin Nutr 74:714–722
11. Campbell S, Stone W, Whaley S, Krishnan K (2003) Development of gamma-tocopherol as a colorectal cancer chemopreventive agent. Crit Rev Oncol Hematol 47:249–259
12. Hensley K, Benaksas EJ, Bolli R, Comp P, Grammas P, Hamdheydari L, Mou S, Pye QN, Stoddard MF, Wallis G, Williamson KS, West M, Wechter WJ, Floyd RA (2004) New perspectives on vitamin E: gamma-tocopherol and carboxyethylhydroxychroman metabolites in biology and medicine. Free Radic Biol Med 36:1–15

13. Lu G, Xiao H, Li G, Chen K-Y, Hao J, Loy S, Yang CS (2010) γ-Tocopherols-rich mixture of tocopherols inhibits chemically-induced lung tumorigenesis in A/J mice and xenograft tumor growth. Carcinogenesis 31:687–694
14. Li GX, Lee MJ, Liu AB, Yang Z, Lin Y, Shih WJ, Yang CS (2011) Delta-tocopherol is more active than alpha- or gamma-tocopherol in inhibiting lung tumorigenesis in vivo. Cancer Prev Res (Phila) 4:404–413
15. Lambert JD, Lu G, Lee MJ, Hu J, Yang CS (2009) Inhibition of lung cancer growth in mice by dietary mixed tocopherols. Mol Nutr Food Res 53:1030–1035
16. Ju J, Hao X, Lee MJ, Lambert JD, Lu G, Xiao H, Newmark HL, Yang CS (2009) A gamma-tocopherol-rich mixture of tocopherols inhibits colon inflammation and carcinogenesis in azoxymethane and dextran sulfate sodium-treated mice. Cancer Prev Res (Phila Pa) 2:143–152
17. Lee HJ, Ju J, Paul S, So JY, DeCastro A, Smolarek AK, Lee MJ, Yang CS, Newmark HL, Suh N (2009) Mixed tocopherols prevent mammary tumorigenesis by inhibiting estrogen action and activating PPAR-g. Clin Cancer Res 15:4242–4249
18. Suh N, Paul S, Lee HJ, Ji Y, Lee MJ, Yang CS, Reddy BS, Newmark HL (2007) Mixed tocopherols inhibit N-methyl-N-nitrosourea-induced mammary tumor growth in rats. Nutr Cancer 59:76–81
19. Zheng X, Cui X-X, Khor TO, Huang Y, DiPaola RS, Goodin S, Lee M-J, Yang CS, Kong A-N, Conney AH (2012) Inhibitory effect of a γ-tocophenol-rich mixture of tocopherols on the formation and growth of LNCaP prostate tumors in immunodeficient mice. Cancers 3:3762–3772
20. Barve A, Khor TO, Nair S, Reuhl K, Suh N, Reddy B, Newmark H, Kong AN (2009) Gamma-tocopherol-enriched mixed tocopherol diet inhibits prostate carcinogenesis in TRAMP mice. Int J Cancer 124:1693–1699
21. Woodson K, Albanes D, Tangrea JA, Rautalahti M, Virtamo J, Taylor PR (1999) Association between alcohol and lung cancer in the alpha-tocopherol, beta-carotene cancer prevention study in Finland. Cancer Causes Control 10:219–226
22. Yong LC, Brown CC, Schatzkin A, Dresser CM, Slesinski MJ, Cox CS, Taylor PR (1997) Intake of vitamins E, C, and A and risk of lung cancer. The NHANES I epidemiologic followup study. First National Health and Nutrition Examination Survey. Am J Epidemiol 146:231–243
23. Shibata A, Paganini-Hill A, Ross RK, Henderson BE (1992) Intake of vegetables, fruits, beta-carotene, vitamin C and vitamin supplements and cancer incidence among the elderly: a prospective study. Br J Cancer 66:673–679
24. Mahabir S, Schendel K, Dong YQ, Barrers SL, Spitz MR, Forman MR (2008) Dietary alpha-, beta-, gamma- and delta-tocopherols in lung cancer risk. Int J Cancer 123:1173–1180
25. Bostick RM, Potter JD, McKenzie DR, Sellers TA, Kushi LH, Steinmetz KA, Folsom AR (1993) Reduced risk of colon cancer with high intake of vitamin E: the Iowa Women's Health Study. Cancer Res 53:4230–4237
26. Ghadirian P, Lacroix A, Maisonneuve P, Perret C, Potvin C, Gravel D, Bernard D, Boyle P (1997) Nutritional factors and colon carcinoma: a case-control study involving French Canadians in Montreal, Quebec, Canada. Cancer 80:858–864
27. White E, Shannon JS, Patterson RE (1997) Relationship between vitamin and calcium supplement use and colon cancer. Cancer Epidemiol Biomarkers Prev 6:769–774
28. Ingles SA, Bird CL, Shikany JM, Frankl HD, Lee ER, Haile RW (1998) Plasma tocopherol and prevalence of colorectal adenomas in a multiethnic population. Cancer Res 58:661–666
29. Helzlsouer KJ, Huang HY, Alberg AJ, Hoffman S, Burke A, Norkus EP, Morris JS, Comstock GW (2000) Association between alpha-tocopherol, gamma-tocopherol, selenium, and subsequent prostate cancer. J Natl Cancer Inst 92:2018–2023
30. Huang HY, Alberg AJ, Norkus EP, Hoffman SC, Comstock GW, Helzlsouer KJ (2003) Prospective study of antioxidant micronutrients in the blood and the risk of developing prostate cancer. Am J Epidemiol 157:335–344

31. Wright ME, Weinstein SJ, Lawson KA, Albanes D, Subar AF, Dixon LB, Mouw T, Schatzkin A, Leitzmann MF (2007) Supplemental and dietary vitamin E intakes and risk of prostate cancer in a large prospective study. Cancer Epidemiol Biomarkers Prev 16:1128–1135
32. Smolarek AK, Suh N (2011) Chemopreventive activity of vitamin E in breast cancer: a focus on gamma-and delta-tocopherol. Nutrients 3:962–986
33. Dorjgochoo T, Shrubsole MJ, Shu XO, Lu W, Ruan Z, Zheng Y, Cai H, Dai Q, Gu K, Gao YT, Zheng W (2008) Vitamin supplement use and risk for breast cancer: the Shanghai Breast Cancer Study. Breast Cancer Res Treat 111:269–278
34. Nagel G, Linseisen J, van Gils CH, Peeters PH, Boutron-Ruault MC, Clavel-Chapelon F, Romieu I, Tjonneland A, Olsen A, Roswall N, Witt PM, Overvad K, Rohrmann S, Kaaks R, Drogan D, Boeing H, Trichopoulou A, Stratigakou V, Zylis D, Engeset D, Lund E, Skeie G, Berrino F, Grioni S, Mattiello A, Masala G, Tumino R, Zanetti R, Ros MM, Bueno-de-Mesquita HB, Ardanaz E, Sanchez MJ, Huerta JM, Amiano P, Rodriguez L, Manjer J, Wirfalt E, Lenner P, Hallmans G, Spencer EA, Key TJ, Bingham S, Khaw KT, Rinaldi S, Slimani N, Boffetta P, Gallo V, Norat T, Riboli E (2010) Dietary beta-carotene, vitamin C and E intake and breast cancer risk in the European prospective investigation into cancer and nutrition (EPIC). Breast Cancer Res Treat 119:753–765
35. Schwenke DC (2002) Does lack of tocopherols and tocotrienols put women at increased risk of breast cancer? J Nutr Biochem 13:2–20
36. Kimmick GG, Bell RA, Bostick RM (1997) Vitamin E and breast cancer: a review. Nutr Cancer 27:109–117
37. Fulan H, Changxing J, Baina WY, Wencui Z, Chunqing L, Fan W, Dandan L, Dianjun S, Tong W, Da P, Yashuang Z (2011) Retinol, vitamins A, C, and E and breast cancer risk: a meta-analysis and meta-regression. Cancer Causes Control 22:1383–1396
38. Lee IM, Cook NR, Gaziano JM, Gordon D, Ridker PM, Manson JE, Hennekens CH, Buring JE (2005) Vitamin E in the primary prevention of cardiovascular disease and cancer: the Women's Health Study: a randomized controlled trial. JAMA 294:56–65
39. Gaziano JM, Glynn RJ, Christen WG, Kurth T, Belanger C, MacFadyen J, Bubes V, Manson JE, Sesso HD, Buring JE (2009) Vitamins E and C in the prevention of prostate and total cancer in men: the Physicians' Health Study II randomized controlled trial. JAMA 301:52–62
40. Lippman SM, Klein EA, Goodman PJ, Lucia MS, Thompson IM, Ford LG, Parnes HL, Minasian LM, Gaziano JM, Hartline JA, Parsons JK, Bearden JD 3rd, Crawford ED, Goodman GE, Claudio J, Winquist E, Cook ED, Karp DD, Walther P, Lieber MM, Kristal AR, Darke AK, Arnold KB, Ganz PA, Santella RM, Albanes D, Taylor PR, Probstfield JL, Jagpal TJ, Crowley JJ, Meyskens FL Jr, Baker LH, Coltman CA Jr (2009) Effect of selenium and vitamin E on risk of prostate cancer and other cancers: the selenium and vitamin E cancer prevention trial (SELECT). JAMA 301:39–51
41. Klein EA, Thompson IM Jr, Tangen CM, Crowley JJ, Lucia MS, Goodman PJ, Minasian LM, Ford LG, Parnes HL, Gaziano JM, Karp DD, Lieber MM, Walther PJ, Klotz L, Parsons JK, Chin JL, Darke AK, Lippman SM, Goodman GE, Meyskens FL Jr, Baker LH (2011) Vitamin E and the risk of prostate cancer: the selenium and vitamin E cancer prevention trial (SELECT). JAMA 306:1549–1556
42. Albanes D, Heinonen OP, Taylor PR, Virtamo J, Edwards BK, Rautalahti M, Hartman AM, Palmgren J, Freedman LS, Haapakoski J, Barrett MJ, Pietinen P, Malila N, Tala E, Liippo K, Salomaa ER, Tangrea JA, Teppo L, Askin FB, Taskinen E, Erozan Y, Greenwald P, Huttunen JK (1996) Alpha-tocopherol and beta-carotene supplements and lung cancer incidence in the alpha-tocopherol, beta-carotene cancer prevention study: effects of base-line characteristics and study compliance. J Natl Cancer Inst 88:1560–1570
43. Weinstein SJ, Wright ME, Lawson KA, Snyder K, Mannisto S, Taylor PR, Virtamo J, Albanes D (2007) Serum and dietary vitamin E in relation to prostate cancer risk. Cancer Epidemiol Biomarkers Prev 16:1253–1259

44. Virtamo J, Pietinen P, Huttunen JK, Korhonen P, Malila N, Virtanen MJ, Albanes D, Taylor PR, Albert P (2003) Incidence of cancer and mortality following alpha-tocopherol and beta-carotene supplementation: a postintervention follow-up. JAMA 290:476–485
45. Heinonen OP, Albanes D, Virtamo J, Taylor PR, Huttunen JK, Hartman AM, Haapakoski J, Malila N, Rautalahti M, Ripatti S, Maenpaa H, Teerenhovi L, Koss L, Virolainen M, Edwards BK (1998) Prostate cancer and supplementation with alpha-tocopherol and beta-carotene: incidence and mortality in a controlled trial. J Natl Cancer Inst 90:440–446
46. Wagner KH, Kamal-Eldin A, Elmadfa I (2004) Gamma-tocopherol – an underestimated vitamin? Ann Nutr Metab 48:169–188
47. Reiter E, Jiang Q, Christen S (2007) Anti-inflammatory properties of alpha- and gamma-tocopherol. Mol Aspects Med 28:668–691
48. McNeil C (2011) Vitamin e and prostate cancer: research focus turns to biologic mechanisms. J Natl Cancer Inst 103:1731–1734
49. Hagiwara A, Boonyaphiphat P, Tanaka H, Kawabe M, Tamano S, Kaneko H, Matsui M, Hirose M, Ito N, Shirai T (1999) Organ-dependent modifying effects of caffeine, and two naturally occurring antioxidants alpha-tocopherol and n-tritriacontane-16,18-dione, on 2-amino-1-methyl-6-phenylimidazo[4,5-b]pyridine (PhIP)-induced mammary and colonic carcinogenesis in female F344 rats. Jpn J Cancer Res 90:399–405
50. Dias MF, Sousa E, Cabrita S, Patrício J, Oliveira CF (2000) Chemoprevention of DMBA-induced mammary tumors in rats by a combined regimen of alpha-tocopherol, selenium, and ascorbic acid. Breast J 6:14–19
51. Hirose M, Nishikawa A, Shibutani M, Imai T, Shirai T (2002) Chemoprevention of heterocyclic amine-induced mammary carcinogenesis in rats. Environ Mol Mutagen 39:271–278
52. Hirose M, Takahashi S, Ogawa K, Futakuchi M, Shirai T (1999) Phenolics: blocking agents for heterocyclic amine-induced carcinogenesis. Food Chem Toxicol 37:985–992
53. Gould MN, Haag JD, Kennan WS, Tanner MA, Elson CE (1991) A comparison of tocopherol and tocotrienol for the chemoprevention of chemically induced rat mammary tumors. Am J Clin Nutr 53:1068S–1070S
54. Bonofiglio D, Gabriele S, Aquila S, Catalano S, Gentile M, Middea E, Giordano F, Ando S (2005) Estrogen receptor alpha binds to peroxisome proliferator-activated receptor response element and negatively interferes with peroxisome proliferator-activated receptor gamma signaling in breast cancer cells. Clin Cancer Res 11:6139–6147
55. Yu W, Simmons-Menchaca M, Gapor A, Sanders BG, Kline K (1999) Induction of apoptosis in human breast cancer cells by tocopherols and tocotrienols. Nutr Cancer 33:26–32
56. Yu W, Jia L, Wang P, Lawson KA, Simmons-Menchaca M, Park SK, Sun L, Sanders BG, Kline K (2008) In vitro and in vivo evaluation of anticancer actions of natural and synthetic vitamin E forms. Mol Nutr Food Res 52:447–456
57. Yu W, Jia L, Park SK, Li J, Gopalan A, Simmons-Menchaca M, Sanders BG, Kline K (2009) Anticancer actions of natural and synthetic vitamin E forms: RRR-alpha-tocopherol blocks the anticancer actions of gamma-tocopherol. Mol Nutr Food Res 53:1573–1581
58. Takahashi S, Takeshita K, Seeni A, Sugiura S, Tang M, Sato SY, Kuriyama H, Nakadate M, Abe K, Maeno Y, Nagao M, Shirai T (2009) Suppression of prostate cancer in a transgenic rat model via gamma-tocopherol activation of caspase signaling. Prostate 69:644–651
59. Jiang Q, Yin X, Lill MA, Danielson ML, Freiser H, Huang J (2008) Long-chain carboxychromanols, metabolites of vitamin E, are potent inhibitors of cyclooxygenases. Proc Natl Acad Sci USA 105:20464–20469
60. Lin J, Cook NR, Albert C, Zaharris E, Gaziano JM, Van Denburgh M, Buring JE, Manson JE (2009) Vitamins C and E and beta carotene supplementation and cancer risk: a randomized controlled trial. J Natl Cancer Inst 101:14–23

Cancer Chemopreventive and Therapeutic Potential of Guggulsterone

Inas Almazari and Young-Joon Surh

Abstract Guggulsterone (GS) is a phytosterol derived from the gum resin of guggul plants that have been used traditionally to treat various disorders such as burns, wounds, gastric ulcer, cough, gum diseases, urinary complaints, diarrhea, stomach cramps, fascioliasis, and intestinal worms. It has anti-inflammatory and antioxidative properties and has recently attracted substantial attention due to its cancer chemopreventive and therapeutic potential exemplified by its antiproliferative, antimetastatic, and proapoptotic properties in many cancer cell lines and animal models. This review highlights some of the cancer chemopreventive/therapeutic targets of GS and the underlying molecular mechanisms.

Keywords Guggulsterone · Cancer chemoprevention · Guggul plants · Phytochemicals

Contents

1 Introduction 36
2 Biological Properties of GS 37
   2.1 Hypolipidemic and Hypocholestremic Properties of GS 38
   2.2 Antioxidant Properties of GS 38
   2.3 Anti-Inflammatory Properties of GS 38

I. Almazari
Tumor Microenvironment Global Core Research Center, College of Pharmacy, Seoul National University, 599 Gwanak-ro, Gwanak-gu, Seoul 151-742, South Korea

Y.-J. Surh (✉)
Tumor Microenvironment Global Core Research Center, College of Pharmacy, Seoul National University, 599 Gwanak-ro, Gwanak-gu, Seoul 151-742, South Korea

WCU Department of Molecular Medicine and Biopharmaceutical Sciences, Seoul National University, Seoul, South Korea

Cancer Research Institute, Seoul National University, Seoul, South Korea
e-mail: surh@plaza.snu.ac.kr

3	Chemopreventive and Chemotherapeutic Potential of GS	39
	3.1 Effect on Colon Cancer	42
	3.2 Effect on Breast Cancer	43
	3.3 Effect on Prostate Cancer	43
	3.4 Effect on Head and Neck Cancer	44
	3.5 Effect on Gastrointestinal Cancer	45
	3.6 Effect on Liver Cancer	46
	3.7 Effect on Lung Cancer	46
4	Signaling Molecules Modulated by GS	46
	4.1 MAPK	46
	4.2 PI3K/Akt	47
	4.3 Nrf2	47
	4.4 STATs	50
	4.5 NF-κB	50
	4.6 14-3-3ζ	51
	4.7 P-Glycoprotein	51
	4.8 iNOS	52
	4.9 Growth Factors	52
5	Conclusion	52
References		54

1 Introduction

The incidence of cancer, in general, is still high all over the world despite the huge international efforts to control it, and this provokes the need for more active action to minimize its spread [1]. Carcinogenesis is a multistage process, composed of at least three stages – initiation, promotion, and progression [2]. Initiation involves DNA damage, which is often caused by carcinogens, and this process is usually rapid and irreversible [3]. In contrast, promotion involves epigenetic changes of the cells, and is a relatively slow and reversible process [3]. The last stage is the progression, which involves the transformation of the cells into malignant ones [3, 4].

As cancer treatment with conventional synthetic anticancer drugs is generally mono-targeted, expensive, and time consuming, which often results in side effects, it is more realistic to control cancer by intervening in the transformation from a preneoplastic state to the cancerous state than to treat it after the malignancy manifests [3]. This is how the concept of chemoprevention was first coined by Michael Sporn in 1976 [5]. Chemoprevention refers to the use of nontoxic chemical substances, such as those present in edible plants, to delay, prevent, or even reverse the carcinogenesis stages [4, 5]. Interrupting promotion, rather than initiation or progression, seems to be a feasible strategy since this carcinogenic step is slow and reversible [3].

Accumulating evidence indicates that many edible phytochemicals have cancer chemopreventive potential [6, 7]. Examples are guggul plants, such as *Commiphora kataf*, *Commiphora erythraea*, *Commiphora wightii* [8], *Commiphora mukul* [9], *Commiphora myrrha*, *Commiphora molmol*, and *Balsamodendron mukul* [10]. The guggul plants are members of the *Burseraceae* family that are found mainly in India, Kenya, China, Bangladesh, Pakistan, Middle East, and Arabia [10–12]. The

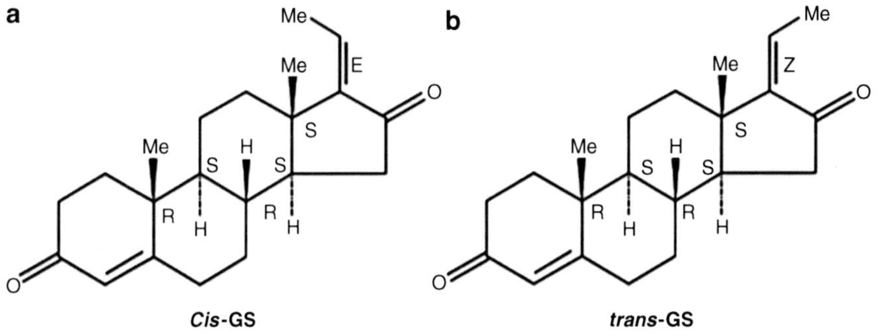

Fig. 1 The chemical structures of GS isomers. (**a**) [4,17(20)-(*cis*)-Pregnadiene-3,16-dione] is the *E* form, and called alternatively *cis*-GS, and (**b**) [4,17(20)-(*trans*)-pregnadiene-3,16-dione] is the Z form, and called alternatively *trans*-GS

guggulu gum resin has been used in making perfumes, especially in the Arabian Peninsula because of its characteristic aromatic odor and has also been prescribed as Ayurvedic medicine to treat hypercholesterolemia, obesity, bone fractures, inflammation and rheumatism, atherosclerosis, urinary complaints, and abdominal disorders [11, 12]. Guggulsterone (GS) is the major constituent of the gum resin and has two stereo-isomers (Fig. 1), *E*-GS (*cis*-GS) and *Z*-GS (*trans*-GS) [10–12].

Inflammation and oxidative stress are two major culprits that are implicated in the pathogenesis of the majority of human malignancies [13]. Oxidative stress induced by reactive oxygen species (ROS) causes not only genetic alterations, such as DNA mutation, but also epigenetic changes that facilitate carcinogenesis [14]. In addition, cancer develops due to chronic inflammation [15]. Many pathophysiological conditions including obesity, diabetes, and infections are associated with chronic inflammation, and are somehow linked to carcinogenesis [16]. Recently, much attention has been focused on GS as a potential chemopreventive phytochemical because it possesses strong antioxidant [11, 17–20] and anti-inflammatory [21–27] properties. The following sections deal with the scientific progress that has been accomplished regarding the mechanisms of action and molecular targets of GS with special focus on its role in the prevention of cancer.

2 Biological Properties of GS

GS has anti-inflammatory, antioxidant, hypolipidemic [11], hypocholesterolemic, and hypoglycemic activities [28, 29]. The aforementioned properties make GS a good choice for treatment or prevention of some metabolic disorders, such as diabetes mellitus, artherosclerosis, and obesity [29]. As oxidative stress and inflammation are implicated in the pathogenesis of diabetes and obesity which, in turn, can increase the risk of cancer, it is anticipated that this medicinal phytochemical has cancer chemopreventive potential as well.

2.1 Hypolipidemic and Hypocholestremic Properties of GS

GS enhances lipid accumulation in hepatocytes and hepatoma cells [30]. It has been proposed that GS-mediated hypolipidemic activity is attributable to its ability to antagonize the bile acid receptor farnesoid X receptor (FXR) and to upregulate the bile salt export pump that plays a role in the excretion of cholesterol metabolites from the liver [11, 31]. Thus, GS interferes with abnormal cellular accumulation of fatty acids, cholesterol, and bile acid [32]. Considering a strong association between FXR and carcinogenesis [33, 34], GS acting as a powerful FXR antagonist is likely to exert anticancer activities.

2.2 Antioxidant Properties of GS

GS inhibits nitric oxide (NO)-nitrosative and H_2O_2-induced oxidative stresses [20, 35, 36]. Oxidative/nitrosative stress is implicated in many disorders, such as rheumatoid arthritis, cardiovascular disorders, diabetes, and neurodegenerative diseases [11, 12]. GS inhibits oxidation of low-density lipoprotein and thereby lowers the cholesterol levels, conferring cardio-protection against artherosclerosis [18, 19, 37]. In addition, GS and the plant gum resin protect cells from cytokine-, endotoxin-, and lipopolysaccharide (LPS)-mediated oxidative stress and toxicity in vivo and in vitro [20, 35, 36, 38, 39]. Interestingly, GS fortifies cellular defense against oxidative stress by inducing the de novo synthesis of the powerful antioxidant enzyme heme oxygenase-1 (HO-1) in human mammary epithelial (MCF-10A) cells [40].

2.3 Anti-Inflammatory Properties of GS

GS [12, 21, 26] and two of its derivatives [41] have been shown to exert anti-inflammatory effects through suppression of nuclear factor-κB (NF-κB), which plays a crucial role in the inflammatory processes by regulating the expression of diverse proinflammatory proteins, including cyclooxygenase-2 (COX-2) [42]. GS exerts potent protective effects against the inflammatory conditions by inhibiting NF-κB signaling in several different types of cells including pancreatic beta cells [35], intestinal epithelial cells [27, 41], fibroblast-like synoviocytes [24], and human primary nonpigment ciliary epithelial cells [26]. Besides GS, guggulipid (GL) was reported to abrogate LPS-induced expression of COX-2, tumor necrosis factor-alpha (TNF-α), and glial fibrillary acidic protein in rat astrocytoma cells [39].

E-GS, unlike its stereoisomer, suppressed the development of inflammatory bowel disease in BALB/c and SCID mice in two different murine colitis models [21]. 2,4,6-Trinitrobenzenesulphonic acid (TNBS)-induced colitis is a model of the T helper (Th)1-mediated disease that results in Crohn's disease, while oxazolone-induced colitis is a model of the Th2-mediated disease that mimics human ulcerative colitis [43, 44]. *E*-GS abrogated TNBS- and oxazolone-induced expression of inflammatory mediators, such as interferon-(IFN)-γ, TNF-α, transforming growth factor-(TGF)-β and interleukin (IL)-2, IL-4, and IL-6 [21].

GS abrogated LPS-induced NF-κB activation through inhibition of IκB kinase (IKK) activity [23, 27]. GS also suppressed LPS-induced NF-κB activation by blocking TIR-domain-containing adapter-inducing interferon-β-dependent signaling of the toll-like receptor (TLR)3 and TLR4 [45, 46]. The GS-mediated inhibition of TLR3 and TLR4 was associated with the inhibition of the expression of COX-2 and IFN-β and the phosphorylation of IRF3 [46].

3 Chemopreventive and Chemotherapeutic Potential of GS

While GS has preventive/therapeutic potential for the management of diabetes, obesity, inflammation, and other human disorders [11, 12, 29, 47, 48], much attention has been paid in the last few years to the ability of this molecule to treat cancer [49–51]. Some of the anticarcinogenic effects of GS reported in the literature are summarized in Table 1.

One common mechanism of the anticarcinogenic activity of GS is its proapoptotic effect [23, 51, 62, 65, 66]. Thus GS significantly induces apoptosis in several different types of cancer cells while it has a minimal effect on normal cell viability [40, 65]. Apoptosis, also known as programmed cell death, accompanies specific morphological and biochemical changes such as DNA fragmentation and cell shrinkage [67]. These characteristics are associated with the expression of a distinct set of stress-inducible signaling molecules [67]. GS delays skin tumor growth in SENCAR mice through inhibition of NF-κB and mitogen activated protein kinase (MAPK) signaling [38].

Apoptosis can be induced via the mitochondrial (intrinsic) pathway or the death receptor (extrinsic) pathway [68]. A common step in both pathways is caspases activation [69]. Caspases are cysteine proteases that are synthesized as inactive procaspases and activated by their cleavage at the post-translational level [70]. They are classified into two categories, initiator (apical) caspases (e.g., caspase-2, caspase-8, caspase-9, and caspase-10) and effector (executioner) caspases, such as caspase-3, caspase-6, and caspase-7 [67]. Initiator caspases directly bind to death-inducing signaling complexes and possess a longer prodomain consisting of a small and a large subunits that contains either CARD domain as in the cases of caspase-2 and caspase-9 or death effector domain as seen in caspase-8 and caspase-10 [71]. Upon activation, the initiators activate effectors, and this in turn cleaves and activates other cytoplasmic and nuclear proteins [70, 71]. GS-induced

Table 1 Anticarcinogenic effects of GS assessed in different experimental models and underlying mechanisms

Form	Cell/tissue type	Models used	Mechanism of action	References
GS	Esophagus	Barrett's esophagus Esophageal adenocarcinoma	Induction of apoptosis by antagonizing FXR	[33]
	Leukemia	Doxorubicin-resistant myelogenous leukemia (K562/DOX)	Reversal of MDR by inhibiting the expression and function of P-glycoprotein	[52]
	Skin	Female SENCAR mice	Inhibition of inflammation by blocking TPA-induced COX-2, and activation of iNOS, NF-κB, IKK, and MAPK	[38]
	Liver	Hepatocellular carcinoma (Hep3B, HepG2)	Enhancement of TRAIL-induced apoptosis through DR5-mediated induction of ROS-dependent ER-stress	[53]
	Osteoclast	Mouse macrophage (RAW 264.7)	Abrogation of RANKL-induced NF-κB activation	[54]
	Breast	Gut-derived adenocarcinoma (Bic-1)	Suppression of bile acid-induced and constitutive CdX2 expression	[55]
		Doxorubicin-resistant breast carcinoma (MCF-7/DOX)	Reversal of MDR and sensitization of cancer cells to doxorubicin by inhibiting P-glycoprotein function and expression	[56]
Z-GS E-GS		Mammary epithelial cells (MCF-10A)	Induction of HO-1 expression through Akt-mediated activation of Nrf2	[40]
Z-GS E-GS	Blood	Monocytic leukemia (U937) Promyelocytic leukemia (HL60) Bone marrow or peripheral blood samples	Induction of apoptosis and differentiation	[57]
Z-GS GS	Bone marrow	Multiple myeloma (U266, MM1)	Induction of apoptosis via inhibition of STAT3 phosphorylation	[58]
Z-GS	Head and neck	HNSC (SCC4 and HSC2)	Suppression of proliferation and induction of apoptosis by blocking the ST-induced PI3K/Akt pathway	[59]
			Inhibition of ST-induced and nicotine-induced activation of NF-κB and STAT3	[60]
			Induction of apoptosis via 14-3-3ζ-mediated induction of intrinsic and extrinsic pathways	[51]
E-GS Z-GS GL		HNSC (PCI-37a, UM-22b, 1483) SV40-immortalized esophageal epithelial (Het-1a) cells HNSC mouse xenografts	Inhibition of STAT-3 signaling	[61]

Z-GS	Lung	Non-small cell lung carcinoma (H1299) Lung epithelial cell carcinoma (A549) T lymphocytic leukemia (Jurkat) Myeloid leukemia (KBM-5)	Abrogation of TNF-, IL-1β, TPA-, H_2O_2-, cigarette smoke condensate-, and okadaic acid-induced activation of IKK-NF-κB signaling	[23]
Z-GS	Blood Bone marrow Skin Lung, Head and neck Breast, Ovarian	Chronic myelogenous leukemia (KBM-5, K562) Monocytic leukemia (U937) T lymphocytic leukemia (Jurkat) Multiple myeloma (U266, MM1) Melanoma (A375, WM35) Non-small cell lung carcinoma (H1299) Normal bronchial epithelial (BEAS-2B) Head and neck squamous carcinoma (HN5, SCC4, FADU) Breast cancer (MCF-7) Ovarian cancer (HEY8, SKOV3)	Inhibition of proliferation, induction of S-phase arrest and apoptosis through suppression of Akt and JNK signaling	[62]
GL	Prostate	Prostate carcinoma (LNCaP, LNCaP-C81)	ROS-dependent apoptosis via c-JNK activation	[63]
Z-GS		Prostate carcinoma (DU145)	Suppression of VEGF and VEGF-R2, and inactivation of Akt leading to inhibition of angiogenesis.	[9]
		Prostate carcinoma (PC-3, LNCaP, DU145)	Generation of ROS which induces JNK-mediated apoptosis	[49]
		Prostate carcinoma (PC-3) Prostate carcinoma (PrEC) PC-3/neo and PC-3/Bcl-2 MEF from Bax or Bak single knock-out mice MEF from Bax–Bak double knock-out mice	Caspase-dependent apoptosis mediated by Bax and Bak	[64]
Z-GS	Colon	Colon adenocarcinoma (HT-29) HT-29 xenograft Normal intestinal (IEC-18) cells	Induction of apoptosis via JNK-mediated inhibition of caspase-3/ caspase-8 and Fas activation	[65]
		Colon adenocarcinoma (HT-29) HT-29 xenograft	Abrogation of angiogenesis and metastasis by inhibiting STAT3, MMP-2 and MMP-9 activities.	[50]

caspase-dependent apoptosis in human prostate cancer cells was found to be mediated by Bax and Bak [64].

Bcl-2 family proteins play an essential role in the mitochondrial apoptotic pathway [68]. They are divided into antiapoptotic proteins, such as Bcl-2, Bcl-X_L, and Mcl-l, and proapoptotic proteins including Bax, Bek, and Bad [67]. GS and GL induce apoptosis in different cells by increasing the expression of proapoptotic proteins, while decreasing the levels of antiapoptotic proteins (e.g., IAP1, XIAP, Bfl-1/A1, Bcl-2, cFLIP, Survivin, etc.) [51, 59, 63, 65].

FXR is involved in cell migration and invasion, and GS inhibits cancer cell metastasis by acting as a potent FXR antagonist [34]. In addition, GS exerts antimetastatic and antiangiogenic effects by inhibiting the activation of NF-κB and signal transducer and activator 3 (STAT3) and the expression of vascular endothelial growth factor (VEGF) [9, 34, 50, 60].

3.1 Effect on Colon Cancer

GS possesses potent anti-inflammatory properties as evidenced by its ability to inhibit NF-κB activation [27], making it a good preventive/therapeutic agent against inflammation-associated cancer [23]. Z-GS exerts a potent antitumor activity in human colon cancer (HT-29) cells by inducing apoptosis and inhibiting angiogenesis and metastasis [50, 65]. STAT3 plays a pivotal role in angiogenesis via modulation of aryl hydrocarbon receptor nuclear translocator (ARNT)-mediated VEGF expression, which is responsible for endothelial proliferation and degradation of extracellular matrix [50, 72]. Z-GS inhibited STAT3 activation and the subsequent ARNT-induced VEGF expression in HT-29 cells [50]. In addition, the antimetastatic properties of Z-GS were confirmed by its suppression of the capillary tube formation and migration of human umbilical vein endothelial cells (HUVECs), and also inhibition of matrix metalloproteinase (MMP)-2 and MMP-9 activities in HT-29 cells [50].

An and colleagues have shown that Z-GS induces apoptosis in HT-29 cells [65]. Interestingly, Z-GS failed to induce apoptosis in the normal intestinal (IEC-18) cells under the same treatment conditions [65]. Activation of the mitochondrial-apoptotic pathway by Z-GS was characterized by the enhancement of caspase-3, and caspase-8 activities, elevated levels of cleaved caspases, and decreased levels of the counterpart procaspases [65]. Z-GS attenuated the expression of the inhibitor of apoptosis proteins (IAP) including cIAP-1 and cIAP-2, suppressed Bcl-2 protein expression, and increased the levels of the truncated-Bid, while it had no effect on Bak expression in HT-29 cells [65]. In addition, Z-GS stimulated the extrinsic apoptosis pathway characterized by caspase-8 activation [65]. This effect was mediated by Fas, which activates Fas receptor-associated death domain protein responsible for the activation of caspase-8 [65]. Z-GS treatment induced Fas expression in H-29 cells by phosphorylating the upstream c-Jun-N-terminal kinase (JNK) and subsequently c-Jun [65]. The anticarcinogenic effect of Z-GS was also investigated in vivo using a HT-29 xenograft model, and it was found that

Z-GS-mediated cell growth suppression was associated with the down-regulation of Bcl-2 protein expression [65].

3.2 Effect on Breast Cancer

GS inhibited cells proliferation and induced apoptosis in human mammary carcinoma (MCF-7) cells and doxorubicin-resistant breast cancer cells [62], while it augmented antioxidative potential in immortalized normal human mammary epithelial (MCF-10A) cells [40]. Z-GS significantly suppressed the activation of Akt and the subsequent phosphorylation of glycogen synthase kinase 3-bata (GSK3β) in some cancerous cells [62].

Bone is the most common site to which breast cancer cells metastasize. The bile acid salt sodium deoxycholate (DC) released from osteoblast-like MG63 cells or bone tissue promotes cell survival and induces the migration of metastatic human breast cancer MDA-MB-231 cells [73]. DC increases the expression and nuclear translocation of FXR in MDA-MB-231 cells, thereby mediating the migration of breast cancer cells. The FXR antagonist Z-GS prevents the migration of MDA-MB-231 cells and induces apoptosis [73]. In another study, Z-GS was found to obliterate the antiapoptotic effect of DC in murine mammary carcinoma 4T1 cells, again by antagonizing FXR [74].

3.3 Effect on Prostate Cancer

Z-GS induced apoptosis in human prostate cancer (PC-3) cells but not in normal human epithelial prostate (PrEC) cells [64]. Z-GS-induced apoptosis in different human prostate cancer lines (PC-3, LNCaP, and DU145) was related to ROS-dependent activation of JNK [49]. Z-GS generated ROS in human prostate cancer cells but not in human prostate epithelial cells (PrEC) that were resistant to Z-GS-induced JNK activation [49]. Activation of JNK induced the expression of proapoptotic Bcl-2 family member proteins, such as Bax and Bak in PC-3 cells [64]. SV40-immortalized mouse embryofibroblasts (MEFs) from Bax–Bak double knock-out mice were resistant to Z-GS-induced apoptosis compared with wild-type cells [64]. SV40-immortalized MEFs derived from Bax–Bak double knock-out mice were also more resistant to Z-GS-induced apoptosis than that observed in Bax or Bak single knock-out MEFs [64]. The proapoptotic activity of Z-GS was caspase-dependent as demonstrated by the cleavage of caspase-8 and caspase-9 [64].

Although PC-3 cells are androgen-independent and lack functional p53, whereas LNCaP cells are androgen-dependent and express functional p53 [75], Z-GS treatment increased Thr^{183}/Tyr^{185} phosphorylation of JNK1/2 and Tyr^{182} p38 MAPK in both cell lines without affecting the total protein levels of these two kinases [49]. However, phosphorylation of extracellular signal-regulated kinase (ERK)1/2 was distinct between those cell lines, as it decreased in PC-3 cells, while both phosphorylated and total protein levels of ERK1/2 increased in LNCaP cells [49]. Z-GS-mediated phosphorylation of JNK and the subsequent DNA fragmentation were attenuated in the presence of SP600125 (pharmacological inhibitor of JNK) in both PC-3 and LNCaP cells, but independent of ERK1/2 and p38 MAPK [49].

Z-GS and E-GS equivalently inhibit capillary-like tube formation in HUVEC, indicative of their antiangiogenic potential [9]. In addition, Z-GS inhibited migration of HUVEC and human prostate cancer (DU145) cells through suppression of Akt phosphorylation [9]. This Z-GS-mediated suppression of cell migration was more pronounced in the presence of the Akt1/2 inhibitor [1,3-dihydro-1-1(1-((4-(6-phenyl-1H-imidazo(4,5-g)quinoxalin-7-yl)phenyl)methyl)-4-piperidinyl)-2H-benzimidazol-2 one] [9]. In addition, migration of DU145 cells transfected with constitutively active Akt was not affected by Z-GS compared to cells transfected with a control vector [9]. Z-GS also suppressed the secretion of the proangiogenic growth factors such as VEGF, granulocyte-colony stimulating factor, IL-17, and MMP-2 in both HUVEC and DU145 cells [76]. VEGF induces cancer cell survival by interacting with one of its receptors VEGF-R1, VEGF-R2, or VEGF-R3, but VEGF-R2 is mainly involved in the regulation of angiogenesis [76]. Z-GS suppressed the VEGF-R2 expression in HUVEC and DU145 cells [9]. Z-GS-mediated inhibition of angiogenesis was also determined in vivo using a DU145-Matrigel plug assay in male nude mice [9]. The tumor volume and the net weight were markedly reduced in mice upon administration of Z-GS five times per week compared with vehicle-treated control mice [9]. Immunohistochemical data in sections taken from DU145-Matrigel plugs showed that Z-GS-treated mice expressed VEGF-R2, factor VIII, and CD31 to a lesser extent than vehicle-treated control mice [9].

3.4 Effect on Head and Neck Cancer

GS exerts antiproliferative and proapoptotic effects in head and neck squamous carcinoma (HNSC) cells (SCC4 and HSC2) [51, 59–62]. Z-GS inhibited smokeless tobacco (ST)-induced and nicotine-induced phosphorylation of Akt (at serine 473 and threonine 308), Bad, Bax, and GSK3β in SCC4 and HSC2 cell lines without affecting the total protein expression levels [59]. The Z-GS-mediated inhibition of Akt, Bad, and Bax was equivalent to that achieved with LY294002, the PI3K/Akt specific inhibitor [51, 59]. In addition, Z-GS inhibited proliferation and induced apoptosis in different human HNSC (HN5, SCC4, and FADU) cell lines through inhibition of Akt signaling [62]. Notably, Z-GS inhibited the activation of the upstream kinases, PI3K and PDK1 and the downstream proteins, RAF, GSK3β, and S6, involved in the Akt pathway [59].

Under survival conditions, the proapoptotic proteins are phosphorylated and localized predominantly in the cytoplasm, but during apoptosis, mitochondrial localization of dephosphorylated Bax and Bad is essential to induce the release of cytochrome c [77]. Z-GS inhibited the ST-induced and nicotine-induced phosphorylation of Bad (at serine 136) and Bax (at serine 184), and their cytoplasmic retention by restoring their mitochondrial relocalization without affecting the total levels of both proteins [59].

ST and nicotine pretreatment induced phosphorylation of Bax and Bad and increased their association with 14-3-3ζ protein, leading to the sequestration of Bad in the cytoplasm [59, 78, 79]. Z-GS treatment induced the PP2A phosphatase-mediated dephosphorylation of pBad (serine 136) and its dissociation from 14-3-3ζ to undergo mitochondrial relocalization in SCC4 cells [51]. In the mitochondrial outer membrane, Bad facilitates Bax release that increases the permeability of the mitochondrial membrane to release cytochrome c and subsequently activate caspase-3 and caspase-9 [51]. Akt and Bax were primarily co-localized in the cytoplasm of SCC4 cells, suggesting that phosphorylation and concurrent inactivation of Bax are mediated via Akt-induced phosphorylation [59].

ST and nicotine treatment induced the phosphorylation of STAT3, activation of NF-κB, and VEGF expression [60]. Z-GS induces apoptosis by inducing MKK4-mediated activation of JNK, which leads to inhibition of STAT3 [61, 62]. Interestingly, Z-GS, but not E-GS, reduced the levels of total and phosphorylated STAT3 in HNSC cells [61]. Z-GS-mediated activation of JNK and suppression of Akt seem to be NF-κB-dependent and related to each other [60, 62]. GS treatment suppressed ST-induced and nicotine-induced NF-κB activation and COX-2 expression in HNSC (SCC4) cells by inhibiting IκBα phosphorylation and degradation [60].

3.5 Effect on Gastrointestinal Cancer

Gastrointestinal carcinogenesis is related to the incomplete differentiation and development of mucosal cells, or occurs as a result of chronic inflammation [80]. In humans there are three CdX (Caudal-related homeobox) proteins: CdX1, CdX2, and CdX4 [81]. Only CdX1 and CdX2 are important in intestinal epithelial development [81]. It is speculated that CdX proteins, particularly Cdx2, may act as a tumor suppressor because $Cdx2^{+/-}$ heterozygous mice develop more polyps than do wild type mice [82–84]. However, other studies have shown that CdX2 possesses an oncogenic potential in intestinal and colon cancer cells [85]. NF-κB is one of the major transcription factors involved in the regulation of CdX2 expression [81], and its activation is suppressed by GS treatment in different cell lines [23, 24, 35, 38, 60]. GS suppresses both bile acid (CDCA and DCA)-induced and constitutive expression of CdX2 in gut-derived adenocarcinoma (Bic-1) cells [55]. While bile acids induced CdX2 expression via NF-κB in esophageal cells [86], GS reduced CdX2 expression at low concentrations without affecting the cell viability or bile acid-induced NF-κB activation, indicating that GS-mediated suppression of CdX2 is not likely to be mediated through NF-κB inactivation [55].

3.6 Effect on Liver Cancer

Hepatocellular carcinoma (HCC) is a tumor of the liver and it can be treated with chemotherapeutic agents such as TRAIL [87, 88]. However, many tumors possess TRAIL resistance that can be reversed by combination therapy [89–92]. Several agents sensitize HCC cells to TRAIL-induced apoptosis via a STAT3-mediated DR5-dependent mechanism [93, 94] or through NF-κB-dependent inhibition of COX-2 expression [95].

GS enhanced TRAIL-induced apoptosis via eIF2α- and CHOP-mediated induction of DR5 expression in HCC (Hep3B, HepG2) cell lines [53]. GS also induced endoplasmic reticulum (ER) stress which accompanies upregulation of ER stress proteins including IRE, JNK, BiP, protein kinase-like endoplasmic reticulum kinase (PERK), eIF2α, and activating transcription factor-4 [53]. GS treatment generated ROS in HCC, which accounts for the activation of PERK and eIF2α, upregulation of CHOP/DR5 expression, and the cleavage of procaspase-3 and PARP [53]. Notably, co-treatment of HCC with GS and TRAIL augmented the apoptosis via ROS-dependent induction of ER-stress [53].

3.7 Effect on Lung Cancer

It has been reported that Z-GS significantly suppresses proliferation and induces apoptosis in nonsmall cell lung (H1299) and other cell types via Akt-dependent and JNK-dependent mechanisms [62]. Z-GS inhibited the survival pathway and induced apoptosis by inhibiting Akt phosphorylation at Ser 473 and Thr 308 residues [62].

4 Signaling Molecules Modulated by GS

Although the mechanism of the anticarcinogenic action of GS is unclear yet, many studies reported its ability to modulate distinct signaling pathways. Some of the identified molecular targets of GS are listed (Table 2).

4.1 MAPK

Multiple lines of evidence support the notion that abnormal regulation of MAPKs is implicated in inflammation-associated carcinogenesis [96]. In mammals, at least six groups of MAPKs have been identified so far: ERK1/2, ERK3/4, ERK5, ERK7/8, JNK1/2/3, and p38 isoforms [97]. ERK regulates cellular proliferation, angiogenesis, and differentiation, depending on the cell type [98, 99], while JNK signaling regulates cellular proliferation and transformation [97]. However, p38 is a stress-responsive or inflammatory-responsive kinase that is involved in the regulation of cellular apoptosis, growth, cell cycle progression, and differentiation [97].

Z-GS-induced apoptosis in human prostate cancer (PC-3 or LNCaP) cells was not mediated by ERK1/2 or p38, but by JNK [49]. Interestingly, Z-GS has weak ability to activate JNK in normal prostate (PrEC) cells [49, 63]. Topical application of GS on mouse skin abrogated 12-O-tetradecanoylphorbol-13-acetate (TPA)-induced phosphorylation of MAPKs including ERK1/2, JNK1/2, and p38 [38]. In addition, GS inhibited the TNF-α-induced activation of JNK and p38, but not ERK in vascular cells [22].

4.2 PI3K/Akt

The activation of the PI3K pathway is essential for cell survival, growth and proliferation [100–102]. Thus, one of the common mechanisms underlying inhibition of cell proliferation and induction of apoptosis in cancer cells by antineoplastic agents involves blockade of abnormally amplified PI3K/Akt signaling [103–105]. Z-GS exerts antiproliferative and proapoptotic effects in many cancer cell lines [52, 59, 62] including leukemia, HNSC, multiple myeloma, lung carcinoma, melanoma, breast carcinoma, and ovarian carcinoma through suppression of Akt phosphorylation. Z-GS-induced apoptosis via PI3K/Akt inhibition was associated with the activation of JNK signaling [62]. In addition, the antiangiogenic activity of Z-GS in prostate cancer is linked to its suppression of Akt signaling [9]. GS-induced Akt inactivation was associated with down-regulation of VEGF and its receptor VEGF-R2 [9].

4.3 Nrf2

NF-E2-related factor2 (Nrf2) is a master regulator in activating the antioxidant response element (ARE) that is located in the promoter regions of majority of antioxidant enzymes and other cytoprotective proteins including HO-1 [106]. GS induced Nrf2 activation with concurrent expression of HO-1 in human mammary epithelial (MCF-10A) cells [40]. E-GS-induced Nrf2 activation appears to be mediated through PTEN inactivation and subsequent activation of PI3K–Akt signaling. PTEN is a negative regulator of the PI3K/Akt signaling. PTEN has essential adjacent cysteine residues (Cys 71 and Cys 124) in its catalytic domain. The oxidation of these two adjacent cysteine residues renders PTEN catalytically inactive [40]. A preliminary study in our laboratory has shown that E-GS treatment generates moderate amounts of ROS in MCF-10A cells, but ROS is unlikely to oxidize PTEN (Fig. 2). It is speculated that E-GS may rather covalently modify a critical cysteine residue of PTEN, thereby activating the PI3K/Akt axis and

Table 2 Molecular targets of GS in different types of cancer

Form	Model used	Molecular targets affected	References
Z-GS	HNSC (SCC4, HSC2)	↑ p21$^{WAF1/CIP1}$, p27, cyclin D1, Bax/Bcl-2, cytosolic cytochrome c, caspase-3, caspase-9, caspase-8, cleavage of PARP, Fas/CD95, and tBid	[51]
		↓ xIAP, cyclin D1, Xiap, Mcl-1, c-Myc, pBad, and Survivin	
		↓ pAkt, pPI3K, pPDK1, pRaf, GSK3β, pS6, pBax, and pBad	[59]
		↓ pp65 (NF-κB), pIκBα, COX-2, IL-6, pSTAT3, and VEGF ↑ IκBα	[60]
	Prostate cancer (PC-3)	↑ Bax, Bak, Bcl-xL and Bcl-2 (initially), DNA fragmentation, caspase-8, caspase-9, and caspase-3 ↓ Bcl-xL and Bcl-2 (delayed), NF-κB	[64]
	Prostate cancer (PC-3, LNCaP)	↑ JNK, p38 MAPK, DNA fragmentation, and ROS generation	[49]
GL	Prostate cancer (LNCaP, LNCaP-C81)	↑ DNA fragmentation, ROS, Bax, Bak, cleavage of PARP, Bcl-2, JNK, c-Jun, Akt, and pAkt	[9]
GS	Esophageal adenocarcinoma	↑ Caspase-3 activity	[33]
Z-GS	Chronic myelogenous leukemia (KBM-5) Monocytic leukemia (U937) Melanoma (A375, WM35) Lymphoblastic leukemia (Jurkat) Chronic myelogenous leukemia (K562) Non-small cell lung carcinoma (H1299) Bronchial epithelial cells (BEAS-2B) Multiple myeloma (U266, MM1) HNSC (HN5, SCC4, FADU) Breast cancer (MCF-7) Ovarian cancer (HEY8, SKOV3)	↑ Caspase-8, caspase-9, caspase-3, bid and cPARP, cytochrome c release, and JNK ↓ Cyclin D1, cdc2, Akt activation, antiapoptotic gene products (Bfl-1, xIAP, cFLIP, Bcl-XL, Bcl-2, and Survivin), c-Myc, COX-2, IL-1β, IL-6, TNF, pAkt, pPDK1, pPI3K, GSK3β, and JNK	[62]
	Colon cancer (HT-29) HT-29 xenograft	↑ Caspase-3, caspase-8, truncated Bid, Fas, p-JNK, and p-c-Jun ↓ cIAP-1, cIAP-2, Bcl-2, caspase-9, and tBid	[65]
		↓ STAT3, ARNT, VEGF, MMP-2, and MMP-9	[50]

(continued)

Table 2 (continued)

Form	Model used	Molecular targets affected	References
E-GS Z-GS	Prostate cancer (DU145)	↓ VEGF, FGF, G-CSF, MMP-2, and IL-17, VEGF-R2, and pAkt	[9]
E-GS Z-GS	Mammary epithelial (MCF-10A) cells	↑ HO-1, Nrf2, ROS, and pAkt ↓ PTEN	[40]
Z-GS E-GS	Monocytic leukemia (U937) Promyelocytic leukemia (HL60) Bone marrow or peripheral blood	↑ ROS and HO-1 ↓ pERK	[57]
E-GS Z-GS GL	HNSC (PCI-37a, UM-22b, 1483) SV40-immortalized esophageal epithelial (Het-1a) cells	↓ pSTAT3 and STAT3	[61]
GS	Hepatocellular carcinoma (Hep3B and HepG2)	↑ ER-stress (IRE, JNK, BiP, PERK, eIF2α, ATF4), CHOP, DR5, ROS, caspase-8, caspase-9, caspase-3, Bid and PARP cleavage, cytochrome c release, and Bad ↓ cIAP-1 and XIAP	[53]
GS	Doxorubicin-resistant myelogenous leukemia (K562/DOX)	↓ P-glycoprotein	[52]
GS	Doxorubicin-resistant breast carcinoma (MCF-7/DOX)	↓ P-glycoprotein	[56]
GS	Female SENCAR mouse skin	↓ COX-2, iNOS, pMAPK (ERK1/2, P38, JNK1/2), NF-κB, pIKKα, pIκBα, and ornithine decarboxylase	[38]
GS	Gut-derived adenocarcinoma (Bic-1)	↓ CdX2 and NF-κB	[55]
GS	Murine macrophage (RAW 264.7)	↓ Activation of NF-κB and IKK	[54]
Z-GS	Nonsmall cell lung carcinoma (H1299) Lung epithelial cell carcinoma (A549) T cell leukemia (Jurkat) Myeloid leukemia (KBM-5)	↓ Activation of NF-κB and IKK, pp65, pIκBα, COX-2, cIAP1, xIAP, Bfl-1, Bcl-2, TRAF1, Cflip, Survivin	[23]
Z-GS GS	Multiple myeloma (U266, MM1)	↓ pSTAT3 ↓ pJAK2, p-c-Src, SHP-1, STAT3, Bcl-2, Mcl-1, cyclin D1, VEGF, and Bcl-xl ↑ Caspase-3 and PARP cleavage	[58]

facilitating nuclear translocation of Nrf2, most likely through phosphorylation of this transcription factor at serine and/or threonine residues [40].

4.4 STATs

STAT3 is a transcription factor that forms a complex with hypoxia-inducible factor (HIF)-1 to activate VEGF gene expression [107]. Inhibition of angiogenesis is achieved by suppressing STAT3-mediated activation of VEGF expression in hypoxic PC-3 cells [108]. Moreover, GS inhibited angiogenesis and metastasis in colon cancer cells by preventing STAT3 and ARNT from binding to VEGF promoter [50]. In addition, Z-isomer but not E-isomer of GS inhibited the constitutive and IL-6-induced phosphorylation of STAT3 via inhibition of Janus kinase (JAK)2 phosphorylation in human multiple myeloma cells [58]. Z-GS suppressed the expression of the antiapoptotic proteins (Bcl-2, Bcl-xl, and Mcl-1), the proliferative protein (cyclin D1), and the angiogenic protein (VEGF) [58]. Notably, Z-GS suppressed the phosphorylation of STAT3 without affecting the total STAT3 levels in human multiple myeloma (U266) cells [58], while it decreased both the total and tyrosine phosphorylated STAT3 in HNSC [61] and HT-29 cell lines [50].

4.5 NF-κB

NF-κB is linked strongly to inflammation and cancer, and NF-κB suppression is considered one of the rational strategies in treating and preventing carcinogenesis [109]. GS inhibits NF-κB activation induced by many inflammatory signals in different cell lines such as RANKL in mouse macrophage (RAW 264.7) cells [54], LPS or IL-1β in human colon cancer (Caco-2) cells and rat nontransformed small (IEC-18) cells [27], IL-1β in fibroblast-like synoviocytes (FLS) [24], and IL-1β and IFN-γ in rat pancreatic β-cells (RINm5F) [35]. GS suppresses LPS-induced and TNF-induced COX-2 expression by inhibiting NF-κB binding to the prompter regions of the inflammatory genes [23, 41]. This is mediated by inhibition of IKK and the subsequent phosphorylation and degradation of IκBα, resulting in suppression of inflammation in normal cells [41] and induction of apoptosis in cancer cells [23]. GS suppressed TNF-induced NF-κB-activation in human lung epithelial cell carcinoma (A549), human myelogenous leukemia [23], and mouse peritoneal macrophage RAW 264.7 cells [54]. In addition, GS suppresses the constitutive and nicotine-induced NF-κB activation in multiple myeloma and HNSC cells [23]. GS, when topically applied onto SENCAR mouse skin, reversed the TPA-induced expression of COX-2 and inducible nitric oxide synthase (iNOS) [38].

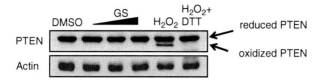

Fig. 2 Identification of reduced and oxidized forms of PTEN by immunoblot analysis. MCF-10A cells were treated with GS (5, 10, or 25 μM), DTT (0.5 mM) or H_2O_2 (5 mM) for 1 h. Cells were then washed with 1× ice-cold phosphate-buffered saline, and lysed with 10% trichloroacetic acid. The cell suspensions were vortexed and centrifuged at 12,000 × g for 5 min. The pellets were washed with acetone and then solubilized in 100 mM Tris–HCl (pH 6.8) containing 2% SDS and 40 mM NEM. The samples (50 μg) were loaded on SDS-PAGE under nonreducing conditions to separate oxidized and reduced forms of PTEN. While addition of the positive reference oxidant H_2O_2 produced the oxidized PTEN which was abolished by the reducing agent dithiothreitol (DTT), GS treatment failed to induce oxidation of PTEN

4.6 14-3-3ζ

14-3-3 is a family of eukaryotic regulatory proteins involved in regulation of cell survival and death [110]. 14-3-3 proteins are capable of binding to distinct phosphorylated ligands including Bad and Bax and sequester them in the cytoplasm leading to loss of their proapoptotic function [110]. Among these, 14-3-3ζ is involved in head and neck cancer progression [111] and in insulin regulation [112, 113].

Under normal conditions, pBad (Ser 136) is sequestered in the cytoplasm as a complex with 14-3-3ζ, thereby inducing proliferation and inhibiting apoptosis [110]. Z-GS-induced apoptosis was mediated via the intrinsic mitochondrial pathway and characterized by reduced expression of antiapoptotic proteins (e.g., xIAP, Mcl1, c-Myc, and Survivin) and dephosphorylation of Bad and its dissociation from 14-3-3ζ, resulting in the release of cytochrome c from mitochondria [51]. In addition, Z-GS induced apoptosis in SCC4 that was medicated by suppression of the cell cycle regulatory protein cyclin D1 and induction of the expression of the cyclin-dependent kinase inhibitor p21$^{WAF1/CIP1}$ and p27 [51].

4.7 P-Glycoprotein

One of the serious problems in cancer therapy is multidrug resistance (MDR) [114]. MDR enables cancer cells to resist structurally unrelated chemotherapeutic agents [115]. One of the main mechanisms of MDR development involves the elevated influx of drugs via energy-dependent transporters as a consequence of increased expression of the P-glycoprotein that belongs to ATP-binding cassette (ABC) transporter superfamily of membrane transport proteins [116]. Some phytochemicals such as curcumin [117], piperine, capsaicin, and sesamin reverse MDR and sensitize cancer cells to chemotherapeutic drugs [118]. GS was reported to reverse MDR in human doxorubicin-resistant cell lines, such as breast carcinoma (MCF-7/DOX) [56] and myelogenous leukemia (K562/DOX) cells [52], and also gleevac-resistant (K562) and dexamethasone-resistant multiple mylenoma (MM1) [62] cells. The GS-induced apoptosis in anticancer drug-resistant cells is likely to be mediated by inhibition of the expression and function of the P-glycoprotein [52].

4.8 iNOS

Constitutive activation of NF-κB was found in many different types of cancer [119]. NO generation induced by toxins is mediated by activation of NF-κB, and it causes cell toxicity and damage that leads to chronic inflammation and carcinogenesis [120]. Z-GS is a potent anti-inflammatory agent that inhibits the expression or production of inflammatory mediators including MMP-2, iNOS, prostaglandin E_2, and COX-2 through inhibition of NF-κB phosphorylation [26, 35]. Z-GS exerts protective effects against various inflammatory and cytotoxic stimuli by targeting of iNOS as exemplified by its ability to reverse LPS-induced inflammation in Lewis rats [26], and cytokine (IL-1β and IFN-γ)-induced toxicity in pancreatic β-cells and rat insulinoma (RIN) cells [35]. Thus, GS treatment gives β-cells the acquired protection against induction of iNOS expression and maintains an appropriate function of insulin in the case of diabetes [35]. The anti-inflammatory and antidiabetic properties of GS, mediated via inhibition of iNOS, make it a good choice for chemoprevention as both inflammation and diabetes are linked directly or indirectly to carcinogenesis [121].

4.9 Growth Factors

Growth factors are the subject of many studies due to their roles in cell proliferation and/or differentiation [122]. Some of the growth factors involved in carcinogenesis include fibroblast growth factor [122, 123], TGF [124], insulin-like growth factor [125], IFN-γ [126–128], IL-1 [129], platelet-derived growth factor [130], and TNF [131]. GS exerts antiproliferative, antiangiogenic, and antimetastatic agent effects by inhibiting cellular signaling mediated by one of the aforementioned growth factors [50].

5 Conclusion

Numerous bioactive substances have been isolated from a vast variety of medicinal plants, and many of them possess strong anticancer activities. GS, a biologically active ingredient of guggul plants, has substantial anti-inflammatory and

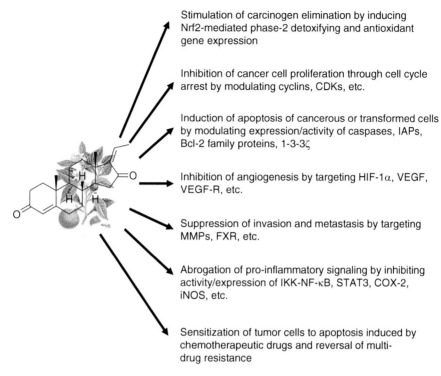

Fig. 3 Multiple mechanisms underlying cancer chemopreventive/therapeutic effects of GS. For simplification, only the structure of Z-isoform of GS is shown

antioxidant properties, which contribute to its cancer thermopreventive and therapeutic potential. GS exerts anticarcinogenic effects by modulating distinct signaling molecules involved in carcinogen detoxification, cell proliferation/cell cycle control, apoptosis, angiogenesis, metastasis, inflammation, MDR, etc. (Fig. 3). These include transcription factors such as Nrf2, NF-κB, STAT3, and AP-1, kinases, such as MAPKs, PI3K/Akt, and 14-3-3 zeta, P-glycoprotein, etc. GS is an electrophilic compound due to its α,β-unsaturated carbonyl functional group, and hence acts as a Michael acceptor. Many of the aforementioned transcription factors and their regulators contain cysteine and other nucleophilic amino acids, such as lysine and histidine that often function as a redox sensor. Direct modification of these signaling molecules by GS represents an important mechanism underlying cancer chemopreventive and therapeutic functions of GS. Further studies will be necessary to identify the bona fide amino acid(s) modified by GS in its modulating the activities of redox-sensitive signaling molecules.

Acknowledgments This work was supported by the grant (No. 2012-0001184) for the Global Core Research Center (GCRC) from the National Research Foundation, Ministry of Education, Science and Technology, Republic of Korea

References

1. Jemal A, Bray F, Center MM, Ferlay J, Ward E, Forman D (2011) Global cancer statistics. CA Cancer J Clin 61:69–90
2. Pitot HC (1993) Multistage carcinogenesis – genetic and epigenetic mechanisms in relation to cancer prevention. Cancer Detect Prev 17:567–573
3. De Flora S, La Maestra S, Micale RT (2011) [Mechanistic issues and prevention strategies targeting occupational carcinogenesis]. G Ital Med Lav Ergon 33:294–299
4. Tsao AS, Kim ES, Hong WK (2004) Chemoprevention of cancer. CA Cancer J Clin 54:150–180
5. Sporn MB (1976) Approaches to prevention of epithelial cancer during the preneoplastic period. Cancer Res 36:2699–2702
6. Surh YJ (2003) Cancer chemoprevention with dietary phytochemicals. Nat Rev Cancer 3:768–780
7. Aggarwal BB, Takada Y, Oommen OV (2004) From chemoprevention to chemotherapy: common targets and common goals. Expert Opin Investig Drugs 13:1327–1338
8. Mathur M, Ramawat KG (2007) Guggulsterone production in cell suspension cultures of the guggul tree, Commiphora wightii, grown in shake-flasks and bioreactors. Biotechnol Lett 29:979–982
9. Xiao D, Singh SV (2008) z-Guggulsterone, a constituent of Ayurvedic medicinal plant Commiphora mukul, inhibits angiogenesis in vitro and in vivo. Mol Cancer Ther 7:171–180
10. Hanus LO, Rezanka T, Dembitsky VM, Moussaieff A (2005) Myrrh – Commiphora chemistry. Biomed Pap Med Fac Univ Palacky Olomouc Czech Repub 149:3–27
11. Deng R (2007) Therapeutic effects of guggul and its constituent guggulsterone: cardiovascular benefits. Cardiovasc Drug Rev 25:375–390
12. Shishodia S, Harikumar KB, Dass S, Ramawat KG, Aggarwal BB (2008) The guggul for chronic diseases: ancient medicine, modern targets. Anticancer Res 28:3647–3664
13. Klaunig JE, Wang Z, Pu X, Zhou S (2011) Oxidative stress and oxidative damage in chemical carcinogenesis. Toxicol Appl Pharmacol 254:86–99
14. Pan JS, Hong MZ, Ren JL (2009) Reactive oxygen species: a double-edged sword in oncogenesis. World J Gastroenterol 15:1702–1707
15. Morrison WB (2012) Inflammation and cancer: a comparative view. J Vet Intern Med 26:18–31
16. Shacter E, Weitzman SA (2002) Chronic inflammation and cancer. Oncology (Williston Park) 16:217–226, 229; discussion 230–212
17. Saxena G, Singh SP, Pal R, Singh S, Pratap R, Nath C (2007) Gugulipid, an extract of Commiphora whighitii with lipid-lowering properties, has protective effects against streptozotocin-induced memory deficits in mice. Pharmacol Biochem Behav 86:797–805
18. Wang X, Greilberger J, Ledinski G, Kager G, Paigen B, Jurgens G (2004) The hypolipidemic natural product Commiphora mukul and its component guggulsterone inhibit oxidative modification of LDL. Atherosclerosis 172:239–246
19. Singh K, Chander R, Kapoor NK (1997) Guggulsterone, a potent hypolipidaemic, prevents oxidation of low density lipoprotein. Phytother Res 11:291–294
20. Xu HB, Li L, Liu GQ (2008) Protection against hydrogen peroxide-induced cytotoxicity in PC12 cells by guggulsterone. Yao Xue Xue Bao 43:1190–1197
21. Mencarelli A, Renga B, Palladino G, Distrutti E, Fiorucci S (2009) The plant sterol guggulsterone attenuates inflammation and immune dysfunction in murine models of inflammatory bowel disease. Biochem Pharmacol 78:1214–1223
22. Gebhard C, Stampfli SF, Gebhard CE, Akhmedov A, Breitenstein A, Camici GG, Holy EW, Luscher TF, Tanner FC (2009) Guggulsterone, an anti-inflammatory phytosterol, inhibits tissue factor and arterial thrombosis. Basic Res Cardiol 104:285–294

23. Shishodia S, Aggarwal BB (2004) Guggulsterone inhibits NF-kappaB and IkappaBalpha kinase activation, suppresses expression of anti-apoptotic gene products, and enhances apoptosis. J Biol Chem 279:47148–47158
24. Lee YR, Lee JH, Noh EM, Kim EK, Song MY, Jung WS, Park SJ, Kim JS, Park JW, Kwon KB et al (2008) Guggulsterone blocks IL-1beta-mediated inflammatory responses by suppressing NF-kappaB activation in fibroblast-like synoviocytes. Life Sci 82:1203–1209
25. Khanna D, Sethi G, Ahn KS, Pandey MK, Kunnumakkara AB, Sung B, Aggarwal A, Aggarwal BB (2007) Natural products as a gold mine for arthritis treatment. Curr Opin Pharmacol 7:344–351
26. Kalariya NM, Shoeb M, Reddy AB, Zhang M, van Kuijk FJ, Ramana KV (2010) Prevention of endotoxin-induced uveitis in rats by plant sterol guggulsterone. Invest Ophthalmol Vis Sci 51:5105–5113
27. Cheon JH, Kim JS, Kim JM, Kim N, Jung HC, Song IS (2006) Plant sterol guggulsterone inhibits nuclear factor-kappaB signaling in intestinal epithelial cells by blocking IkappaB kinase and ameliorates acute murine colitis. Inflamm Bowel Dis 12:1152–1161
28. Burris TP, Montrose C, Houck KA, Osborne HE, Bocchinfuso WP, Yaden BC, Cheng CC, Zink RW, Barr RJ, Hepler CD et al (2005) The hypolipidemic natural product guggulsterone is a promiscuous steroid receptor ligand. Mol Pharmacol 67:948–954
29. Sharma B, Salunke R, Srivastava S, Majumder C, Roy P (2009) Effects of guggulsterone isolated from Commiphora mukul in high fat diet induced diabetic rats. Food Chem Toxicol 47:2631–2639
30. Moya M, Gomez-Lechon MJ, Castell JV, Jover R (2010) Enhanced steatosis by nuclear receptor ligands: a study in cultured human hepatocytes and hepatoma cells with a characterized nuclear receptor expression profile. Chem Biol Interact 184:376–387
31. Owsley E, Chiang JY (2003) Guggulsterone antagonizes farnesoid X receptor induction of bile salt export pump but activates pregnane X receptor to inhibit cholesterol 7alpha-hydroxylase gene. Biochem Biophys Res Commun 304:191–195
32. Wu J, Xia C, Meier J, Li S, Hu X, Lala DS (2002) The hypolipidemic natural product guggulsterone acts as an antagonist of the bile acid receptor. Mol Endocrinol 16:1590–1597
33. De Gottardi A, Dumonceau JM, Bruttin F, Vonlaufen A, Morard I, Spahr L, Rubbia-Brandt L, Frossard JL, Dinjens WN, Rabinovitch PS et al (2006) Expression of the bile acid receptor FXR in Barrett's esophagus and enhancement of apoptosis by guggulsterone in vitro. Mol Cancer 5:48
34. Lee JY, Lee KT, Lee JK, Lee KH, Jang KT, Heo JS, Choi SH, Kim Y, Rhee JC (2011) Farnesoid X receptor, overexpressed in pancreatic cancer with lymph node metastasis promotes cell migration and invasion. Br J Cancer 104:1027–1037
35. Lv N, Song MY, Kim EK, Park JW, Kwon KB, Park BH (2008) Guggulsterone, a plant sterol, inhibits NF-kappaB activation and protects pancreatic beta cells from cytokine toxicity. Mol Cell Endocrinol 289:49–59
36. Meselhy MR (2003) Inhibition of LPS-induced NO production by the oleogum resin of *Commiphora wightii* and its constituents. Phytochemistry 62:213–218
37. Chander R, Khanna AK, Kapoor NK (2002) Antioxidant activity of guggulsterone, the active principle of guggulipid from Commiphora mukul. J Med Aromatic Plant Sci 24:371–375
38. Sarfaraz S, Siddiqui IA, Syed DN, Afaq F, Mukhtar H (2008) Guggulsterone modulates MAPK and NF-kappaB pathways and inhibits skin tumorigenesis in SENCAR mice. Carcinogenesis 29:2011–2018
39. Niranjan R, Kamat PK, Nath C, Shukla R (2010) Evaluation of guggulipid and nimesulide on production of inflammatory mediators and GFAP expression in LPS stimulated rat astrocytoma, cell line (C6). J Ethnopharmacol 127:625–630
40. Almazari I, Park JM, Park SA, Suh JY, Na HK, Cha YN, Surh YJ (2012) Guggulsterone induces heme oxygenase-1 expression through activation of Nrf2 in human mammary epithelial cells: PTEN as a putative target. Carcinogenesis 33:368–376

41. Kim JM, Kang HW, Cha MY, Yoo D, Kim N, Kim IK, Ku J, Kim S, Ma SH, Jung HC et al (2010) Novel guggulsterone derivative GG-52 inhibits NF-kappaB signaling in intestinal epithelial cells and attenuates acute murine colitis. Lab Invest 90:1004–1015
42. Rivest S (1999) Activation of the nuclear factor kappa B (NF-kappaB) and cyclooxygenase-2 (COX-2) genes in cerebral blood vessels in response to systemic inflammation. Mol Psychiatry 4:500
43. Caruso R, Sarra M, Stolfi C, Rizzo A, Fina D, Fantini MC, Pallone F, MacDonald TT, Monteleone G (2009) Interleukin-25 inhibits interleukin-12 production and Th1 cell-driven inflammation in the gut. Gastroenterology 136:2270–2279
44. Boirivant M, Fuss IJ, Chu A, Strober W (1998) Oxazolone colitis: a murine model of T helper cell type 2 colitis treatable with antibodies to interleukin 4. J Exp Med 188:1929–1939
45. Youn HS, Ahn SI, Lee BY (2009) Guggulsterone suppresses the activation of transcription factor IRF3 induced by TLR3 or TLR4 agonists. Int Immunopharmacol 9:108–112
46. Ahn SI, Youn HS (2008) Guggulsterone suppresses the activation of NF-κB and expression of COX-2 induced by toll-like receptor 2, 3, and 4 agonists. Food Sci Biotechnol 17:1294–1298
47. Francis JA, Raja SN, Nair MG (2004) Bioactive terpenoids and guggulusteroids from Commiphora mukul gum resin of potential anti-inflammatory interest. Chem Biodivers 1:1842–1853
48. Rayalam S, Yang JY, Della-Fera MA, Park HJ, Ambati S, Baile CA (2009) Anti-obesity effects of xanthohumol plus guggulsterone in 3T3-L1 adipocytes. J Med Food 12:846–853
49. Singh SV, Choi S, Zeng Y, Hahm ER, Xiao D (2007) Guggulsterone-induced apoptosis in human prostate cancer cells is caused by reactive oxygen intermediate dependent activation of c-Jun NH2-terminal kinase. Cancer Res 67:7439–7449
50. Kim ES, Hong SY, Lee HK, Kim SW, An MJ, Kim TI, Lee KR, Kim WH, Cheon JH (2008) Guggulsterone inhibits angiogenesis by blocking STAT3 and VEGF expression in colon cancer cells. Oncol Rep 20:1321–1327
51. Macha MA, Matta A, Chauhan S, Siu KM, Ralhan R (2010) 14-3-3 Zeta is a molecular target in guggulsterone induced apoptosis in head and neck cancer cells. BMC Cancer 10:655
52. Xu HB, Li L, Liu GQ (2009) Reversal of P-glycoprotein-mediated multidrug resistance by guggulsterone in doxorubicin-resistant human myelogenous leukemia (K562/DOX) cells. Pharmazie 64:660–665
53. Moon DO, Park SY, Choi YH, Ahn JS, Kim GY (2011) Guggulsterone sensitizes hepatoma cells to TRAIL-induced apoptosis through the induction of CHOP-dependent DR5: involvement of ROS-dependent ER-stress. Biochem Pharmacol 82:1641–1650
54. Ichikawa H, Aggarwal BB (2006) Guggulsterone inhibits osteoclastogenesis induced by receptor activator of nuclear factor-kappaB ligand and by tumor cells by suppressing nuclear factor-kappaB activation. Clin Cancer Res 12:662–668
55. Yamada T, Osawa S, Hamaya Y, Furuta T, Hishida A, Kajimura M, Ikuma M (2010) Guggulsterone suppresses bile acid-induced and constitutive caudal-related homeobox 2 expression in gut-derived adenocarcinoma cells. Anticancer Res 30:1953–1960
56. Xu HB, Li L, Liu GQ (2011) Reversal of multidrug resistance by guggulsterone in drug-resistant MCF-7 cell lines. Chemotherapy 57:62–70
57. Samudio I, Konopleva M, Safe S, McQueen T, Andreeff M (2005) Guggulsterones induce apoptosis and differentiation in acute myeloid leukemia: identification of isomer-specific antileukemic activities of the pregnadienedione structure. Mol Cancer Ther 4:1982–1992
58. Ahn KS, Sethi G, Sung B, Goel A, Ralhan R, Aggarwal BB (2008) Guggulsterone, a farnesoid X receptor antagonist, inhibits constitutive and inducible STAT3 activation through induction of a protein tyrosine phosphatase SHP-1. Cancer Res 68:4406–4415
59. Macha MA, Matta A, Chauhan SS, Siu KW, Ralhan R (2011) Guggulsterone targets smokeless tobacco induced PI3K/Akt pathway in head and neck cancer cells. PLoS One 6: e14728

60. Macha MA, Matta A, Chauhan SS, Siu KW, Ralhan R (2011) Guggulsterone (GS) inhibits smokeless tobacco and nicotine-induced NF-kappaB and STAT3 pathways in head and neck cancer cells. Carcinogenesis 32:368–380
61. Leeman-Neill RJ, Wheeler SE, Singh SV, Thomas SM, Seethala RR, Neill DB, Panahandeh MC, Hahm ER, Joyce SC, Sen M et al (2009) Guggulsterone enhances head and neck cancer therapies via inhibition of signal transducer and activator of transcription-3. Carcinogenesis 30:1848–1856
62. Shishodia S, Sethi G, Ahn KS, Aggarwal BB (2007) Guggulsterone inhibits tumor cell proliferation, induces S-phase arrest, and promotes apoptosis through activation of c-Jun N-terminal kinase, suppression of Akt pathway, and downregulation of antiapoptotic gene products. Biochem Pharmacol 74:118–130
63. Xiao D, Zeng Y, Prakash L, Badmaev V, Majeed M, Singh SV (2011) Reactive oxygen species-dependent apoptosis by gugulipid extract of ayurvedic medicine plant Commiphora mukul in human prostate cancer cells is regulated by c-Jun N-terminal kinase. Mol Pharmacol 79:499–507
64. Singh SV, Zeng Y, Xiao D, Vogel VG, Nelson JB, Dhir R, Tripathi YB (2005) Caspase-dependent apoptosis induction by guggulsterone, a constituent of ayurvedic medicinal plant Commiphora mukul, in PC-3 human prostate cancer cells is mediated by Bax and Bak. Mol Cancer Ther 4:1747–1754
65. An MJ, Cheon JH, Kim SW, Kim ES, Kim TI, Kim WH (2009) Guggulsterone induces apoptosis in colon cancer cells and inhibits tumor growth in murine colorectal cancer xenografts. Cancer Lett 279:93–100
66. Yang JY, Della-Fera MA, Baile CA (2008) Guggulsterone inhibits adipocyte differentiation and induces apoptosis in 3T3-L1 cells. Obesity (Silver Spring) 16:16–22
67. Debatin KM (2004) Apoptosis pathways in cancer and cancer therapy. Cancer Immunol Immunother 53:153–159
68. Sprick MR, Walczak H (2004) The interplay between the Bcl-2 family and death receptor-mediated apoptosis. Biochim Biophys Acta 1644:125–132
69. Ricci MS, Zong WX (2006) Chemotherapeutic approaches for targeting cell death pathways. Oncologist 11:342–357
70. Thornberry NA (1997) The caspase family of cysteine proteases. Br Med Bull 53:478–490
71. Danial NN, Korsmeyer SJ (2004) Cell death: critical control points. Cell 116:205–219
72. Zachary I, Gliki G (2001) Signaling transduction mechanisms mediating biological actions of the vascular endothelial growth factor family. Cardiovasc Res 49:568–581
73. Silva J, Dasgupta S, Wang G, Krishnamurthy K, Ritter E, Bieberich E (2006) Lipids isolated from bone induce the migration of human breast cancer cells. J Lipid Res 47:724–733
74. Krishnamurthy K, Wang G, Rokhfeld D, Bieberich E (2008) Deoxycholate promotes survival of breast cancer cells by reducing the level of pro-apoptotic ceramide. Breast Cancer Res 10:R106
75. Chung LW, Kao C, Sikes RA, Zhau HE (1997) Human prostate cancer progression models and therapeutic intervention. Hinyokika Kiyo 43:815–820
76. Kambhampati S, Ray G, Sengupta K, Reddy VP, Banerjee SK, Van Veldhuizen PJ (2005) Growth factors involved in prostate carcinogenesis. Front Biosci 10:1355–1367
77. Datta SR, Katsov A, Hu L, Petros A, Fesik SW, Yaffe MB, Greenberg ME (2000) 14-3-3 Proteins and survival kinases cooperate to inactivate BAD by BH3 domain phosphorylation. Mol Cell 6:41–51
78. Zha J, Harada H, Yang E, Jockel J, Korsmeyer SJ (1996) Serine phosphorylation of death agonist BAD in response to survival factor results in binding to 14-3-3 not BCL-X(L). Cell 87:619–628
79. Nomura M, Shimizu S, Sugiyama T, Narita M, Ito T, Matsuda H, Tsujimoto Y (2003) 14-3-3 Interacts directly with and negatively regulates pro-apoptotic Bax. J Biol Chem 278:2058–2065
80. Yuasa Y (2003) Control of gut differentiation and intestinal-type gastric carcinogenesis. Nat Rev Cancer 3:592–600

81. Guo RJ, Suh ER, Lynch JP (2004) The role of Cdx proteins in intestinal development and cancer. Cancer Biol Ther 3:593–601
82. Beck F, Stringer EJ (2010) The role of Cdx genes in the gut and in axial development. Biochem Soc Trans 38:353–357
83. Aoki K, Tamai Y, Horiike S, Oshima M, Taketo MM (2003) Colonic polyposis caused by mTOR-mediated chromosomal instability in Apc+/Delta716 Cdx2+/− compound mutant mice. Nat Genet 35:323–330
84. Bonhomme C, Duluc I, Martin E, Chawengsaksophak K, Chenard MP, Kedinger M, Beck F, Freund JN, Domon-Dell C (2003) The Cdx2 homeobox gene has a tumour suppressor function in the distal colon in addition to a homeotic role during gut development. Gut 52:1465–1471
85. Dang LH, Chen F, Ying C, Chun SY, Knock SA, Appelman HD, Dang DT (2006) CDX2 has tumorigenic potential in the human colon cancer cell lines LOVO and SW48. Oncogene 25:2264–2272
86. Debruyne PR, Witek M, Gong L, Birbe R, Chervoneva I, Jin T, Domon-Cell C, Palazzo JP, Freund JN, Li P et al (2006) Bile acids induce ectopic expression of intestinal guanylyl cyclase C through nuclear factor-kappaB and Cdx2 in human esophageal cells. Gastroenterology 130:1191–1206
87. Anan A, Gores GJ (2005) A new TRAIL to therapy of hepatocellular carcinoma: blocking the proteasome. Hepatology 42:527–529
88. He SQ, Chen Y, Chen XP, Zhang WG, Wang HP, Zhao YZ, Wang SF (2003) [Antitumor effects of soluble TRAIL in human hepatocellular carcinoma]. Zhonghua Zhong Liu Za Zhi 25:116–119
89. Yang JF, Cao JG, Tian L, Liu F (2012) 5,7-Dimethoxyflavone sensitizes TRAIL-induced apoptosis through DR5 upregulation in hepatocellular carcinoma cells. Cancer Chemother Pharmacol 69:195–206
90. Wang W, Gallant JN, Katz SI, Dolloff NG, Smith CD, Abdulghani J, Allen JE, Dicker DT, Hong B, Navaraj A et al (2011) Quinacrine sensitizes hepatocellular carcinoma cells to TRAIL and chemotherapeutic agents. Cancer Biol Ther 12:229–238
91. Koehler BC, Urbanik T, Vick B, Boger RJ, Heeger S, Galle PR, Schuchmann M, Schulze-Bergkamen H (2009) TRAIL-induced apoptosis of hepatocellular carcinoma cells is augmented by targeted therapies. World J Gastroenterol 15:5924–5935
92. Jin CY, Park C, Moon SK, Kim GY, Kwon TK, Lee SJ, Kim WJ, Choi YH (2009) Genistein sensitizes human hepatocellular carcinoma cells to TRAIL-mediated apoptosis by enhancing Bid cleavage. Anticancer Drugs 20:713–722
93. Chen KF, Chen HL, Liu CY, Tai WT, Ichikawa K, Chen PJ, Cheng AL (2012) Dovitinib sensitizes hepatocellular carcinoma cells to TRAIL and tigatuzumab, a novel anti-DR5 antibody, through SHP-1-dependent inhibition of STAT3. Biochem Pharmacol 83:769–777
94. Carlisi D, D'Anneo A, Angileri L, Lauricella M, Emanuele S, Santulli A, Vento R, Tesoriere G (2011) Parthenolide sensitizes hepatocellular carcinoma cells to TRAIL by inducing the expression of death receptors through inhibition of STAT3 activation. J Cell Physiol 226:1632–1641
95. Yamanaka Y, Shiraki K, Inoue T, Miyashita K, Fuke H, Yamaguchi Y, Yamamoto N, Ito K, Sugimoto K, Nakano T (2006) COX-2 inhibitors sensitize human hepatocellular carcinoma cells to TRAIL-induced apoptosis. Int J Mol Med 18:41–47
96. Huang P, Han J, Hui L (2010) MAPK signaling in inflammation-associated cancer development. Protein Cell 1:218–226
97. Dhillon AS, Hagan S, Rath O, Kolch W (2007) MAP kinase signalling pathways in cancer. Oncogene 26:3279–3290
98. Ren Y, Chan HM, Li Z, Lin C, Nicholls J, Chen CF, Lee PY, Lui V, Bacher M, Tam PK (2004) Upregulation of macrophage migration inhibitory factor contributes to induced N-Myc expression by the activation of ERK signaling pathway and increased expression of interleukin-8 and VEGF in neuroblastoma. Oncogene 23:4146–4154

99. Yang B, Cao DJ, Sainz I, Colman RW, Guo YL (2004) Different roles of ERK and p38 MAP kinases during tube formation from endothelial cells cultured in 3-dimensional collagen matrices. J Cell Physiol 200:360–369
100. Liu Y, Mei C, Sun L, Li X, Liu M, Wang L, Li Z, Yin P, Zhao C, Shi Y et al (2011) The PI3K-Akt pathway regulates calpain 6 expression, proliferation, and apoptosis. Cell Signal 23:827–836
101. Huang JG, Xia C, Zheng XP, Yi TT, Wang XY, Song G, Zhang B (2011) 17beta-Estradiol promotes cell proliferation in rat osteoarthritis model chondrocytes via PI3K/Akt pathway. Cell Mol Biol Lett 16:564–575
102. Yang J, Zhang X, Wang W, Liu J (2010) Insulin stimulates osteoblast proliferation and differentiation through ERK and PI3K in MG-63 cells. Cell Biochem Funct 28:334–341
103. Vazquez de la Torre A, Junyent F, Folch J, Pelegri C, Vilaplana J, Auladell C, Beas-Zarate C, Pallas M, Camins A, Verdaguer E (2011) Study of the pathways involved in apoptosis induced by PI3K inhibition in cerebellar granule neurons. Neurochem Int 59:159–167
104. Jeong SJ, Dasgupta A, Jung KJ, Um JH, Burke A, Park HU, Brady JN (2008) PI3K/AKT inhibition induces caspase-dependent apoptosis in HTLV-1-transformed cells. Virology 370:264–272
105. Gunadharini DN, Elumalai P, Arunkumar R, Senthilkumar K, Arunakaran J (2011) Induction of apoptosis and inhibition of PI3K/Akt pathway in PC-3 and LNCaP prostate cancer cells by ethanolic neem leaf extract. J Ethnopharmacol 134:644–650
106. Paine A, Eiz-Vesper B, Blasczyk R, Immenschuh S (2010) Signaling to heme oxygenase-1 and its anti-inflammatory therapeutic potential. Biochem Pharmacol 80:1895–1903
107. Jung JE, Lee HG, Cho IH, Chung DH, Yoon SH, Yang YM, Lee JW, Choi S, Park JW, Ye SK et al (2005) STAT3 is a potential modulator of HIF-1-mediated VEGF expression in human renal carcinoma cells. FASEB J 19:1296–1298
108. Shin J, Lee HJ, Jung DB, Jung JH, Lee EO, Lee SG, Shim BS, Choi SH, Ko SG, Ahn KS et al (2011) Suppression of STAT3 and HIF-1 alpha mediates anti-angiogenic activity of betulinic acid in hypoxic PC-3 prostate cancer cells. PLoS One 6:e21492
109. Dolcet X, Llobet D, Pallares J, Matias-Guiu X (2005) NF-kB in development and progression of human cancer. Virchows Arch 446:475–482
110. Masters SC, Fu H (2001) 14-3-3 Proteins mediate an essential anti-apoptotic signal. J Biol Chem 276:45193–45200
111. Matta A, DeSouza LV, Ralhan R, Siu KW (2010) Small interfering RNA targeting 14-3-3zeta increases efficacy of chemotherapeutic agents in head and neck cancer cells. Mol Cancer Ther 9:2676–2688
112. Chen S, Synowsky S, Tinti M, MacKintosh C (2011) The capture of phosphoproteins by 14-3-3 proteins mediates actions of insulin. Trends Endocrinol Metab 22:429–436
113. Xiang X, Yuan M, Song Y, Ruderman N, Wen R, Luo Z (2002) 14-3-3 Facilitates insulin-stimulated intracellular trafficking of insulin receptor substrate 1. Mol Endocrinol 16:552–562
114. Baguley BC (2010) Multidrug resistance in cancer. Methods Mol Biol 596:1–14
115. Ullah MF (2008) Cancer multidrug resistance (MDR): a major impediment to effective chemotherapy. Asian Pac J Cancer Prev 9:1–6
116. Chen KG, Valencia JC, Gillet JP, Hearing VJ, Gottesman MM (2009) Involvement of ABC transporters in melanogenesis and the development of multidrug resistance of melanoma. Pigment Cell Melanoma Res 22:740–749
117. Li Y, Revalde JL, Reid G, Paxton JW (2011) Modulatory effects of curcumin on multi-drug resistance-associated protein 5 in pancreatic cancer cells. Cancer Chemother Pharmacol 68:603–610
118. Okura T, Ibe M, Umegaki K, Shinozuka K, Yamada S (2010) Effects of dietary ingredients on function and expression of P-glycoprotein in human intestinal epithelial cells. Biol Pharm Bull 33:255–259
119. Prasad S, Ravindran J, Aggarwal BB (2010) NF-kappaB and cancer: how intimate is this relationship. Mol Cell Biochem 336:25–37

120. Nomura Y (2001) NF-kappaB activation and IkappaB alpha dynamism involved in iNOS and chemokine induction in astroglial cells. Life Sci 68:1695–1701
121. Pandey A, Forte V, Abdallah M, Alickaj A, Mahmud S, Asad S, McFarlane SI (2011) Diabetes mellitus and the risk of cancer. Minerva Endocrinol 36:187–209
122. Wesche J, Haglund K, Haugsten EM (2011) Fibroblast growth factors and their receptors in cancer. Biochem J 437:199–213
123. Turner N, Grose R (2010) Fibroblast growth factor signalling: from development to cancer. Nat Rev Cancer 10:116–129
124. Jakowlew SB (2006) Transforming growth factor-beta in cancer and metastasis. Cancer Metastasis Rev 25:435–457
125. Ibrahim YH, Yee D (2004) Insulin-like growth factor-I and cancer risk. Growth Horm IGF Res 14:261–269
126. Conti G, De Pol A, Scarpini E, Vaccina F, De Riz M, Baron P, Tiriticco M, Scarlato G (2002) Interleukin-1 beta and interferon-gamma induce proliferation and apoptosis in cultured Schwann cells. J Neuroimmunol 124:29–35
127. Drexler HG, Zaborski M, Quentmeier H (1997) Interferon-gamma induced proliferation of human myeloid leukaemia cell lines. Br J Haematol 98:699–710
128. Murohashi I, Hoang T (1991) Interferon-gamma enhances growth factor-dependent proliferation of clonogenic cells in acute myeloblastic leukemia. Blood 78:1085–1095
129. Wu X, Xu K (2010) [Interleukin-1 and cancer]. Zhongguo Fei Ai Za Zhi 13:1145–1148
130. Wang Z, Kong D, Li Y, Sarkar FH (2009) PDGF-D signaling: a novel target in cancer therapy. Curr Drug Targets 10:38–41
131. Wajant H (2009) The role of TNF in cancer. Results Probl Cell Differ 49:1–15

# Inhibition of UVB-Induced Nonmelanoma Skin Cancer: A Path from Tea to Caffeine to Exercise to Decreased Tissue Fat

Allan H. Conney, You-Rong Lou, Paul Nghiem, Jamie J. Bernard, George C. Wagner, and Yao-Ping Lu

Abstract Oral administration of green tea, black tea, or caffeine (but not the decaffeinated teas) inhibited ultraviolet B radiation (UVB)-induced skin carcinogenesis in SKH-1 mice. Studies with caffeine indicated that its inhibitory effect on the ATR/Chk1 pathway is an important mechanism for caffeine's inhibition of UVB-induced carcinogenesis. The regular teas or caffeine increased locomotor activity and decreased tissue fat. In these studies, decreased dermal fat thickness was associated with a decrease in the number of tumors per mouse. Administration of caffeine, voluntary exercise, and removal of the parametrial fat pads all stimulated UVB-induced apoptosis, inhibited UVB-induced carcinogenesis, and stimulated apoptosis in UVB-induced tumors. These results suggest that caffeine administration, voluntary exercise, and removal of the parametrial fat pads inhibit UVB-induced carcinogenesis by stimulating UVB-induced apoptosis and by enhancing apoptosis in DNA-damaged precancer cells and in cancer cells. We hypothesize that tissue fat secretes antiapoptotic adipokines that have a tumor promoting effect.

Keywords Adipokines · Sunlight-induced skin cancer

A.H. Conney (✉), Y.-R. Lou, J.J. Bernard and Y.-P. Lu
Susan Lehman Cullman Laboratory for Cancer Research, Department of Chemical Biology, Ernest Mario School of Pharmacy, Rutgers, The State University of New Jersey, 164 Frelinghuysen Road, Piscataway, NJ 08854, USA
e-mail: aconney@pharmacy.rutgers.edu

P. Nghiem
Division of Dermatology, Department of Medicine, Fred Hutchinson Cancer Research Center, University of Washington Medical School, 815 Mercer Street, Box 358050, Seattle, WA 98109, USA

G.C. Wagner
Department of Psychology, Rutgers, The State University of New Jersey, 152 Frelinghuysen Road, Piscataway, NJ 08854, USA

Contents

1	Introduction	62
2	Inhibitory Effects of Green Tea and Caffeine on UVB-Induced Carcinogenesis	62
3	Mechanism Studies	64
4	Effects of Oral Administration of Tea, Decaffeinated Tea, and Caffeine on Tissue Fat and Skin Carcinogenesis in UVB-Pretreated High-Risk Mice	65
5	Relationship Between the Thickness of the Dermal Fat Layer Away from Tumors and Tumor Multiplicity	66
6	Effects of Topical Applications of Caffeine on Apoptosis in Tumors During Carcinogenesis in UVB-Pretreated High Risk Mice	67
7	Effects of Running Wheel Exercise on UVB-Induced Apoptosis, UVB-Induced Carcinogenesis, and Apoptosis in Tumors	67
8	Effects of a Combination of Running Wheel Exercise Together with Oral Caffeine on Tissue Fat and UVB-Induced Apoptosis	68
9	Stimulatory Effect of Fat Removal (Partial Lipectomy) on UVB-Induced Apoptosis in the Epidermis of SKH-1 Mice	69
10	Surgical Removal of the Parametrial Fat Pads Decreases Serum Levels of TIMP1 and Other Adipokines	69
11	Surgical Removal of the Parametrial Fat Pads Inhibits UVB-Induced Formation of Skin Tumors in Mice Fed a High Fat Diet	70
References		71

1 Introduction

Sunlight-induced nonmelanoma skin cancer is the most prevalent cancer in the United States with more than two million cases per year (more than the number of cases for all of the other cancers combined) [1], and the number of nonmelanoma skin cancer cases has been increasing in recent years [2, 3]. Possible reasons for the increasing incidence of nonmelanoma skin cancer are increased recreational exposure to sunlight, increased use of "sun tanning salons," and depletion of the ozone layer. We also wonder whether the increasing incidence may be related to the use of certain moisturizing creams [4].

2 Inhibitory Effects of Green Tea and Caffeine on UVB-Induced Carcinogenesis

In an early study we found that oral administration of green tea inhibited the formation of ultraviolet B radiation (UVB)-induced nonmelanoma skin cancer in SKH-1 mice, but decaffeinated green tea was inactive [5] (Table 1). Oral administration of caffeine had a strong inhibitory effect on UVB-induced carcinogenesis, and adding caffeine to the decaffeinated green tea restored its inhibitory activity [5] (Table 1). Similar observations were made with black tea [5]. Our results indicate that caffeine is a biologically important component of tea.

Table 1 Effect of oral administration of green tea, decaffeinated green tea or caffeine on UVB-induced complete carcinogenesis

Treatment	Number of keratoacanthomas per mouse	Number of squamous cell carcinomas per mouse
Water	5.75 ± 1.04	1.17 ± 0.27
Green tea	2.21 ± 0.46*	0.52 ± 0.18*
Decaf. green tea	4.58 ± 0.64	1.35 ± 0.29
Caffeine	1.81 ± 0.44*	0.63 ± 0.14*
Decaf. green tea + caffeine	2.53 ± 0.43*	0.47 ± 0.11*

Female SKH-1 mice were treated with UVB (30 mJ/cm^2) twice weekly for 44 weeks. Tea leaf extracts (1.25 g tea leaf/100 mL hot water; ~4 mg tea solids/mL) or caffeine (0.36 mg/mL) were administered as the drinking fluid. Each value is the mean ± S.E. from 24–30 mice
*$p < 0.05$ (Taken from [5])

Table 2 Inhibitory effect of oral administration or topical applications of caffeine on tumor formation in UVB-pretreated high risk mice

		Keratoacanthomas		Squamous cell carcinomas	
Exp.	Treatment	Tumors per mouse	Percent decrease	Tumors per mouse	Percent decrease
1	Water	4.00 ± 0.47	–	1.82 ± 0.30	–
	Oral caffeine	1.70 ± 0.48*	57	0.63 ± 0.31*	65
2	Acetone	7.07 ± 1.27	–	1.18 ± 0.25	–
	Topical caffeine	3.93 ± 0.74*	44	0.33 ± 0.12*	72

In Experiment 1, UVB-pretreated high risk SKH-1 mice (30/group) with no observable tumors were given caffeine (0.44 mg/mL) as their sole source of drinking fluid for 23 weeks. The number of tumors per mouse is expressed as the mean ± S.E. In Experiment 2, high risk UVB-pretreated SKH-1 mice (30/group) were treated topically with 100 μL acetone or caffeine (6.2 μmol) in 100 μL acetone once daily 5 days a week for 18 weeks. Each value represents the mean ± S.E.
*$p < 0.01$ (Taken from [6, 7])

In additional studies, we irradiated SKH-1 mice with UVB (30 mJ/cm^2) twice a week for 20 weeks and then stopped UVB irradiation. These UVB-pretreated mice have no tumors but develop tumors over the next several months in the absence of further UVB irradiation (high risk mice) [6]. Treatment of these UVB-pretreated high risk mice with oral or topical administration of caffeine inhibited tumor formation (Table 2) [6, 7]. These results parallel epidemiological studies indicating that people ingesting regular coffee had a decreased risk of nonmelanoma skin cancer, and decaffeinated coffee was inactive [8, 9].

Oral administration of green tea (6 mg tea solids/mL) or caffeine (0.4 mg/mL) as the sole source of drinking fluid during irradiation of SKH-1 mice with UVB twice a week for 20 weeks inhibited UVB-induced formation of mutant p53 positive patches in the epidermis by ~40% [10]. Oral administration of green tea (6 mg tea solids/mL) as the sole source of drinking fluid or topical applications of caffeine (6.2 μmol) once a day 5 days a week starting immediately after discontinuation of UVB treatment enhanced the rate and extent of disappearance of the mutant

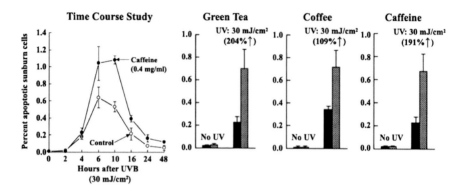

Fig. 1 *Stimulatory effect of oral administration of green tea, coffee or caffeine on UVB-induced apoptosis.* The time course for the effect of oral caffeine (0.4 mg/mL) for 2 weeks on UVB-induced apoptosis in female SKH-1 mice is shown in the *first panel*. In additional studies, SKH-1 female mice were treated with green tea (6 mg tea solids/mL), coffee (10 mg coffee solids/mL), or caffeine (0.4 mg/mL) as their sole source of drinking fluid for 2 weeks. The mice were irradiated with UVB (30 mJ/cm^2) and killed 10 h later. Apoptotic sunburn cells in the epidermis were determined. The *solid bars* represent control animals treated with water. The *dashed bars* indicate treatment with green tea, coffee, or caffeine as indicated. (Taken from [12, 14])

p53-positive patches [10]. Topical applications of caffeine to the dorsal skin of mice pretreated with UVB for 20 weeks resulted in enhanced apoptosis selectively in focal basal cell hyperplastic areas of the epidermis (putative precancerous lesions), but not in areas of the epidermis that only had diffuse hyperplasia [10]. These studies indicate that the chemopreventive effect of caffeine or green tea may occur by a proapoptotic effect, preferentially in early precancerous lesions.

3 Mechanism Studies

Mechanistic studies indicated that caffeine has a sunscreen effect [11] and also enhances UVB-induced apoptosis [12, 13]. The stimulatory effect of oral administration of green tea, coffee, and caffeine on UVB-induced apoptosis is shown in Fig. 1 [14]. In other studies, topical application of caffeine immediately after UVB irradiation also enhanced UVB-induced apoptosis [13], and the stimulatory effect of topical caffeine on UVB-induced apoptosis occurred by p53-dependent and p53-independent mechanisms [12, 15]. Application of caffeine after UVB irradiation avoided the potential sunscreen effect of caffeine. Studies on the p53-independent pathway suggested that oral or topical caffeine administration enhanced lethal mitosis in UVB irradiated mice by inhibiting the ATR/Chk-1 pathway in the epidermis [16] and in tumors from UVB-treated mice [17]. In addition, inhibition of the ATR/Chk-1 pathway by caffeine was associated with enhanced UVB-induced apoptosis in primary human keratinocytes [18].

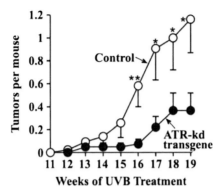

Fig. 2 *ATR-kd transgene delays tumor onset and suppresses UV tumorigenesis.* ATR-kd transgene suppresses UV-induced tumor development. Mean number of tumors per mouse is shown up to 19 weeks when some mice with advanced tumors were sacrificed and the cohort was no longer complete. *Error bars* represent SEM. Statistical significance in mean number of tumors per mouse between the groups was as shown at the indicated time points: *$P \leq 0.05$, **$P < 0.01$ (Taken from [19])

Additional evidence for the importance of blocking the ATR/Chk-1 pathway for inhibition of UVB-induced carcinogenesis came from finding that genetic inhibition of epidermal ATR kinase resulted in inhibition of UVB-induced carcinogenesis [19]. To test the effect of genetic inhibition of the ATR-Chk-1 pathway on UVB carcinogenesis, transgenic FVB mice were prepared that expressed a kinase dead form of human ATR (ATR-kd) under a human keratin-14 promoter. These mice were crossed into $Xpc^{-/-}$ mice with a global repair deficiency. UVB-induced carcinogenesis was determined in ATR-kd transgenic mice and transgene-negative littermate controls. Formation of UVB-induced skin tumors was markedly decreased in ATR-kd transgenic mice when compared with UVB-induced tumor formation in transgene-negative controls, indicating that genetic inhibition of the ATR/Chk-1 pathway inhibits UVB-induced carcinogenesis (Fig. 2) [19].

The results of mechanistic studies indicate that caffeine can inhibit UVB-induced carcinogenesis by exerting a sunscreen effect, by stimulating UVB-induced upregulation of wild-type p53, and by inhibition of the ATR/Chk-1 pathway.

4 Effects of Oral Administration of Tea, Decaffeinated Tea, and Caffeine on Tissue Fat and Skin Carcinogenesis in UVB-Pretreated High-Risk Mice

We found that oral administration of green tea or black tea (6 mg tea solids/mL) for 23 weeks to UVB-pretreated high risk mice in the absence of continued treatment with UVB decreased the number of tumors per mouse by 66–68%, the size of the parametrial fat pads by 32–54%, and the thickness of the dermal fat layer by 39–53% [20]. Administration of the decaffeinated teas had little or no effect on any of these parameters, and adding caffeine (equivalent to the amount in the regular teas) to the decaffeinated teas restored their inhibitory effects [20]. Administration of caffeine alone (0.4 mg/mL) decreased the number of tumors per mouse by 61%, decreased the average size of the parametrial fat pads by 56%, and caused a substantial decrease in the thickness of the dermal fat layer [20].

We observed that the dermal fat layer was much thinner under tumors than away from tumors in all experimental groups [20]. For instance, in UVB-pretreated high risk mice given only water as their drinking fluid for 23 weeks, the thickness of the dermal fat layer away from tumors was 162 μm but was only 60 μm directly under tumors. In high risk mice given 0.6% green tea for 23 weeks, the average thickness of the dermal fat layer away from tumors was 100 μm but was only 28 μm directly under tumors. Administration of caffeinated beverages decreased the average thickness of the dermal fat layer directly under tumors by 36% for small tumors (≤ 0.5 mm diameter), by 57% for tumors 0.5–1 mm in diameter, by 70% for tumors 1–2 mm in diameter, by 90% for tumors 2–3 mm in diameter, and by 97% for tumors >3 mm in diameter. In addition to the effect of caffeine to decrease the thickness of the dermal fat layer under tumors, our results suggest that tumors may be utilizing dermal fat as a source of energy or that tumors are secreting substances that enhance lipolysis.

5 Relationship Between the Thickness of the Dermal Fat Layer Away from Tumors and Tumor Multiplicity

In the above study with UVB-pretreated high risk mice treated with water, green tea, black tea, decaffeinated green tea, decaffeinated black tea, decaffeinated green tea plus caffeine, decaffeinated black tea plus caffeine, or caffeine alone, all mice at the end of the study were analyzed histologically for tumors, and 152 of these mice had a total of 689 tumors and 27 mice had no tumors. The relationship between the thickness of the dermal fat layer away from tumors (possible surrogate for total body fat levels) in individual mice and the number of tumors per mouse in all 179 mice was evaluated [20] (Table 3). Fourteen mice with a very thin dermal fat layer (≤ 50 μm) away from tumors had an average of only 1.6 ± 0.7 tumors/mouse whereas 7 mice with a thick dermal fat layer (>250 μm) away from tumors had 7.4 ± 1.8 tumors/mouse. Regression analysis was performed with data from all 179 mice to assess the relationship between the thickness of the dermal fat layer away from tumors for each mouse and the number of tumors per mouse. There was a highly significant positive linear association between the number of tumors per mouse and the thickness of the dermal fat layer away from tumors ($p = 0.0001$).

Table 3 Relationship between the thickness of the dermal fat layer (away from tumors) and tumor multiplicity

Thickness of dermal fat layer (μm)	Number of mice	Number of tumors per mouse
≤50	14	1.6 ± 0.7
50–100	63	2.9 ± 0.4
100–150	68	3.8 ± 0.6
150–200	17	5.5 ± 1.0
200–250	10	7.8 ± 1.4
>250	7	7.4 ± 1.8

UVB-pretreated high risk SKH-1 mice were given water, green tea, black tea, decaffeinated green tea, decaffeinated black tea, caffeine, decaffeinated green tea + caffeine or decaffeinated black tea + caffeine for 23 weeks. The thickness of the dermal fat layer in areas away from tumors or in mice with no tumors was determined. Each value represents the mean ± S.E. $p = 0.0001$ (from the Pearson correlation coefficient) for the thickness of the dermal fat layer away from tumors vs the number of tumors/mouse for all 179 mice. (Taken from [20])

6 Effects of Topical Applications of Caffeine on Apoptosis in Tumors During Carcinogenesis in UVB-Pretreated High Risk Mice

Tumor-free high risk mice (30 mice per group) were treated topically with 100 μL of acetone or with caffeine (6.2 μmoles) in 100 μL of acetone once a day 5 days a week for 18 weeks, and all tumors in the treated areas of the mice were counted and characterized by histological examination. The treatments with caffeine decreased the number of nonmalignant tumors (mostly keratoacanthomas) and squamous cell carcinomas by 44 and 72%, respectively (Table 2), and tumor volume per mouse was decreased by 72 and 79%, respectively [7].

The results of immunohistochemical staining of tumors described in the above study indicated that topical applications of caffeine to high risk mice enhanced apoptosis in the tumors but not in areas away from the tumors (Table 4) [7]. These results suggest that the inhibitory effect of caffeine administration on tumorigenesis in high risk mice may be caused in part by enhanced apoptosis in small tumors during their formation and growth.

7 Effects of Running Wheel Exercise on UVB-Induced Apoptosis, UVB-Induced Carcinogenesis, and Apoptosis in Tumors

During the course of our studies we observed that mice treated orally with green tea or caffeine had increased locomotor activity and decreased tissue fat [21]. Because of these observations, we studied the effect of voluntary exercise (running

Table 4 Stimulatory effect of topical applications of caffeine on apoptosis in tumors

Treatment	Number of tumors examined	Percent caspase 3 positive cells	Percent increase
Nontumor areas			
Control	–	0.159 ± 0.015	–
Caffeine	–	0.165 ± 0.027	4
Keratoacanthomas			
Control	198	0.229 ± 0.017	–
Caffeine	118	0.430 ± 0.034*	88
Carcinomas			
Control	33	0.196 ± 0.022	–
Caffeine	10	0.376 ± 0.056*	92

High risk mice (30 per group) were treated topically with acetone (100 μL) or with caffeine (6.2 μmol) in 100 μL acetone once daily 5 days a week for 18 weeks. Each value for the percent of caspase 3 positive cells represents the mean ± S.E.
*$p < 0.01$. (Animals are from Table 2, Exp. 2.) (Taken from [7])

wheel in the cage) on UVB-induced apoptosis, UVB-induced carcinogenesis, and apoptosis in UVB-induced tumors. An inhibitory effect of voluntary exercise on UVB-induced tumor formation and a stimulatory effect of voluntary exercise on UVB-induced apoptosis and apoptosis in tumors were observed [22, 23]. These results are similar to those observed for animals treated with caffeine.

8 Effects of a Combination of Running Wheel Exercise Together with Oral Caffeine on Tissue Fat and UVB-Induced Apoptosis

Treatment of SKH-1 mice orally with caffeine (0.1 mg/mL in the drinking water), voluntary running wheel exercise, or a combination of caffeine and exercise for 2 weeks (1) decreased the weight of the parametrial fat pads by 35, 62, and 77%, respectively, (2) decreased the thickness of the dermal fat layer by 38, 42, and 68%, respectively, and (3) stimulated the formation of UVB-induced caspase 3 (active form) positive cells in the epidermis by 92, 120, and 389%, respectively [23]. No effects of voluntary exercise or oral caffeine administration (alone or together) on apoptosis in the epidermis were observed in the absence of UVB irradiation. The plasma concentration of caffeine in mice ingesting caffeine (0.1 mg/mL drinking water) is similar to that in the plasma of most coffee drinkers (1–2 cups/day). The results of our studies indicate a greater than additive stimulatory effect of combined voluntary exercise and oral administration of a low dose of caffeine on UVB-induced apoptosis. In an additional study, oral administration of caffeine (0.1 mg/mL in the drinking water), voluntary running wheel exercise or the combination to SKH-1 mice irradiated with UVB (30 mJ/cm^2) twice a week for 34 weeks inhibited the formation of tumors (tumors/mouse) by 25, 35, and 62%, respectively.

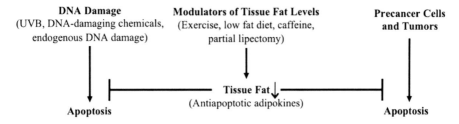

Fig. 3 Proposed inhibitory effect of tissue fat on DNA damage-induced apoptosis in precancer cells and in tumors

9 Stimulatory Effect of Fat Removal (Partial Lipectomy) on UVB-Induced Apoptosis in the Epidermis of SKH-1 Mice

Since administration of caffeine or running wheel exercise decreased tissue fat and enhanced UVB-induced apoptosis, we evaluated the effect of removal of tissue fat on UVB-induced apoptosis. Surgical removal of the two parametrial fat pads 2 weeks before UVB irradiation enhanced UVB-induced apoptosis in the epidermis by 107% at 6 h after irradiation when compared with the effect of UVB on apoptosis in sham-operated control mice [24]. In control studies with mice that did not receive UVB irradiation, partial lipectomy had no effect on the small number of apoptotic cells in the epidermis. Our results suggest that tissue fat may secrete antiapoptotic substances that enhance carcinogenesis by inhibiting the death of DNA-damaged precancer cells and cancer cells as hypothesized in Fig. 3. According to this hypothesis, factors that decrease tissue fat will decrease cancer risk by decreasing the amount of antiapoptotic adipokines, thereby enhancing apoptosis in DNA-damaged precancer cells and in cancer cells. Antiapoptotic adipokines associated with tissue fat may help explain why obese individuals have an increased risk of cancer.

10 Surgical Removal of the Parametrial Fat Pads Decreases Serum Levels of TIMP1 and Other Adipokines

Feeding SKH-1 mice, a 40% kcal high fat diet rich in omega-6 fatty acids as described earlier [25] or a 60% kcal very high fat diet for 2 weeks increased the serum levels of TIMP1 (tissue inhibitor of metalloproteinase 1) and several other adipokines. TIMP1 was reported to enhance cell proliferation and to inhibit apoptosis [26], suggesting that it has tumor promoting activity. TIMP1 was also reported to be a useful indicator of cutaneous cancer invasion and progression [27]. Removal of the parametrial fat pads from mice on a high fat diet resulted in a marked decrease in the serum level of TIMP1 and other adipokines when compared with the sham-operated control mice. Our results suggest that a high fat diet increases adipokines that have tumor promoting properties and that partial lipectomy decreases the serum levels of these adipokines.

11 Surgical Removal of the Parametrial Fat Pads Inhibits UVB-Induced Formation of Skin Tumors in Mice Fed a High Fat Diet

Our previous studies showed that a 40% kcal high fat diet rich in omega-6 fatty acids enhanced UVB-induced skin tumor formation when compared with mice fed a diet rich in omega-3 fatty acids [25]. We investigated the effect of lipectomy on UVB-induced skin tumorigenesis in mice fed either a high fat diet rich in omega-6 fatty acids or a low fat Chow diet.

SKH-1 mice were given a high fat diet and other mice were given a low fat Chow diet for 2 weeks. Mice on each diet were then divided into two groups. One group of mice had their parametrial fat pads removed and the other group of mice was a sham-operated control. The average weight of the removed parametrial fat pads from the mice that were fed a Chow diet or the high fat diet was about 15% of total body fat. All animals were treated with UVB (30 mJ/cm^2) once a day, twice a week for 33 weeks.

Surgical removal of the parametrial fat pads markedly inhibited UVB-induced skin tumorigenesis in mice fed the high fat diet, but this effect was not observed in mice fed the low fat Chow diet. Although there was no difference in body weight between lipectomized mice and sham-operated control animals fed the high fat diet, histopathology examination indicated that removal of the parametrial fat pads decreased the number of keratoacanthomas and squamous cell carcinomas per mouse by 75–80% when compared to the sham-operated controls. Partial lipectomy decreased the tumor volume per mouse for keratoacanthomas and carcinomas by ~90% when compared to the sham-operated controls.

Immunohistochemical analysis of the tumor samples indicated that lipectomy increased the percentage of caspase 3 (active form) positive cells in areas away from the tumors by 48%, in keratoacanthomas by 68%, and in carcinomas by 224%, respectively, and proliferation was also inhibited in lipectomized mice when compared with sham-operated mice. These results indicate that inhibition of UVB-induced carcinogenesis may have resulted from an increase in apoptosis and an inhibition of proliferation in tumors and in precancerous areas away from tumors. Our proposed effect of caffeine administration, exercise, low fat diet, and partial lipectomy to decrease tissue fat and associated antiapoptotic adipokines is shown in Fig. 3.

It was of considerable interest that compensatory fat appeared in the peritoneal cavity of partially lipectomized mice near where the parametrial fat pads had been removed. Biochemical properties of the compensatory fat in lipectomized mice at the end of the above tumor study were compared with the biochemical properties of the parametrial fat pads in sham-operated control mice at the end of the tumor study in mice fed the 40% high fat diet. It was found by RT-PCR that mRNAs for TIMP1, Serpin E1, and MCP1 were 50- to 80-fold higher in the parametrial fat pads than in the compensatory fat. Our results suggest that the parametrial fat pads secrete pro-inflammatory/tumor promoting adipokines that are not secreted in appreciable amounts by the compensatory fat.

Acknowledgement We thank Ms. Annette Dionisio for her excellent help in the preparation of this manuscript.

Grant Support: Research described in this chapter was supported in part by NIH grants CA 49756, CA 80759, CA 88961, CA 114442, CA 130857, CA 128997, 5T32 ES007148, and AR 49832.

References

1. Siegal R, Ward E, Brawley O, Jemal A (2011) Cancer statistics, 2011. CA Cancer J Clin 61:212–236
2. Rogers HW, Weinstock MA, Harris AR et al (2010) Incidence estimate of nonmelanoma skin cancer in the United States. Arch Dermatol 146:283–287
3. Athas WF, Hunt WC, Key CR (2003) Changes in nonmelanoma skin cancer incidence between 1977–1978 and 1998–1999 in northcentral New Mexico. Cancer Epidemiol Biomarkers Prev 12:1105–1108
4. Lu Y-P, Lou Y-R, Xie J-G et al (2009) Tumorigenic effect of some commonly used moisturizing creams when applied topically to UVB-pretreated high risk mice. J Invest Dermatol 129:468–475
5. Huang M-T, Xie J-G, Wang Z-Y et al (1997) Effects of tea, decaffeinated tea, and caffeine on UVB light-induced complete carcinogenesis in SKH-1 mice: demonstration of caffeine as a biologically important constituent of tea. Cancer Res 57:2623–2629
6. Lou Y-R, Lu Y-P, Xie J-G, Huang M-T, Conney AH (1999) Effects of oral administration of tea, decaffeinated tea, and caffeine on the formation and growth of tumors in high-risk SKH-1 mice previously treated with ultraviolet-B light. Nutr Cancer 33:146–153
7. Lu Y-P, Lou Y-R, Xie J-G et al (2002) Topical applications of caffeine or (−)-epigallocatechin gallate (EGCG) inhibit carcinogenesis and selectively increase apoptosis in UVB-induced skin tumors in mice. Proc Natl Acad Sci USA 99:12455–12460
8. Jacobsen BK, Bjelke E, Kvåle G, Heuch I (1986) Coffee drinking, mortality, and cancer incidence: results from a Norwegian prospective study. J Natl Cancer Inst 76:823–831
9. Abel EL, Hendrix SO, McNeeley SG et al (2007) Daily coffee consumption and prevalence of nonmelanoma skin cancer in Caucasian women. Eur J Cancer Prev 16:446–452
10. Lu Y-P, Lou Y-R, Liao J et al (2005) Administration of green tea or caffeine enhances the disappearance of UVB-induced patches of mutant p53 positive epidermal cells in SKH-1 mice. Carcinogenesis 26:1465–1472
11. Lu Y-P, Lou Y-R, Xie J-G et al (2007) Caffeine and caffeine sodium benzoate have a sunscreen effect, enhance UVB-induced apoptosis, and inhibit UVB-induced skin carcinogenesis in SKH-1 mice. Carcinogenesis 28:199–206
12. Lu Y-P, Lou Y-R, Li XH et al (2000) Stimulatory effect of oral administration of green tea or caffeine on ultraviolet light-induced increases in epidermal wild-type p53, p21(WAF1/CIP1) and apoptotic sunburn cells in SKH-1 mice. Cancer Res 60:4785–4791
13. Lu Y-P, Lou Y-R, Li X-H et al (2002) Stimulatory effect of topical application of caffeine on UVB-induced apoptosis in mouse skin. Oncol Res 13:61–70

14. Conney AH, Zhou S, Lee M-J et al (2007) Stimulatory effect of oral administration of tea, coffee or caffeine on UVB-induced apoptosis in the epidermis of SKH-1 mice. Toxicol Appl Pharmacol 224:209–213
15. Lu Y-P, Lou Y-R, Peng Q-Y, Xie J-G, Conney AH (2004) Stimulatory effect of topical application of caffeine on UVB-induced apoptosis in the epidermis of p53 and Bax knockout mice. Cancer Res 64:5020–5027
16. Lu Y-P, Lou Y-R, Peng Q-Y, Xie J-G, Nghiem P, Conney AH (2008) Effect of caffeine on the ATR/Chk1 pathway in the epidermis of UVB-irradiated mice. Cancer Res 68:2523–2529
17. Lu Y-P, Lou Y-R, Peng Q-Y, Nghiem P, Conney AH (2011) Caffeine decreases phospho-Chk1 (Ser317) and increases mitotic cells with cyclin B1 and caspase 3 in tumors from UVB-treated mice. Cancer Prev Res 4:1118–1125
18. Heffernan TP, Kawasumi M, Blasina A, Anderes K, Conney AH, Nghiem P (2009) ATR-Chk1 pathway inhibition promotes apoptosis after UV treatment in primary human keratinocytes: potential basis for the UV protective effects of caffeine. J Invest Dermatol 129:1805–1815
19. Kawasumi M, Lemos B, Bradner JE et al (2011) Protection from UV-induced skin carcinogenesis by genetic inhibition of the ataxia telangiectasia and Rad3-related (ATR) kinase. Proc Natl Acad Sci USA 108:13716–13721
20. Lu Y-P, Lou Y-R, Lin Y et al (2001) Inhibitory effects of orally administered green tea, black tea and caffeine on skin carcinogenesis in mice previously treated with ultraviolet B light (high risk mice): relationship to decreased tissue fat. Cancer Res 61:5002–5009
21. Michna L, Lu Y-P, Lou Y-R, Wagner GC, Conney AH (2003) Stimulatory effect of oral administration of green tea and caffeine on locomotor activity in SKH-1 mice. Life Sci 73:1383–1392
22. Michna L, Wagner GC, Lou Y-R et al (2006) Inhibitory effects of voluntary running wheel exercise on UVB-induced skin carcinogenesis in SKH-1 mice. Carcinogenesis 27:2108–2115
23. Lu Y-P, Nolan B, Lou Y-R, Peng Q-Y, Wagner GC, Conney AH (2007) Voluntary exercise together with oral caffeine markedly stimulates UVB light-induced apoptosis and decreases tissue fat in SKH-1 mice. Proc Natl Acad Sci USA 104:12936–12941
24. Lu Y-P, Lou Y-R, Nolan B et al (2006) Stimulatory effect of voluntary exercise or fat removal (partial lipectomy) on apoptosis in the skin of UVB light-irradiated mice. Proc Natl Acad Sci USA 103:16301–16306
25. Lou Y-R, Peng Q-Y, Li T et al (2011) Effects of high-fat diets rich in either omega-3 or omega-6 fatty acids on UVB-induced skin carcinogenesis in SKH-1 mice. Carcinogenesis 32:1078–1084
26. Chirco R, Liu XW, Jung KK, Kim HR (2006) Novel functions of TIMPs in cell signaling. Cancer Metastasis Rev 25:99–113
27. O'Grady A, Dunne C, O'Kelly P, Murphy GM, Leader M, Kay E (2007) Differential expression of matrix metalloproteinase (MMP)-2, MMP-9 and tissue inhibitor of metalloproteinase (TIMP)-1 and TIMP-2 in non-melanoma skin cancer: implications for tumour progression. Histopathology 51:793–804

Cancer Chemoprevention and Nutri-Epigenetics: State of the Art and Future Challenges

Clarissa Gerhauser

Abstract The term "epigenetics" refers to modifications in gene expression caused by heritable, but potentially reversible, changes in DNA methylation and chromatin structure. Epigenetic alterations have been identified as promising new targets for cancer prevention strategies as they occur early during carcinogenesis and represent potentially initiating events for cancer development. Over the past few years, nutri-epigenetics – the influence of dietary components on mechanisms influencing the epigenome – has emerged as an exciting new field in current epigenetic research. During carcinogenesis, major cellular functions and pathways, including drug metabolism, cell cycle regulation, potential to repair DNA damage or to induce apoptosis, response to inflammatory stimuli, cell signalling, and cell growth control and differentiation become deregulated. Recent evidence now indicates that epigenetic alterations contribute to these cellular defects, for example epigenetic silencing of detoxifying enzymes, tumor suppressor genes, cell cycle regulators, apoptosis-inducing and DNA repair genes, nuclear receptors, signal transducers and transcription factors by promoter methylation, and modifications of histones and non-histone proteins such as *p53*, *NF-κB*, and the chaperone *HSP90* by acetylation or methylation.

The present review will summarize the potential of natural chemopreventive agents to counteract these cancer-related epigenetic alterations by influencing the activity or expression of DNA methyltransferases and histone modifying enzymes. Chemopreventive agents that target the epigenome include micronutrients (folate, retinoic acid, and selenium compounds), butyrate, polyphenols from green tea, apples, coffee, black raspberries, and other dietary sources, genistein and soy isoflavones, curcumin, resveratrol, dihydrocoumarin, nordihydroguaiaretic acid (NDGA), lycopene, anacardic acid, garcinol, constituents of *Allium* species and cruciferous vegetables, including indol-3-carbinol (I3C), diindolylmethane (DIM),

C. Gerhauser (✉)
Division Epigenomics and Cancer Risk Factors, German Cancer Research Center,
Im Neuenheimer Feld 280, 69120 Heidelberg, Germany
e-mail: c.gerhauser@dkfz.de

sulforaphane, phenylethyl isothiocyanate (PEITC), phenylhexyl isothiocyanate (PHI), diallyldisulfide (DADS) and its metabolite allyl mercaptan (AM), cambinol, and relatively unexplored modulators of histone lysine methylation (chaetocin, polyamine analogs). So far, data are still mainly derived from in vitro investigations, and results of animal models or human intervention studies are limited that demonstrate the functional relevance of epigenetic mechanisms for health promoting or cancer preventive efficacy of natural products. Also, most studies have focused on single candidate genes or mechanisms. With the emergence of novel technologies such as next-generation sequencing, future research has the potential to explore nutri-epigenomics at a genome-wide level to understand better the importance of epigenetic mechanisms for gene regulation in cancer chemoprevention.

Keywords Cancer chemoprevention • Dietary compounds • DNA methylation • Histone modifications • Nutri-epigenetics

Contents

1	Introduction	74
2	DNA Methylation	75
3	Histone Modifications	77
4	MicroRNAs	78
5	Interplay Between Chemopreventive and Epigenetic Mechanisms and Natural Products Effects	79
6	Detoxification	81
7	Cell Cycle Regulation	83
8	Apoptosis	89
9	DNA Repair	91
10	Inflammation and Regulation of NF-κB	92
11	Cell Signaling and Cell Growth	94
12	Cell Differentiation	98
13	Summary and Conclusions	100
References		116

1 Introduction

The term "epigenetics" refers to modifications in gene expression caused by heritable, but potentially reversible, changes in DNA methylation and chromatin structure [1]. Given the fact that epigenetic modifications are reversible and occur early during carcinogenesis as potentially initiating events for cancer development, they have been identified as promising new targets for cancer prevention strategies. Major epigenetic mechanisms of gene regulation include DNA methylation, modifications of the chromatin structure by histone tail acetylation and methylation, and small non-coding microRNAs, that affect gene expression by targeted degradation of mRNAs or inhibition of their translation (overview in Fig. 1) [3, 4].

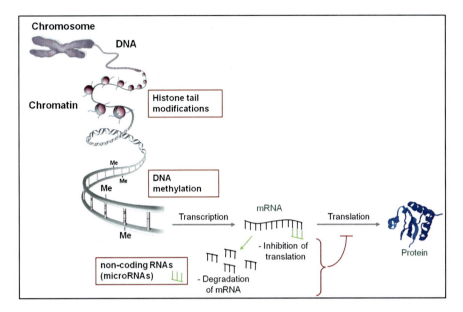

Fig. 1 Overview of epigenetic mechanisms including DNA methylation, histone tail modifications and non-coding (micro) RNAs, targeting DNA, N-terminal histone tails and mRNA (modified from [2], with permission of Nature Publishing Group)

Epigenetic mechanisms are essential to control normal cellular functions and they play an important role during development. Distinct patterns of DNA methylation regulate tissue specific gene expression and are involved in X-chromosome inactivation and genomic imprinting [5–7]. Histone modifications are critical for memory formation [4, 8]. Interestingly, epigenetic profiles can be modified to adapt to changes in the environment (e.g., nutrition, chemical exposure, smoking, radiation, etc.) [3, 9] as has been exemplified in studies with monozygotic twins and inbred animals [10, 11]. Consequently, alterations in DNA methylation and histone marks eventually contribute to the development of age-related and lifestyle-related diseases, such as metabolic syndrome, Alzheimer's disease, and cancer [8, 12, 13].

2 DNA Methylation

DNA methylation is mediated by DNA methyltransferases (*DNMT*) that transfer methyl groups from *S*-adenosyl-L-methionine (SAM) to the 5′-position of cytosines. This reaction mainly takes place at cytosines when positioned next to a guanine (CpG dinucleotides) and creates 5-methylcytosine (5mC) and *S*-adenosyl-L-homocysteine (SAH). Three active mammalian DNMTs have been identified so far, i.e., *DNMT1*, *3a*, and *3b*. *DNMT1* is a maintenance methyltransferase that maintains DNA methylation during DNA replication. It preferentially methylates the newly synthesized,

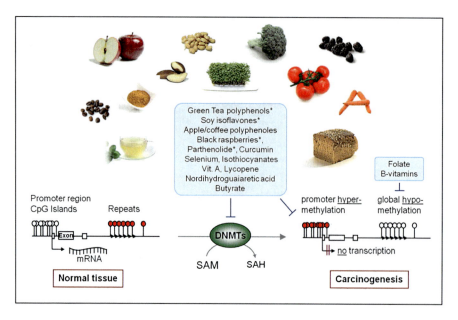

Fig. 2 Overview of DNA methylation changes during carcinogenesis and cancer chemopreventive agents inhibiting the activity of expression of DNMTs, thereby preventing aberrant (promoter) hypermethylation or genome wide hypomethylation. DNA methylation is catalyzed by DNA methyltransferases (DNMTs) using S-adenosylmethionine (SAM) as a substrate. See text and Table 1 (Appendix) for further details. *Asterisks* indicate epigenetic activity in vivo. Empty circles: unmethylated CpG dinucleotide; red circles: methylated CpG site

unmethylated DNA strand after replication and thus assures transmission of DNA methylation patterns to daughter cells. *DNMT3a* and *DNMT3b* are "de novo" methyltransferases that catalyze methylation of previously unmethylated sequences. *DNMT3b* is believed to play an important role during tumorigenesis [14, 15].

In normal cells, CpG-rich sequences (so-called CpG islands, CGIs) in gene promoter regions are generally unmethylated, with the exception of about 6–8% CGIs methylated in a tissue-specific manner [7]. Conversely, the majority of CpG sites in repetitive sequences such as ribosomal DNA repeats, satellite repeats, or centromeric repeats are often heavily methylated, thereby contributing to chromosomal stability by limiting accessibility to the transcription machinery [16]. This controlled pattern of DNA methylation is disrupted during ageing, carcinogenesis, or development of chronic diseases. Increased methylation (DNA hypermethylation) of promoter CGIs leads to transcriptional silencing of tumor suppressors and other genes with important biological functions [12, 16, 17]. In contrast, global loss of DNA methylation at repetitive genomic sequences (DNA hypomethylation) during carcinogenesis has been associated with genomic instability and chromosomal aberrations and was first described about 30 years ago [18, 19] (Fig. 2). Different from irreversible gene inactivation by genetic deletions or nonsense mutations, genes silenced by epigenetic modifications are still intact and can potentially be reactivated by small molecules acting as modifiers of epigenetic

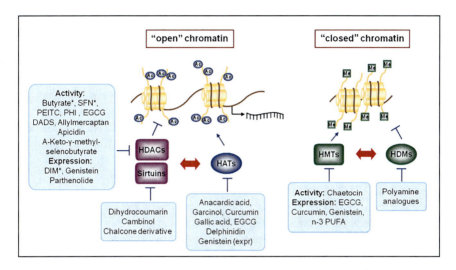

Fig. 3 Simplified overview of histone modifying enzymes with a focus on histone deacetylases (HDACs), histone acetyltransferases (HATs), histone methyl transferases (HTMs), and histone demethylases (HDM), and their influence on chromatin structure. Sirtuins represent a NAD$^+$-dependent subclass of HDACs (class III). Also indicated is the inhibitory potential of chemopreventive agents. See text and Tables 2 and 3 (Appendix) for further details. *Asterisks* indicate epigenetic activity in vivo

mechanisms. Consequently, development of agents or food components that prevent or reverse methylation-induced inactivation of gene expression is a new promising approach for cancer prevention [20].

3 Histone Modifications

Epigenetic regulation of gene expression is also mediated by post-translational modifications at the N-terminal tails of histones. These include acetylation, methylation, phosphorylation, ubiquitinylation, sumoylation, and ADP ribosylation and contribute to genomic stability, DNA damage response, and cell cycle checkpoint integrity [118–120]. Histones can be modified through sequence-specific transcription factors or on a more global scale through histone-modifying enzymes [120]. So far, histone acetylation and histone methylation have been investigated the most and disturbance of their balance has been associated with neoplastic transformation (Fig. 3).

Histone acetylation is maintained by the interplay of *histone acetyltransferases* (*HATs*) and *histone deacetylases* (*HDACs*). HATs transfer acetyl groups from acetyl-CoA to the ε-amino group of lysine (K) residues in histone tails, whereas HDACs remove histone acetyl groups by catalyzing their transfer to Coenzyme A (CoA). Acetylation of histone tails opens up the chromatin structure, allowing transcription factors to access the DNA. Consequently, proteins with *HAT* catalytic activity are often transcriptional coactivators. So far at least 25 *HAT* proteins have been characterized. They are organized into four families based on structure homology [189] and often possess distinct histone specificity. Subgroups include the GNAT (*hGCN5*, *PCAF*), MYST (*MYST*, *Tip60*), p300/CBP (*p300/CBP*), SRC (*SRC-1*), and TAFII250 families (*TAFII250*) [119, 190]. In contrast to histone acetylation, histone deacetylation generally leads to chromatin condensation and transcriptional repression. So far, 18 proteins with *HDAC* activity have been classified [191, 192]. *HDACs* 1–11 are subdivided into three classes – I, II, and IV – based on homology, size, sub-cellular expression, and number of enzymatic domains. Class III is comprised of *sirtuins* 1–7, which are structurally unrelated to class I and II HDACs and require NAD^+ as a cofactor for activity [191, 192]. Interestingly, *HDAC* substrates are not limited to histones. As further outlined below, several important regulatory proteins and transcription factors such as *p53*, *E2F*, and *nuclear factor-κB* (*NF-κB*) involved in stress response, inflammation, and apoptosis have been shown to be regulated by acetylation [193–195].

Histone methylation takes place at lysine and arginine residues. Histone lysine methylation has activating or repressive effects on gene expression. This is dependent on the lysine residue that is methylated (e.g., K4, K9, K27, K36, K79 in H3), the methylation status (mono-, di-, or tri-methylation), and the location (interaction with promoter vs gene coding regions) [118, 119, 196]. Methylation at H3K4, H3K36, and H3K79 is generally associated with transcriptional active chromatin (euchromatin), whereas methylation at H3K9, H3K27, and H4K20 is frequently associated with transcriptional inactive heterochromatin [190, 197]. Histone lysine methylation is mediated by *histone lysine methyltransferases* (*HMTs*) that transfer a methyl group from SAM to the lysine residue. *HMTs* can be classified as *Dot1* protein family and proteins containing a so-called SET domain, based on sequence similarity with Drosophila proteins suppressor of variegation (*SUV*), enhancer of zeste (*EZH*), and homeobox gene regulator *Trithorax* (*TRX*). So far, more than 50 SET domain family members have been identified in humans [197]. They are grouped into six subfamilies, SET1, SET2, SUV39, EZH, SMYD, and PRDM, and several SET-containing *HMTs* that do not fall into these groups [197].

Several types of *histone lysine demethylases* (*HDMs*) have been identified so far, for example *lysine specific demethylase 1* (*LSD1*) and the family of about 20 *Jumonji domain-containing* (*JmjC*) *histone demethylases* [118, 119, 197]. Similar to lysine acetylation, lysine methylation is not limited to histone proteins, and several non-histone protein substrates including *p53*, retinoblastoma protein (*RB*), the *NF-κB* subunit *RelA*, and estrogen receptor α (*ERα*) have been identified (summarized in [198–200]).

4 MicroRNAs

MicroRNAs (miRNAs) are small non-coding RNAs of 20–22 nucleotides that inhibit gene expression at the posttranscriptional level. MiRNAs are involved in the regulation of key biological processes, including development, differentiation, apoptosis, and proliferation, and are known to be altered in a variety of chronic degenerative diseases including cancer [201]. MiRNAs are generated from RNA precursor structures by a protein complex system composed of members of the Argonaute protein family, polymerase II-dependent transcription, and the ribonucleases Drosha and Dicer [202]. MiRNAs regulate the transformation of mRNA into proteins, either by imperfect base-pairing to the mRNA 3′-untranslated regions to repress protein synthesis, or by affecting mRNA stability. Each miRNA is expected to control several hundred genes. They have been implicated in cancer initiation and progression, and their expression is often down-regulated during carcinogenesis. Major mechanisms of miRNA deregulation include genetic and epigenetic alterations as well as defects in the miRNA processing machinery [196].

5 Interplay Between Chemopreventive and Epigenetic Mechanisms and Natural Products Effects

Over the last few years, evidence has accumulated that natural products and dietary constituents with chemopreventive potential have an impact on DNA methylation (Fig. 2), histone modifications (Fig. 3), and miRNA expression. The available information on the topic has been summarized in several recent review articles [20–36, 121, 122, 203, 204].

As indicated in Fig. 2, *folate* and *B-vitamins* have a potential impact on DNA hypomethylation. They affect the so called "one-carbon metabolism" which provides methyl groups for methylation reactions. Folate is an important factor for the maintenance of DNA biosynthesis and DNA repair, and folate deficiency leads to global DNA hypomethylation, genomic instability, and chromosomal damage. As an essential micronutrient, folate needs to be taken up from dietary sources, such as citrus fruits, dark green vegetables, whole grains, and dried beans. Alcohol misuse is often associated with folate deficiency. Epidemiological studies have indicated that low folate levels are associated with an increased risk for colorectum, breast, ovary, pancreas, brain, lung, and cervix cancer [66, 76, 205]. Consequently, the relationship between folate status, DNA methylation, and cancer risk has been analyzed in numerous rodent carcinogenesis models and in human intervention studies. Overall, the results are inconclusive and depend on various parameters, for example dose and timing of the intervention, the severity of folate deficiency, and health status (reviewed in [23, 66–68, 76]). Excessive intake of synthetic folic acid (from high-dose supplements or fortified foods) may even increase human cancer risk by accelerating growth of precancerous lesions [66]. Therefore folate supplementation cannot be generally recommended, and deficiencies should be prevented by dietary intake. In a cohort-based observation study with 1,100 participants, Stidley et al. investigated the effect of various dietary factors on promoter methylation levels of eight genes commonly hypermethylated

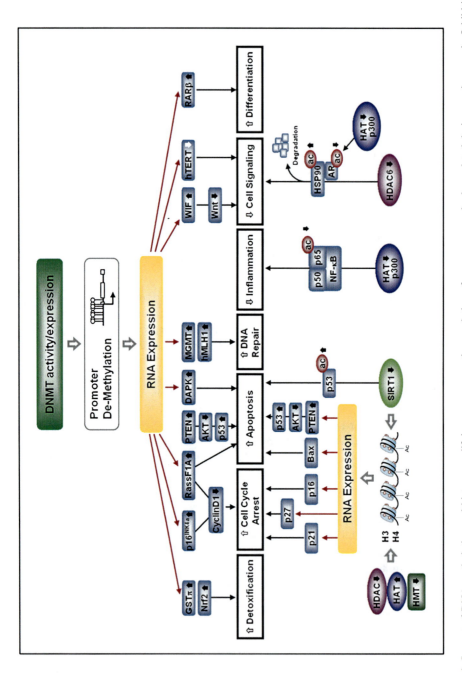

Fig. 4 Impact of DNA methylation and histone-modifying enzymes on the regulation of genes commonly deregulated during carcinogenesis. Inhibition/modulation of the activity or expression of DNMTs or histone-modifying proteins by chemopreventive agents can lead to reactivation of epigenetically silenced genes. See text for a detailed description of the indicated mechanisms and pathways and the influence of chemopreventive agents

in cancer, including *RassF1A*, *p16*, *MGMT*, *DAPK*, *GATA4*, *GATA5*, *PAX5α*, and *PAX5β* in exfoliated aerodigestive tract cells from sputum samples of current and former smokers. Significant protection from DNA methylation (less than two genes methylated) was observed for regular consumption of folate [OR (odds ratio) = 0.84 per 750 μg/day; CI (95% confidence interval), 0.72–0.99], leafy green vegetables (OR, 0.83 per 12 monthly servings; CI, 0.74–0.93), and multivitamin use (OR, 0.57; CI, 0.40–0.83) [77].

The following chapter will focus on pathways which are relevant for chemoprevention and are commonly deregulated by epigenetic mechanisms in cancer cells, including drug detoxification, cell cycle regulation, apoptosis induction, DNA repair, tumor-associated inflammation, cell signaling that promotes cell growth, and cell differentiation (overview in Fig. 4). It will present a summary of natural chemopreventive agents targeting these pathways by affecting DNA methylation and histone tail modifications. Their effect on miRNAs and subsequent gene expression will not be discussed.

Plant compounds which affect DNA methylation and inhibit DNMT enzymatic activity (DNMT inhibitors, DNMTi), revert aberrant DNA promoter methylation, or reactivate genes silenced by promoter hypermethylation, are listed in Table 1 (Appendix). Natural products with influence on histone acetylation and methylation that inhibit the activity or modulate the expression of histone-modifying enzymes including HDACs, SIRTs, HATs, and HMTs are summarized in Tables 2 and 3 (Appendix).

6 Detoxification

GSTP1 is a member of the glutathione *S*-transferase family of isoenzymes that conjugate reactive chemicals and carcinogens with the tripeptide glutathione (GSH) and thus enhance their excretion and detoxification [206]. Induction of GSTs and other enzymes involved in phase 2 of drug metabolism via the *Nrf2-Keap1* pathway is an important mechanism in cancer chemoprevention [207]. Recently, *GSTP1* activity has also been associated with cell-signaling functions critical for survival, for example the regulation of *c-Jun N-terminal kinase* (*JNK*) activity and modulation of protein functions by *S*-glutathionylation [208].

Loss of *GSTP1* expression by CGI hypermethylation is very common in prostate cancer [209]. *GSTP1* is expressed and unmethylated in normal prostate tissue. Hypermethylation increases with increasing prostate carcinogenesis and can be detected in up to 70–100% prostate adenocarcinoma [209]. *GSTP1* hypermethylation is also detectable in plasma, ejaculate, or urine, and is discussed as a promising prostate cancer biomarker. In addition to prostate cancer, *GSTP1* hypermethylation is frequent in ~30% and >80% of breast cancer and hepatocellular carcinoma, respectively [209]. Deletion of *GSTP1* in mice was shown to enhance susceptibility to chemically-induced skin and lung cancer, and to increase adenoma incidence and multiplicity when *mGstp1/p2* knockout mice were crossed with $APC^{Min/+}$ mice [206]. Gene expression studies in these models indicate a protective role of *GSTP1* in inflammation and immune response.

Reexpression of *GSTP1* after treatment with natural products has been tested in prostate and breast cancer cell lines. Ramachandran et al. was unable to detect demethylation and reexpression of *GSTP1* in LNCaP and PC-3 prostate cancer cells after treatment with *seleno-DL-methionine*. More recently, reactivation of *GSTP1* by *sodium selenite* in LNCaP cells was shown to involve a dual effect on both DNA methylation and histone modifications. Incubation with low dose sodium selenite lowered *DNMT1* mRNA and protein expression, reduced global DNA methylation, and led to the reexpression of *GSTP1* associated with reduced *GSTP1* promoter methylation [115]. An earlier study identified sodium selenite and organic seleno-compounds as inhibitors of DNMT activity in vitro [112]. Therefore, direct inhibition of DNMT enzyme activity might contribute to the demethylating potential of sodium selenite. *Phenethylisothiocyanate (PEITC)* derived from the glucosinolate gluconasturtiin from watercress was able to revert epigenetic silencing of *GSTP1* in LNCaP cells. Reduced DNA methylation at specific CpG sites was associated with enhanced protein expression and increased *GSTP1* enzymatic activity [100]. *Green tea polyphenols (GTP)* and *epigallocatechin gallate (EGCG)* inhibited DNMT enzyme activity and DNMT protein expression in LNCaP cells. DNMT inhibition was associated with reduced methylation of the *GSTP1* proximal promoter and reactivation of *GSTP1* expression. Transcription was facilitated by enhanced binding of transcription factor Sp1 to the *GSTP1* promoter [45]. Intervention of prostate cancer cell lines with the *soy phytoestrogens genistein and daidzein* significantly reduced *GSTP1* promoter methylation and resulted in reexpression of *GSTP1* protein, determined by immunocytochemistry and western blotting [83, 84]. The mechanism of inhibition was not further analyzed. King-Batoon et al. investigated the effects of genistein and the tomato-derived carotenoid *lycopene* on DNA methylation in breast cancer cells. A single application of lycopene reactivated *GSTP1* mRNA expression within 1 week, associated with reduced promoter methylation in MDA-MB-468 cells, whereas genistein was weakly effective only after repetitive treatments. Both compounds were ineffective in the MCF7 cell line, and also did not reduced *RARβ* and *HIN1* promoter methylation in both cancer cell lines [79]. Similarly, treatment of MCF7 cells with a series of dietary polyphenols, including *ellagic acid, protocatechuic acid, sinapic acid, syringic acid, rosmarinic acid, betanin,* and *phloretin* did not lead to demethylation and reexpression of *GSTP1, RASSF1A*, and *HIN1*, although all of these compounds at the same concentrations inhibited DNMT activity in vitro by 20–88% [40]. Lack of demethylating activity in cell culture might indicate an unspecific enzyme inhibitory effect.

As mentioned above, transcription factor *Nrf2* (nuclear factor-erythroid 2 p45-related factor 2) plays an important role in phase 2 enzyme induction [207]. Recently, *Nrf2* was shown to be epigenetically silenced by promoter methylation at specific CpG sites during prostate carcinogenesis in tumors of transgenic adenocarcinoma of mouse prostate (TRAMP) mice and tumorigenic TRAMP C1 cells. In contrast, the *Nrf2* promoter CGI was unmethylated in normal prostate tissue and non-tumorigenic TRAMP C3 cells. Methylation led to transcriptional repression by increased binding of methyl binding protein 2 (MBD2) and H3K9me3, and reduced interaction with RNA polymerase II and the activating histone mark acetylated histone 3 (ac-H3) [210]. Treatment of TRAMP C1 cells with *curcumin* significantly reduced *Nrf2* promoter methylation at five specific CpG sites and led to mRNA reexpression of *Nrf2* and NAD(P)H:quinone reductase (*NQO1*) as a downstream target [49]. Curcumin (diferuloyl methane) is a well characterized cancer chemopreventive agent derived from turmeric (*Curcuma longa*) [211].

7 Cell Cycle Regulation

One of the hallmarks of cancer cells is their ability to evade growth-suppressing signals. Various genes affecting cell cycle progression have been identified as tumor suppressor genes, first of all *p53* and *pRB* [212]. Progression through the cell cycle is regulated through activation and inactivation of cyclin-dependent kinase (Cdks) that form sequential complexes with cyclins A–E during the different phases G_1, S, G_2, and M of the cell cycle. During G_1 phase, Cdk2–cyclin E and Cdk4/6–cyclin D1 complexes promote entry into S-phase by phosphorylation of *pRB*, thereby releasing the transcription factor *E2F* [213]. The activity of Cdks is controlled by binding of Cdk inhibitors (CKIs) to Cdk–cyclin complexes. CKIs *p21*, *p27*, and *p57* preferentially interact with Cdk2– and Cdk4–cyclin complexes, whereas CKIs *p15^{INK4B}* and *p16^{INK4A}* are more specific for Cdk4– and Cdk6–cyclin complexes and block their interaction with cyclin D [213].

Interestingly, both DNA methylation and histone acetylation are involved in the regulation of CKI expression, as exemplified with *p16^{INK4A}* and *p21$^{CIP1/WAF1}$*. *p16^{INK4A}* (inhibitor of Cdk4, also known as *CDKN2*, CDK inhibitor 2) is genetically inactivated by point mutations, deletion, or DNA methylation in about 50% of all human cancers [214]. Hypermethylation of the *p16* promoter is frequently observed in all major human malignancies, including hepatocellular carcinoma, primary gastric carcinoma, Barrett's esophagus and esophageal adenocarcinoma [214], breast cancer [215], squamous cell carcinoma of the lung [216], colorectal cancer [217], lymphoma [218], as well as tumors of the ovary, uterus, head and neck, brain, kidney, bladder, and pancreas [219]. Murine *p16* knockout strains are more prone to spontaneous tumorigenesis than wildtype littermates, whereas overexpression of *p16* led to a threefold reduction of spontaneous cancers [220].

Several studies have investigated whether natural products were able to demethylate and reactivate *p16* in a wide variety of cancer cell lines. Fang et al. reported demethylation and re-expression of *p16* in KYSE510 esophageal cancer cells and HCT116 colon cancer cells after treatment with *EGCG* [37, 55]. These results could not be confirmed in a subsequent study by Chuang et al. [56] using T24 bladder cancer cells, HT 29 colon cancer cells, and PC3 prostate cancer cells. In A431 epidermoid carcinoma cells, EGCG decreased global methylation and inhibited DNMT activity as well as expression of *DNMT1*, *3a*, and *3b*, which led to the reexpression of *p16* mRNA and protein [61]. *Genistein* treatment of KYSE510 esophageal cancer cells resulted in dose-dependent and time-dependent demethylation and re-expression of *p16* [78]. In a study by Fini et al., intervention of RKO, SW48, and SW480 colon cancer cells with an *apple polyphenol extract* also resulted in *p16* promoter demethylation and mRNA or protein reexpression. This was explained by downregulation of *DNMT 1* and *DNMT 3b* protein expression in RKO and SW480 cells [38]. *Nordihydroguaiaretic acid (NDGA)* was investigated in RKO and T47D breast cancer cell lines. *p16* promoter demethylation and reactivation was associated with reduced *cyclin D1* expression and *RB* phosphorylation, G_1 cell cycle arrest, and increased senescence [96]. *Phenylhexyl isothiocyanate (PHI)* was initially identified as an HDAC inhibitor, as described below. Lu et al. were able to demonstrate that intervention in RPMI8226 myeloma cells reduced *p16* promoter methylation and induced cell cycle arrest in G_1 phase [102].

p21, also known as CDK-interacting protein 1 (Cip1) or wild-type p53-activated fragment 1 (WAF1), is encoded by the cyclin-dependent kinase inhibitor 1 *CDKN1A* gene locus [221–223]. *p21* directly inhibits the activity of Cdk2/cyclin E and functions as an adaptor protein for Cdk4/6/cyclin D complexes, thereby modulating cell cycle progression at S-phase [224]. Overexpression of *p21* can lead to G_1-phase, G_2-phase, or S-phase arrest, whereas *p21*-deficient cells fail to undergo cell cycle arrest in response to p53 activation after DNA damage [225]. In addition to cell cycle regulation, *p21* is involved in regulation of cell differentiation, senescence, gene transcription, apoptosis, and DNA repair (review in [223]). *p21* knockout mice are prone to development of spontaneous tumors [223]. In contrast to *p16* or *p53*, mutations in *p21* are extremely rare (summarized in [225]). In comparison to other tumor suppressor genes, methylation at the *p21* promoter was not frequently observed in hematological malignancies [226]. *p21* was overexpressed after downregulation of DNMTs, but the mechanism of induction might be independent of changes in promoter methylation and rather involve competing interactions of DNMTs and *p21* with *PCNA* and enhanced stability [224, 227]. *p21* expression is more commonly regulated at the transcriptional level, and chromatin structure controlled by histone acetylation seems to play an important role. The *p21* promoter region contains binding sites for *p53* and Sp1/3, several E-boxes, and can be repressed by the oncogene *c-Myc* [224]. Inhibition of *HDAC* activity, in addition to opening the chromatin structure, has been suggested to lead to a release of *HDAC1* from the *p21* promoter, thereby facilitating binding of Sp1/3 and HATs *p300* or *PCAF*. Indirectly, hyperacetylation of *p53* through HDAC inhibition may promote *p21* transcription by enhancing the affinity of *p53* to the *p21* promoter (summarized in [224]). Alternatively, *p21* expression can be transcriptionally silenced through recruitment of *CTIP2 (COUP-TF-interacting protein 2)* and interactions with HDACs and histone methyltransferases (HMTs) [180].

Butyric acid (its sodium salt being referred to as "butyrate") is a major short-chain fatty acid produced by colonic fermentation of resistant starch and dietary fiber. Butyrate was first described to inhibit *HDAC* activity in vitro and in cell culture models more than 30 years ago. Initial work focused on its anti-proliferative and differentiation-inducing effects in leukemia cell lines [228–230]. Since dietary fiber consumption has been associated with colon cancer prevention [231], Archer et al. established a link between butyrate-mediated *HDAC* inhibition, *p21* induction, and cell growth inhibition in colon cancer cell lines [130]. Induction of *p21* mRNA and protein expression was also associated with histone hyperacetylation and colon cancer prevention in 1,2-dimethylhydrazine-induced tumorigenesis in a mouse model of colorectal cancer [133].

Dietary sources of selenium, such as Se-methyl-Se-cysteine (SMC) and Se-methionine (SM), can be metabolized to *α-methylselenopyruvate* (*MSP*) and *α-keto-γ-methylselenobutyrate* (*KMSB*) with structural similarity to butyrate [156]. Consequently Nian et al. investigate *HDAC*-inhibitory potential of these α-keto acid metabolites. MSP and KMSB caused a dose-dependent inhibition of human *HDAC1* and *HDAC8* activities in vitro. Enzymatic kinetic studies and computational molecular modeling identified MSP as a competitive inhibitor of *HDAC8*, based on reversible interaction with the active site zinc atom. In human colon cancer cells, MSP and KMSB dose-dependently inhibited *HDAC* activity and increased global H3 acetylation and *p21* expression levels, which led to G_2/M cell cycle arrest and apoptosis induction [156]. In a seminal study published in 2004, Myzak et al. first suggested that *sulforaphane* (*SFN*) might possess *HDAC*-inhibitory activity, based on the observation that SFN treatment caused *p21* upregulation and cell cycle arrest, similar to the activities of butyrate. SFN failed to inhibit directly *HDAC* activity in cell-free systems in vitro. Rather, in silico modeling indicated that SFN-Cys, an SFN metabolite, might possess *HDAC* inhibitory potential. Consistently, cell culture media after incubation with SFN contained a metabolite able to inhibit *HDAC* enzymatic activity [169]. Further studies confirmed the *HDAC* inhibitory activity of SFN intervention in various human cancer cell lines [169, 170, 174]. In human prostate cancer cells, SFN treatment increased global histone acetylation, accompanied by locus-specific hyperacetylation of H3, H4, or both at the *p21* promoter [170]. A study of SFN intervention in $APC^{Min/+}$ mice underlined the relevance of HDAC inhibition for chemopreventive activity of SFN. A single dose of SFN lowered *HDAC* activity and transiently increased ac-H3 and ac-H4 levels in colonic mucosa of wild-type mice [176]. Long-term application for 10 weeks produced similar effects in ileum, colon, prostate, and peripheral blood mononuclear cells (PBMC). In $APC^{Min/+}$ mice, SFN treatment reduced tumor multiplicity, increased ac-H3 levels, and ac-H3 occupancy at the *p21* and *Bax* promoter in tumor samples, and induced expression of pro-apoptotic *Bax* [176]. *Bax* is a member of the Bcl-2 protein family of apoptosis regulators which play an important role in mediating the intrinsic, mitochondrial pathway of apoptosis induction [232, 233]. SFN reduced growth of androgen-independent human prostate cancer cells in a xenograft model, and increased global histone acetylation in prostate tissue and in xenografts [177]. In a human pilot study, three healthy volunteers ingested 68 g of broccoli sprouts as a source of SFN. After 3 h and 6 h the intervention transiently induced strong hyperacetylation of H3 and H4 in PBMCs, concomitant with HDAC inhibition. Both acetylation and enzyme activity returned to normal levels by 24 and 48 h [178]. These findings support a role for SFN as an HDAC inhibitor in vivo, with evidence for decreased HDAC activity in various tissues, increased global histone acetylation, as well as enhanced localization of acetylated histones at specific promoters. These findings may also be relevant for human cancer prevention.

Two additional isothiocyanates (ITCs), *PEITC* found in water cress [234, 235], as well as the synthetic *PHI* were also confirmed as inhibitors of HDACs, suggesting that this might be a more common mechanism of ITCs. Exposure of prostate cancer cells to PEITC significantly enhanced histone acetylation, cell cycle arrest, and *p53*-independent up-regulation of CKIs, including *p21* and *p27* [158]. Similar to SFN and PEITC, PHI was first identified as an HDAC inhibitor and inducer of cell cycle arrest, but was also shown to reduce *p16* promoter methylation in myeloma cells [102]. HDAC inhibitory potential and chromatin modifications were confirmed in human prostate and liver cancer, and leukemia and myeloma cells. PHI affected both the expression as well as the activity of *HDAC1* in LNCaP and HL-60 cells [159, 160]. In leukemia cells, PHI treatment increased expression of the HAT *p300/CBP* [161]. Increased levels of ac-H3 and ac-H4 were commonly detected in all cell lines, as well as in bone marrow of AML patients [163]. This was further associated with increased interaction of acetylated histones with the *p21* promoter, *p21* induction, G_0/G_1 cell cycle arrest, and apoptosis induction [160–162].

In addition to sulfur-containing ITCs, dietary organosulfur compounds found in garlic and other *Allium* species such as *diallyldisulfide (DADS)* have been shown to inhibit HDAC activity. After consumption, DADS is converted to the active metabolite S-allylmercaptocysteine (SAMC). Both compounds are further metabolized to *allyl mercaptan (AM)* and other metabolites (reviewed in [121]). Induction of histone acetylation by DADS and SAMC was first described in murine erythroleukemia cells [236]. Interestingly, when testing *HDAC* inhibitory potential in vitro, AM was more potent than the precursor compounds DADS and SAMC. Nian et al. predicted direct binding of AM to the *HDAC* active site by in silico docking studies and confirmed inhibitory potential in vitro and in cell culture. *HDAC* inhibition by AM led to hyperacetylation of H3 and H4, enhanced ac-H3 association with the *p21* promoter, upregulation of *p21*, and cell cycle arrest [123]. DADS treatment induced transient histone hyperacetylation followed by *p21* induction, cell-cycle arrest, and induction of differentiation and apoptosis in various cancer cell lines (reviewed in [141]). Intracecal perfusion or intraperitoneal injection of DADS (200 mg/kg b.w.) to male rats also resulted in histone hyperacetylation in normal hepatocytes and colonocytes [142]. These data indicate that effects on histone acetylation and downstream mechanisms induced by organosulfur compounds may be relevant for preventive efficacy, although the described effects observed both in vitro as well as in vivo require doses that might not be reached by dietary consumption of *Allium* vegetables. Also, inhibition of *HDAC* activity and histone hyperacetylation are transient effects. This may suggest that the compounds or dietary sources have to be consumed regularly to achieve long-term effects in vivo. *Apicidin*, a fungal metabolite, is a cyclic tetrapeptide antibiotic with broad spectrum antiparasitic, antiprotozoal, and potential antimalarial properties [127]. Apicidin treatment at low microgram per milliliter concentrations inhibited cell proliferation in a series of cancer cell lines. Apicidin induced morphological changes, accumulation of ac-H4, and G_1 cell cycle arrest in human cervical cancer cells. This led to induction of *p21* and *gelsolin* involved in cell cycle control and cell morphology, respectively. Decreased phosphorylation of *Rb* protein was indicative of Cdk inhibition. Interestingly, in contrast to the dietary HDAC inhibitors described above, the effects of apicidin on cell morphology, expression of *gelsolin*, and *HDAC1* activity appeared to be irreversible [127]. So far, apicidin has not been tested in animal models for chemopreventive activity.

In addition to these direct effects on HDAC activity, several chemopreventive agents, including the soy isoflavone *genistein*, *3,3'-diindolylmethane* (*DIM*) derived from cruciferous vegetables, *parthenolide*, a sesquiterpene lactone from feverfew, the fungal metabolite *chaetocin*, and *EGCG* have been described to modulate histone acetylation by changing the expression of histone modifying enzymes.

In prostate cancer cell lines, *genistein* treatment caused an upregulation of histone acetyl transferases (HATs) *CREB-binding protein* (*CREBBP*), *p300*, *PCAF*, and *HAT1*. This resulted in hyperacetylation of histones H3 and H4, increased association of acetylated H3K4 with the transcription start sites of *p16* and *p21*, re-expression of *p16* and *p21*, and cell cycle arrest [153]. Indole-3-carbinol (IC3) is the main hydrolysis product of the glucosinolate glucobrassicin [234]. Under low gastric pH conditions I3C is condensed to polycyclic compounds such as *DIM* as the major condensation product [237]. In a study by Li et al., DIM selectively induced proteasomal degradation of the class I histone deacetylases *HDAC1, 2, 3,* and *8* in human colon cancer cells in vitro and in tumor xenografts, without affecting class II HDACs. HDAC depletion resulted in re-expression of *p21* and *p27* and triggered cell cycle arrest in G_2/M phase. Additionally, HDAC depletion was associated with DNA damage and apoptosis induction [144]. *Parthenolide* was described as an HDACi-like compound with ability to induce transient and selective ubiquitination and proteasomal degradation of *HDAC1* in breast cancer and other cancer cell lines, whereas other classes I and II HDACs were not affected. Downstream effects were similar to those of HDACi, with *p53*-independent upregulation of *p21* and global histone hyperacetylation. Downregulation of *HDAC1* involved the phosphoinositide-3-kinase-like kinase *ATM* (ataxia telangiectasia), as siRNA-mediated knockdown of *ATM* severely affected parthenolide-induced degradation of *HDAC1*. However, the exact mechanism how parthenolide induces *HDAC1* degradation via *ATM* is presently unknown [157].

In addition to increased histone acetylation through various mechanisms, inhibition of repressive histone methylation marks also results in upregulation of *p21*. *Chaetocin*, a fungal metabolite, was one of the first identified selective inhibitors for the *SUV39* class of HMTs targeting H3K9 (overview in [238]). H3K9 trimethylation is generally associated with repressed chromatin. Chaetocin treatment of microglial cells transfected with a *p21*-promoter reporter construct repressed H3K9 trimethylation at the *p21* promoter, stimulated *p21* expression, and induced cell cycle arrest [180].

Recent research indicates that *EGCG* may regulate expression of cell cycle regulators *p21* and *p27* and apoptotic proteins by influencing *polycomb group* (*PcG*)-mediated histone modifications [184]. *PcG* proteins, including *BMI-1* and

EZH2, are HMTs that increase H3K27 methylation leading to a repressed chromatin conformation and enhanced cell survival. In skin cancer cells EGCG treatment reduced levels of *BMI-1* and *EZH2*, lowered H3K27me3 levels, and reduced cell survival. This was associated with induction of cell cycle regulators and activation of caspases and *Bcl-2* family proteins. The inhibitory effects of EGCG on *BMI-1* expression were corroborated by overexpression of *BMI-1* [184]. EGCG treatment of human epidermoid carcinoma cells reduced H3K9 methylation and concomitantly increased H3 and H4 acetylation by HDAC inhibition. This was associated with an upregulation of *p16* and *p21* mRNA and protein levels [61].

RassF1A (Ras Association Domain family 1, isoform A) is a candidate tumor suppressor gene located on the chromosome 3p21.3 locus that is frequently inactivated in cancer by loss of heterozygosity. *RassF1A* promoter methylation and silencing have been described as the most frequent epigenetic change observed in human cancers, including lung, breast, pancreas, kidney, liver, cervix, nasopharyngeal, prostate, thyroid, and other cancers [239, 240]. Loss of *RassF1A* is associated with advanced tumor stage and poor prognosis. Since *RassF1A* hypermethylation is detectable in various body fluids including blood, urine, nipple aspirates, sputum, and bronchial alveolar lavages, it may serve as a valuable diagnostic or prognostic marker [239]. *RassF1A* knockout mice are viable and fertile, but prone to spontaneous tumorigenesis [241]. *RassF1A* is involved in two pathways commonly deregulated in cancer – cell cycle regulation and apoptosis [239, 240]. Overexpression of *RassF1A* in vitro was found to inhibit accumulation of *cyclin D1*, thereby blocking G_1/S cell cycle progression [242].

Numerous studies have attempted to demethylate and reexpress *RassF1A* by chemopreventive agents in vitro or dietary intervention in vivo. Most of these studies have reported negative results. As summarized in Table 1 (Appendix), *genistein* and *seleno-D,L-methionine* did not influence the methylation status of *RassF1A* in prostate cancer cell lines in vitro [83, 111]. In a randomized 4-week human intervention study with *cruciferous vegetables* or *soy products* in combination with *green tea*, neither treatments influenced methylation of *RassF1A* and a series of other candidate genes in PMBCs of heavy smokers, whereas methylation of the repetitive element *Line1* (long interspersed nuclear element) was slightly but significantly increased [47]. Also, 4-week dietary intervention in 34 healthy premenopausal women with daily doses of 40 or 140 mg *isoflavones* did not influence *RassF1A* methylation in intraductal specimens [92]. Jagadeesh et al. tested the effect of *mahanine*, a carbazole alkaloid found in some Asian vegetables, in a series of prostate cancer and several other human cancer cell lines. Mahanine treatment at low microgram per milliliter concentrations led to reexpression of *RassF1A*, reduced expression of *cyclin D1* and inhibition of cell proliferation. The authors did not investigate changes in *RassF1A* promoter methylation, but *DNMT* activity in mahanine-treated prostate cancer cell lines was significantly reduced. In a subsequent study, a synthesized mahanine derivative was equally or even more effective as mahanine with respect to inhibition of PC-3 cell proliferation, DNA synthesis, and *DNMT* activity, reactivation of *RassF1A* mRNA expression, and downregulation of *cyclin D1* [94]. The derivative was shown to act by sequestering

DNMT3b, but not *DNMT3a* in the cytoplasm. Consistently, depletion of *DNMT3b* was shown previously to cause *RASSF1A* reactivation, cell growth inhibition, and apoptosis induction in cancer cell lines, but not in normal cells [14]. In Balb/c nude mice, the mahanine derivative was not toxic after oral application at concentrations up to 550 mg/kg. It reduced growth of PC-3 xenografts by 40% when applied at 10 mg/kg body weight every other day for 4 weeks. The influence of epigenetic mechanisms for tumor growth inhibition was however not investigated [94].

8 Apoptosis

Tissue homeostasis is balanced by cell proliferation and cell death. Evading apoptosis (programmed cell death) has been recognized as one of the hallmarks of cancer cells [243]. Apoptosis can be triggered when cells sense abnormalities such as DNA damage, imbalance in signaling by aberrant activation of oncogenes, lack of survival factors, or hypoxia [243]. *p53* is one of the most important pro-apoptotic mediators involved in sensing DNA damage. It is lost or functionally inactivated in more than 50% of all human tumors [243]. p53 activity is also epigenetically controlled: deacetylation of *p53* through *SIRT1* (silent information regulator 1), a member of the sirtuin HDAC class III family, prevents *p53*-mediated transactivation of cell cycle inhibitor *p21* and pro-apoptotic *Bax*, allowing promotion of cell survival after DNA damage and ultimately tumorigenesis [193]. Inhibition of *SIRT1* should therefore lead to induction of apoptosis by counteracting the deacetylation of *p53* and other key factors such as *FOXO3a*. However, despite the fact that SIRT1 can inactivate *p53* and is upregulated in several human cancer types, recent data suggest that SIRT1 is a tumor suppressor in vivo [244].

Two natural products, *cambinol* and *dihydrocoumarin* (*DHC*) have been identified as *SIRT* inhibitors. The β-naphthol compound *cambinol* was identified in a chemical screen and inhibits both *SIRT1* and *SIRT2*, whereas class I and II *HDAC*s were not affected [134]. Cambinol acts as a competitive inhibitor with respect to the histone H4 peptide and as a non-competitive inhibitor with respect to the co-substrate NAD$^+$. In lung cancer cells, cambinol treatment in combination with etoposide to induce DNA damage led to hyperacetylation of *SIRT* target proteins such as *p53, FOXO3a* and *Ku70*. Deacetylation of these later proteins promoted cell survival under stress, which was abrogated by inhibition of *SIRT* with cambinol. *BCL6* is a transcriptional repressor that is also deacetylated by *SIRT*. In *BCL6*-expressing Burkitt lymphoma cells, treatment with cambinol induced apoptosis, accompanied by hyperacetylation of *BCL6* and *p53*. In vivo, cambinol intervention at a dose of 100 mg/kg i.v. or i.p. inhibited growth of Burkitt lymphoma xenografts in SCID mice and was well tolerated [134]. *DHC*, a component of *Melilotus officinalis* (sweet clover), is frequently used in cosmetics or as a flavoring agent. DHC was identified as an inhibitor of yeast *Sir2p* and human *SIRT1* activity. Treatment of human TK6 lymphoblastoid cells with DHC led to a dose-dependent induction of *ac-p53*, cytotoxicity, and apoptosis [143]. Kahyo et al. attempted to identify novel inhibitors of sirtuins (*SIRT*s), also known as class III HDACs. Using acetylated *p53* as a substrate, they identified the synthetic *3,2′,3′,4′-tetrahydroxychalcone* as an inhibitor of *SIRT* activity and *p53* deacetylation in vitro. Treatment of human embryonic kidney cells with the chalcone induced hyperacetylation of endogenous *p53*, increased *p21* expression and suppressed cell growth. Since HDAC inhibitory potential of the compound was not tested, it is difficult to conclude whether *p21* induction is indeed mediated via inhibition of *SIRT1* [135].

An alternative mechanism leading to hyperacetylation of *p53* and apoptosis induction is mediated through the activity of *MTA1/HDAC1* in the nucleosome remodeling deacetylation (NuRD) complex. *MTA1* (metastasis-associated protein 1) expressed in various cancers has been associated with aggressiveness and metastasis [165]. Kai et al. identified that treatment of prostate cancer cells with *resveratrol* resulted in down-regulation of *MTA1*. This functionally blocked the *MTA*/NuRD complex and led to hyperacetylation of *p53*, trans-activation of *p21* and *Bax*, and apoptosis induction. This effect was corroborated by knockdown of *MTA1* and further enhanced by cotreatment with the HDACi suberoylanilide hydroxamic acid (SAHA). These combination effects might present an innovative therapeutic strategy for the management of prostate cancer [165].

The tumor suppressor *PTEN* (phosphatase and tensin homolog deleted on chromosome 10) negatively regulates the *phosphatidylinositol 3-kinase (PI3K)-AKT* pathway that transmits anti-apoptotic survival signals and regulates cell proliferation, growth and motility [245]. Downstream signaling is indirectly mediated via transcription factors such as *NF-κB* and *FOXO* [245, 246]. Somatic *PTEN* deletions and mutations, and epigenetic inactivation of *PTEN* by promoter methylation or miRNA silencing are common in multiple tumor types. Silencing through epigenetic mechanisms frequently occurs in breast, prostate, thyroid, and lung cancer, glioma, and melanoma, whereas mutations and deletions are common in endometrium, bladder, kidney, colorectal cancer, and leukemias. PTEN$^{-/-}$ was shown to lead to early onset of prostate or mammary cancer in mouse models [245, 246].

PTEN is hypermethylated in breast cancer cell lines MCF-7 and MDA-MB-231. Stefanska et al. analyzed whether *PTEN* silencing could be reversed in these cell lines after incubation with the chemopreventive agents *all-trans-retinoic acid* (ATRA), *Vitamin D₃*, and *resveratrol* alone and in combination with nucleoside analogs such as 2-chloro-2′-deoxyadenosine (2CdA), 9-β-D-arabinosyl-2-fluoroadenine (F-ara-A), and 5-aza-2′-deoxycytosine (5-Aza) [104]. In MCF-7 cells with a methylation level of about 30% at the *PTEN* promoter, incubation with all three natural products resulted in demethylation and reexpression of *PTEN*. This was associated with down-regulation of *DNMT1* and upregulation of *p21* after incubation with vitamin D₃ and resveratrol. The effects were further enhanced by co-incubation with 2CdA and F-ara-A. In highly invasive MDA-MB-231 cells, the PTEN promoter was >90% methylated. Only Vitamin D₃ treatment was able to reduce methylation and to enhance concomitantly expression of *PTEN*, whereas the combined treatment with nucleoside analogs did not enhance efficacy [104]. Kikuno et al. investigated whether *genistein* might suppress *AKT* signaling via epigenetic mechanisms. In prostate cancer cell lines, genistein treatment led to reexpression of *PTEN* and consequential inactivation of *AKT*, resulting in induction of *p53* and *FOXO3a*. Genistein treatment also upregulated the endogenous *NF-κB* inhibitor *CYLD* and decreased constitutive *NF-κB* activity. These effects were likely unrelated to inhibition of DNA methylation, as promoter regions of all of these factors were unmethylated in the investigated cell lines. Rather, reexpression was associated with elevated H3K9 acetylation (*PTEN*, *CYLD*, *p53*, and *FOXO3a*) and loss of H3K9 methylation (*PTEN* and *CYLD*). H3K9 hyperacetylation could be associated with reduced expression and nuclear localization of *SIRT1* after genistein treatment [154].

Death-associated protein kinase (*DAPK*) is a pro-apoptotic serine/threonine kinase acting in the extrinsic death receptor-mediated pathway of apoptosis induction [233, 247]. *DAPK* is induced by *p53* activation and in turn elevates *p53* expression, supporting the existence of an autoregulatory feedback loop between *DAPK* and p53 that controls apoptosis. In addition to apoptosis induction, *DAPK* is also involved in the control of autophagy, which can lead to cell survival or cell death depending on the cellular context (review in [247]). *DAPK* expression is reduced in a wide range of cancer types by promoter methylation, including lung, bladder, head and neck, kidney, breast, and B-cell malignancies. Detection of *DAPK* methylation has been suggested as a useful prognostic biomarker for invasive and metastatic potential [247]. *DAPK* is an *NF-κB* regulated gene. Hypermethylation of *DAPK* might be mediated by a targeted recruitment of *DNMTs* to *RelB* (a subunit of NF-κB)-regulated genes via *Daxx*, an apoptosis regulator. *DAPK* function is also lost by deletion and point mutations [247]. In a study by Fang et al. treatment of mouse lung cancer cells with EGCG in combination with trichostatin (TSA) or butyrate synergistically increased mRNA levels of *DAPK* and retinoic acid receptor β (*RARβ*), indicating a reversal of epigenetic silencing. *DAPK* promoter methylation was not investigated in this study.

9 DNA Repair

Cancer genomes are characterized by accumulation of genomic instability and chromosomal aberrations, associated with underlying defects in the DNA repair machinery [248]. Important DNA repair genes, such as the mismatch repair gene *hMLH1* and the DNA-alkyl repair gene *MGMT* (O^6-methylguanine DNA methyltransferase) are commonly inactivated in human cancers by CpG island hypermethylation. Loss of *hMLH1* expression by germ-line mutations and promoter hypermethylation leads to microsatellite instability that is mainly associated with hereditary non-polyposis colorectal cancer (HNPCC), but also observed in endometrial and gastric tumors [249]. *MGMT* repairs promutagenic O^6-methylguanine adducts by transferring the methyl group to a cysteine residue in its active site. Methylated *MGMT* is then degraded by the proteasome. *MGMT* has been shown to be silenced by aberrant methylation in a large spectrum of human tumors, with highest hypermethylation rates in tumors of the testis and colon, in retinoblastoma, glioma, head and neck and cervical cancer, lymphoma, lung, esophageal, gastric and pancreatic cancer, and several further cancer types. It has been suggested that silencing of *MGMT* is associated with 72% of the mutations observed in the *p53* gene, and with 40% of the colon cancer cases induced through *K-ras* mutations [250]. Noteworthy, although loss of *MGMT* expression contributes to tumorigenesis and is a marker of poor prognosis, glioma patients with reduced *MGMT* activity respond better to treatment with alkylating agents [251].

Several studies have investigated the effect of natural products on the methylation status and expression of repair genes. *EGCG* and *genistein* treatment resulted in reduced *MGMT* and *hMLH1* promoter methylation and mRNA/protein reexpression in human esophageal carcinoma cells [37, 55, 78, 252]. Incubation of colon cancer cell lines with *apple polyphenols* also led to reexpression of *hMLH1* by promoter hypomethylation due to reduced *DNMT1* and *DNMT3b* protein expression [241]. This effect on DNA methylation may contribute to the colon cancer preventive efficacy of apple polyphenols (reviewed in [253]). In the transgenic adenocarcinoma of the mouse prostate (TRAMP) model, intervention with *PEITC* given at a dose of 15 µmol daily by gavage for 13 weeks significantly reduced prostate tumor formation and lowered *MGMT* promoter methylation in tumor tissue [101]. In the same model, intervention with *5-aza-2'-deoxycytidine* (5-Aza) at a dose of 0.25 mg/kg twice per week completely prevented prostate cancer development at 24 weeks of age, whereas in 54% of the control mice poorly differentiated prostate cancers were detected upon necropsy. Treatment with 5-Aza also prevented lymph node metastases and dramatically extended survival compared with control-treated mice. In tumor tissue, *MGMT* promoter methylation was reduced by 5-Aza treatment, and *MGMT* mRNA expression was induced [254].

10 Inflammation and Regulation of NF-κB

Epidemiological evidence indicates that chronic infections and subsequent inflammation are causally linked to about 15–20% of all cancer deaths [255, 256]. Examples include chronic infections with *Hepatitis B* and *C* virus and risk for hepatocellular carcinoma, infections with *Helicobacter pylori* and gastric cancer, chronic inflammatory bowel diseases and colorectal cancer, and chronic airway irritations and inflammation caused by tobacco smoke and lung cancer [255]. Chronic inflammatory conditions are characterized by the accumulation of inflammatory cells, which are recruited to the tumor tissue and contribute to the stromal tumor microenvironment and the release of tumor-promoting pro-inflammatory mediators [256]. These factors facilitate evasion from host defense mechanisms, promote genomic instability, regulate growth, migration, and differentiation, alter response to hormones and chemotherapeutic agents, and stimulate angiogenesis and metastasis [256, 257].

One of the most important transcription factors controlling inflammatory conditions is *NF-κB* [258]. *NF-κB* is a homodimer or heterodimer of members of the *NF-κB* subunit family, consisting of *RELA* (also known as *p65*), *RELB*, *REL*, *p50*, and *p52*. All these members contain a REL homology domain that allows DNA-binding and dimerization (for further detailed information refer to [255, 259]). During carcinogenesis, aberrant *NF-κB* activation regulates transcription of anti-apoptotic genes, cyclins, and oncogenes that promote cell proliferation, pro-angiogenic genes, as well as matrix metalloproteinases and cell adhesion genes [259]. Interestingly, *NF-κB* activity is partly controlled by post-translational modifications, including phosphorylation, acetylation, methylation, and ubiquitinylation [259]. Reversible acetylation at lysine 310 mediated by the HAT *p300* is required for full trans-activating activity [260–262].

NF-κB has been extensively studied as a target for chemopreventive agents [263]. Interestingly, recent research now establishes a link between *NF-κB* and chemopreventive agents via an indirect epigenetic mechanism by inhibition of *NF-κB* acetylation mediated by *p300* HAT. *Anacardic acid* (6-nonadecyl salicylic acid) isolated from cashew nut shell liquid was identified as the first natural product inhibitor of *p300* HAT activity. In a natural product screen it was found to inhibit *p300* and *PCAF* activities with IC_{50} values of 8.5 and 5 μM, respectively [124]. In a study by Sung et al., anacardic acid blocked *NF-κB* activation by *TNF-α* and a series of other stimuli and suppressed acetylation and nuclear translocation of the *NF-κB* subunit *p65*. Anacardic acid-mediated effects could be mimicked by down-regulation of *p300* HAT by siRNA, indicating that *p300* is a key mediator of the effects of anacardic acid on *NF-κB* signaling. In cancer cell lines, anacardic acid potentiated *TNF-α*-, cisplatin-, and doxorubicin-mediated apoptosis induction, and strongly suppressed *TNF-α*-mediated upregulation of *NF-κB* target genes, including the anti-apoptotic proteins *Bcl-2*, *Bcl-xL*, *cFLIP*, *cIAP-1*, and *survivin*, as well as *cyclin D1*, *c-Myc*, *Cox-2*, *VEGF*, *ICAM*-1, and *MMP9* involved in invasion and angiogenesis. Based on these results, anacardic acid might be an interesting lead compound for further development in cancer prevention [126]. *Garcinol* is a polyisoprenylated benzophenone isolated from the Mangosteen tree *Garcinia indica* Choisy (*Clusiaceae*) [264]. Garcinol was identified as a cell-permeable inhibitor of *PCAF* and *p300* HAT activities with IC_{50} values of 5 and 7 μM, respectively. In HeLa cells, garcinol treatment repressed general histone acetylation and induced apoptosis [151]. Similar to the activities of anacardic acid, garcinol reduced the expression of various *NF-κB* target proteins, including anti-apoptotic *survivin*, *Bcl-2*, *XIAP*, and *cFLIP* [265]. Although garcinol has previously been reported to inhibit *NF-κB*, acetylation of *p65* was not analyzed in this study. *Curcumin* was identified as a specific inhibitor of *p300/CBP* in vitro and in cell culture, whereas other histone-modifying enzymes, including *PCAF*, *HDAC*, and *HTM* activities were not inhibited by curcumin. *HAT* inhibition was attributed to a structural modification of *p300*, thereby preventing binding of histones or cofactor acetyl-CoA. Curcumin also inhibited acetylation of *p53* as a non-histone target of *p300/CBP* [137, 138]. In Raji cells, curcumin treatment significantly down-regulated levels of *HDAC1* and *p300* protein and mRNA. Reduction was prevented by co-treatment with MG-132, an inhibitor of the 26S proteasome [136]. Although not specifically addressed in these studies, direct inhibition and down-regulation of *p300* might contribute to the well-known inhibition of *NF-κB* by curcumin [266].

In a natural product screen, Choi et al. identified *gallic acid* from rose flowers, a simple polyphenol found in various fruits, tea, and wine, as a novel inhibitor of *p65* acetylation, leading to suppression of lipopolysaccharide (LPS)-induced *NF-κB* signaling [149]. Gallic acid was found to inhibit uncompetitively *p300* HAT activity with an IC_{50} value of 14 μM. Other HATs, such as *PCAF* and *Tip60*, were inhibited to a lesser extent, whereas *SIRT1*, *HDAC*, and *HMT* activities were not affected. In cell culture, gallic acid prevented *p65* acetylation, binding to the *IL-6* promoter, activation of an *NF-κB* reporter construct by LPS, inhibited inflammatory response to various stimuli, and downregulated the expression of *NF-κB*-dependent inflammatory and anti-apoptotic proteins. Inhibition of *p65* acetylation was also confirmed in vivo in macrophages of LPS-stimulated mice [149]. The same group also identified *EGCG* as a *p300* inhibitor with similar effects on *p65* acetylation and downstream pathways as described for gallic acid. Inhibition of *p65* acetylation reduced EBV-induced B-lymphocyte transformation [147]. Recently, they also reported that *delphinidin*, an anthocyanidin plant pigment isolated from pomegranate (*Punica granatum* L.), potently inhibited *p300* HAT activity and suppressed pro-inflammatory signaling through inhibition of *NF-κB* acetylation in synoviocyte cells and in T lymphocytes [140]. Interestingly, all three compounds structurally share a 1,2,3-trihydroxybenzene moiety. The authors did not discuss whether this structural feature might be important for the observed *p300*-inhibitory activity. Overall these data demonstrate that acetylation of NF-κB seems to play an important role in mediating downstream signaling events, and that regulation of *p65* acetylation by inhibition of *p300* might be an interesting target for chemoprevention.

11 Cell Signaling and Cell Growth

Normal cells do not proliferate without mitogenic stimulatory signals. Consequently, "self-sufficiency in growth signals" was defined as one of the hallmarks of cancer cells [243].

Androgen receptor (AR) signaling provides the most important growth stimulus in hormone-dependent prostate cancer. Androgen action is mediated via circulating testosterone levels. Free testosterone enters prostate cells and is converted by 5α-reductase to dihydrotestosterone (DHT) with higher affinity to the *AR* than testosterone. *AR* is sequestered in the cytosol by complexation with heat shock proteins (HSP) such as *HSP90*. After DHT binding, receptor dimerization, phosphorylation, and nuclear translocation, the receptor-ligand complex binds to the androgen-response element in promoter regions of androgen-responsive genes. This leads to recruitment of co-activators, which then facilitate transcription of androgen-sensitive target genes, resulting in increased proliferation and survival [267]. In early stages of prostate cancer, androgen signaling primarily controls cellular growth and proliferation [268], and therefore androgen ablation therapy is carried out as a first line of treatment [269]. An initial response is often followed by an androgen-resistant, lethal disease state. This transition has been attributed to aberrant reactivation of *AR*-signaling that is hypothesized to occur through multiple mechanisms, including *AR* amplification, *AR* mutations, ligand-independent *AR* activation, excessive production of co-activators, and enhanced local production of androgens [270, 271].

Anti-androgen therapy is achieved by compounds binding to the androgen receptor. Alternatively, compounds inhibiting 5α-reductase and the formation of DHT (such as finasteride) are used, but their application in the prevention of prostate cancer is controversial [272].

Chemopreventive agents might indirectly target *AR* signaling via epigenetic mechanisms. *HDAC6* was shown to deacetylate and activate non-histone proteins, including the AR-chaperone *heat shock protein 90* (*HSP90*). Basak et al. reported that *genistein* treatment of LNCaP cells led to enhanced proteosomal degradation of *AR*. Genistein downregulated the expression of *HDAC6*, which resulted in hyperacetylation of *HSP90* and consequent dissociation of the *AR*. Genistein-mediated effects of *HDAC6* downregulation on *AR* were mimicked by *HDAC6* siRNA. These data indicate that prostate cancer preventive potential of genistein may be mediated through modulating the complex of *HDAC6* with *HSP90* and *AR* [152]. Similarly, *SFN* treatment of LNCaP cells induced rapid hyperacetylation of *HSP90* and dissociation of the *AR* by inhibition of *HDAC6* activity. *AR* degradation led to decreased expression of AR target genes such as prostate specific antigen (PSA) and the androgen-regulated fusion of *TMPRSS2* with the oncogene *ERG*. SFN-mediated effects on *AR* were mimicked by *HDAC6* siRNA or treatment with TSA, whereas overexpression of *HDAC6* restored the effects of *HDAC6* inhibition. Therefore, similar to genistein [152], SFN may act as a prostate cancer preventive agent by affecting the complex of *HSP90-AR* through *HDAC6* inhibition [171]. Recently, *EGCG* was shown to affect acetylation of *AR* via inhibition of HAT activity. This was associated with reduced acetylation and nuclear translocation of AR, leading to inhibition of cell proliferation, especially in hormone-dependent prostate cancer cells [148]. In summary, these indirect epigenetic mechanisms might be interesting tools to counteract androgen signaling as a means for prostate cancer prevention.

Wnt signaling plays an important role during embryonic tissue development and tissue homeostasis in adults. Aberrant *Wnt* signaling has been implicated in cancer development in various organs, including colon, skin, liver ovary, breast, and lung [273]. The main function of canonical *Wnt* signaling is controlling the levels of the transcriptional co-activator *β-catenin*. In the absence of Wnt, *β-catenin* levels in the cytosol are regulated through interaction and complex formation with the scaffolding protein *Axin*, *APC* (the gene product of the *adenomatous polyposis coli* gene), *casein kinase* (*CK1*), and *glycogen-synthase kinase 3β* (*GSK3β*). Phosphorylation by *CK1* and *GSK3β* marks *β-catenin* for ubiquitinylation and degradation through the proteasome. Under these conditions, *β-catenin* levels in the nucleus are low, and *Wnt*-target genes are repressed by binding of the *Tcf/Lef* (T cell factor/lymphoid enhancer factor) family of proteins in conjunction with *Groucho* corepressors [274]. Binding of a *Wnt* ligand to the transmembrane receptor *Frizzled* activates the Wnt signaling pathway and ultimately results in the recruitment of *Axin* to the membrane. Consequently, the *CK1/APC/GSK3β* destruction complex gets disrupted, and *β-catenin* is stabilized, accumulates in the cytosol, and finally translocates to the nucleus, where it interacts with *Tcf/Lef* and activates the transcription of *Wnt* target genes, including *c-Myc*, *cyclin D1* and many others [274].

Components of the *Wnt* signaling pathway are mutated or altered in over 90% of human colorectal cancers and in high fractions of other cancer types. In addition to these genetic alterations, endogenous *Wnt* antagonists that inhibit *Wnt* signaling through direct binding to *Wnt* are frequently disrupted by DNA methylation in various cancers. These include *secreted frizzled-related proteins* (*sFRPs*) and *Wnt-inhibitory factor 1* (*WIF-1*) [274].

Several recent studies indicate that the chemopreventive agents *EGCG*, *genistein*, and *black raspberries* reactivate silenced *Wnt* pathway antagonists by promoter demethylation [41, 60, 80]. In lung cancer cell lines treated with *EGCG*, promoter methylation of *WIF-1* was potently reduced, resulting in reexpression of *WIF-1*. This was associated with decreased *β-catenin* levels and reduced *Tcf/Lef* reporter activity, indicating that EGCG can inhibit aberrant *Wnt* signaling in vitro [60]. Wang and Chen reported variable methylation and expression levels of the *Wnt* receptor ligand *Wnt5a* in colon cancer cell lines [80]. In the SW1116 cell line derived from an early stage colorectal cancer, *Wnt5a* promoter methylation correlated with lowest expression compared to cell lines derived from later stage tumors that were not methylated. Treatment with *genistein* reduced SW1116 cell viability by about 80%. Under these conditions, *Wnt5a* mRNA levels increased upon treatment, accompanied by about a 10% decrease in *Wnt5a* promoter methylation [80]. Dose-dependent effects were not analyzed in this study.

Wang et al. performed a small human Phase 1 pilot study with 20 colorectal cancer patients to investigate the effects of intervention with 60 g/day freeze-dried *black raspberries* (*BRB*) for 1–9 weeks on biomarkers of colorectal cancer [41]. Promoter sequences of *Wnt*-inhibitory genes *WIF1*, *sFRP2*, and *sFRP4*, as well as *p16* and the developmental gene *PAX1* were analyzed for methylation changes. Also, expression of downstream *Wnt* target genes, including *β-catenin*, *E-cadherin*, and *c-Myc*, as well as of markers of proliferation, apoptosis, and angiogenesis, was measured in colorectal cancer and adjacent normal tissue. At least a 4 weeks intervention was necessary to detect a significant reduction in promoter methylation of *sFRP2* and *Pax6* in both normal and tumor tissue, comparing samples from before and after intervention. In tumor tissue, promoter methylation of *WIF1* was also significantly lower in the group with higher BRB uptake than in the group with uptake for only about 2 weeks. Reduced methylation levels correlated with lowered expression of *DNMT1* in both normal and tumor tissue in the high BRB dose group. Overall, demethylation of *Wnt* inhibitors led to reduced expression of *β-catenin*, *E-cadherin*, and *Ki67* as a proliferation marker in tumor tissue, and induced apoptosis [41]. This is one of the first studies demonstrating modulation of epigenetic markers and downstream effects in human target tissue after chemopreventive intervention.

Interestingly, a study by Huang et al. indicates that *Wnt* inhibitory genes are repressed not only by DNA methylation but also by histone lysine methylation. As outlined above, histone lysine methylation is regulated by the balance between HMT and HDMs (compare also Fig. 3). *LSD1* is a FAD-dependent amine oxidase which demethylates mono-methylated and di-methylated H3K4 as part of a multiprotein co-repressor complex and thereby broadly represses gene expression ([187] and references cited therein). Since *LSD1* has high homology with monoamine and polyamine oxidases and histone lysine residues resemble polyamines, Huang et al. tested the hypothesis that polyamine analogs might inhibit *LSD1* activity and lead to reexpression of epigenetically silenced genes. Treatment of colon cancer cells with polyamine analogs indeed resulted in re-expression *sFRP1*, *sFRP4*, *sFRP5*s, and transcription factor *GATA5* [186]. This was accompanied by a dose-dependent global increase in H3K4me1 and H3K4me2 levels and enhanced occupancy of these activating histone marks and H3K9ac at the promoters of all re-expressed genes, whereas binding of the repressive marks H3K9me1 and H3K9me2 was reduced. Knockdown of *LSD1* by siRNA recapitulated the effects of the *LSD1* inhibitors on *sFRP* and *GATA5* gene expression [186]. These results were further strengthened by a follow up study that identified two decamine analogs, *PG11144* and *PG11150*, as *LSD1* inhibitors with similar effects on histone methylation and *sFRP* reexpression leading to reduced proliferation and apoptosis induction in colon cancer cell lines. Combined treatment with PG11144 and 5-Aza strongly repressed tumor growth of HCT116 colon cancer xenografts [187]. These data indicate the potential value of *LSD1* inhibitors for the reactivation of silenced genes in cancer prevention or therapy.

hTERT is a catalytic subunit of the enzyme telomerase, which is often upregulated in cancer cells. Telomerase activity is responsible for the maintenance of telomeres which protect chromosome ends from degradation and repair activities to ensure chromosomal stability. Loss of telomeres is associated with ageing, whereas gain of telomerase activity during carcinogenesis enables unlimited cell division [275]. Sequence variations at the *hTERT* locus on chromosome 5 have been associated with many types of cancer, including acute myelogenous leukemia and tumors of the lung, bladder, prostate, cervix, and pancreas (review in [275]). *hTERT* transcription is repressed through binding of the repressor *E2F* to its promoter region. In tumor cells, methylation at the *E2F* binding site prevents *E2F* binding, contributing to elevated expression [54].

ATRA treatment is used in differentiation therapy of leukemia. In human promyelocytic leukemia (HL60) and human teratocarcinoma (HT) cells, ATRA treatment induced cell differentiation and led to progressive histone hypoacetylation. This was coupled with gradual accumulation of *hTERT* promoter methylation, reduced *hTERT* expression, and lower telomerase activity [107]. *hTERT* methylation was not influenced by ATRA treatment in SKBr3 breast cancer cells [276]. In two studies with estrogen receptor (ER)-positive and negative breast cancer cell lines in comparison with an immortalized breast epithelial cell line, treatment with *EGCG* or a prodrug of EGCG with enhanced bioavailability and stability differentially reduced promoter methylation of *hTERT* at selected CpG sites in the cancer cell lines. This allowed enhanced binding of the *E2F* repressor measured by chromatin immunoprecipitation (ChIP), and reduced expression of *hTERT* mRNA. Concomitantly, cell proliferation was reduced in the cancer cell lines by apoptosis induction [54, 62]. Similarly, *genistein* treatment inhibited hTERT transcription by increasing the binding of the repressor *E2F-1* to the *hTERT* core promoter. This was facilitated by site-specific hypomethylation of the *E2F-1* binding site. Reduced methylation was concomitant with genistein-mediated downregulation of *DNMT* expression [81]. Only recently Meeran et al. identified SFN as a DNA demethylating agent. SFN treatment of breast cancer cell lines inhibited telomerase activity and repressed *hTERT* mRNA expression. SFN intervention reduced *DNMT1* and *DNMT3a* protein expression and significantly lowered *hTERT* methylation at CpG sites in exon 1. These sites were identified as binding region for the transcription factor *CTCF* that is also known to act as an *hTERT* repressor. Activating histone marks, including ac-H3, H3K9ac, and ac-H4, were enhanced at the *hTERT* promoter, whereas the inactivating marks H3K9me3 and H3K27me3 were decreased. SFN-induced histone hyperacetylation facilitated binding of *hTERT* repressors *MAD1* and *CTCF* and decreased binding of *c-Myc*. The importance to *CTCF* for SFN-mediated effects was demonstrated by knockdown of *CTCF* that restored *hTERT* expression and decreased the apoptosis-inducing potential of SFN. In addition, SFN treatment inhibited *HDAC* activity and may modulated histone methylation by increased expression of the histone demethylase *RBP2* [173, 178].

12 Cell Differentiation

Retinoid acid receptors (*RAR*) belong to the steroid hormone receptor superfamily of nuclear receptors that play important roles in embryonic development, maintenance of differentiated cellular phenotypes, metabolism, and cell death. Dysfunction of nuclear receptor signaling is implicated in the development of proliferative, reproductive or metabolic diseases such as obesity, diabetes, and cancer [277]. Genetic studies have identified three isoforms of *RAR*, namely *RARα*, *RARβ*, and *RARγ*, that are activated by binding of ATRA and function as heterodimers with a member of the *9-cis retinoic acid receptor* (*RXR*) family represented by *RXRα*, *RXRβ*, and *RXRγ*. *RXR* heterodimerization with *RARs* or other steroid hormone receptors allows fine-tuning of nuclear hormone receptor signaling [277].

Alterations in *RAR* function may contribute to cancer development in two ways.

A fusion of *RARα* with the *promyelocytic leukemia* (*PML*) gene caused by translocation of *RARα* leads to formation of a *PML-RARα fusion protein* that acts as a co-repressor of ATRA-responsive genes and is involved in the development of acute promyelocytic leukemia (APL). This defect is efficiently treated by differentiation therapy with ATRA. Some ATRA-resistant leukemia cells fail to respond to ATRA treatment [278]. Treatment of these ATRA-refractory APL blasts with ATRA plus *HDAC* inhibitors or with demethylating agents restored ATRA sensitivity and cell differentiation [226].

RARβ has been identified as silenced by promoter methylation in various tumor types, including colorectal, breast, prostate, head and neck, stomach, and liver cancer, and lymphoma (overview in [279]). Combination of ATRA with natural or synthetic *DNMT* or *HDAC* inhibitors has been suggested to facilitate reexpression of *RARβ* and may provide beneficial effects for chemoprevention [280]. This was recently demonstrated by the combined intervention with ATRA and *butyrate* as an HDACi in colon cancer cell lines that led to demethylation and reexpression of *RARβ*. Butyrate treatment alone resulted in demethylation of single CpG sites in the *RARβ* promoter. Its effect on *RARβ* reexpression was further enhanced by cotreatment with the soy isoflavone genistein alone or in combination with ATRA [42]. Loss of expression of the *RARβ2* gene is commonly observed during breast carcinogenesis. ATRA therapy failed to induce *RARβ2* in primary breast tumors if the *RARβP2* promoter was methylated. When breast cancer cell lines were treated with ATRA alone or in combination with trichostatin A (TSA) to induce histone acetylation, reactivation of *RARβ2* transcription was facilitated, accompanied by inhibition of cell growth and apoptosis induction [105, 110]. Treatment of APL cells with ATRA reduced *RARβ2* promoter methylation linked with *RARβ2* mRNA reexpression [106]. In the same cell line, Nouzawa et al. were unable to detect ATRA-mediated alterations in *RARβ* CpG island methylation. However, following ATRA-induced differentiation, more than 100 CpG islands within 1 kB of transcription start sites of a known human gene became hyperacetylated [108]. Tang et al. investigated the effect of ATRA at two concentrations alone and in combination with 5-Aza on carcinogen-induced oral cavity carcinogenesis in mice. Both compounds alone and in combination reduced the average number of oral lesions per mouse; combined treatment additionally reduced severity of tongue lesion. Reduction of *RARβ2* mRNA expression in tongue tissue as a consequence of the carcinogen treatment was partly prevented by the combined intervention, whereas carcinogen-induced *Cox-2* and *c-Myc* mRNA expression was inhibited [281].

In studies with natural products, treatment of esophageal cancer cell lines with *EGCG* led to demethylation and reexpression *RARβ2* in a time-dependent and dose-dependent manner [37, 55]. Similar effects were observed with *genistein* in the same cell line [78]. In breast cancer cell lines, Lee et al. reported a slight reduction of *RARβ2* promoter methylation by EGCG intervention [44]. Also, treatment with two coffee polyphenols, *caffeic acid* and *chlorogenic acid*, led to a partial demethylation of the *RARβ2* promoter. Both compounds were potent inhibitors of DNMT activity in vitro [43]. King-Batoon et al. investigated the effects of *lycopene* and *genistein* on *RARβ2* methylation in breast (cancer) cells. A single low dose of *lycopene*, a caroteinoid isolated from tomatoes, reduced *RARβ2* and *HIN1* promoter methylation in immortalized MCF10A human breast cells, but not in MCF-7 breast cancer cells [79]. The mechanism of DNA demethylating activity was not further investigated. In the same study, *genistein* treatment did not result in demethylation of the *RARβ2* promoter in MCF-7 and MDA-MB468 breast cancer cell lines [79]. In a 4-week human intervention trial in 34 healthy premenopausal women, soy isoflavones at two doses led to dose-dependent changes in *RARβ2* and *CCND2* promoter methylation in mammary tissue. Before treatment, methylation levels of both genes were very low. The low dose of isoflavones further reduced methylation, whereas the high dose weakly increased methylation levels of both genes [92].

Jha et al. investigated *RARβ2* promoter methylation in cervical cancer cell lines [51]. Both genistein and curcumin resulted in demethylation of the *RARβ2* promoter and led to the reactivation of the gene, especially after incubation for 6 days. Concomitantly with reduction of *RARβ2* promoter methylation, both compounds induced apoptosis in the cervical cancer cell lines at higher concentrations [51]. Since DNMT bears a cysteine in its active center, Lin et al. speculated that *disulfiram* as a thiol-reactive dithiocarbamate might inhibit DNMT activity. Disulfiram is an inhibitor of aldehyde dehydrogenase currently used clinically for the treatment of alcoholism [282], and has been shown to prevent chemically-induced carcinogenesis in various animal models. Lin et al. demonstrated that disulfiram dose-dependently inhibited *DNMT1* enzyme activity in vitro. In prostate cancer cell lines, global levels of 5me-C decreased upon disulfiram treatment. At the same time, disulfiram intervention decreased *APC* and *RARβ2* promoter methylation and led to reexpression of the genes. Cell growth and clonogenic survival of prostate cancer cell cultures were inhibited in vitro. In vivo, there was a trend for reduced growth of prostate cancer xenografts. So far, a direct causal relationship between tumor growth inhibition and demethylating effects has not been established. Volate et al. analyzed the effect of *green tea* intervention on azoxymethane-induced colon carcinogenesis in the $APC^{Min/+}$ mouse model that is characterized by a defect in *Wnt* signaling due to a mutation in the *APC* gene [64]. Intervention with green tea as a 0.6% solution for 8 weeks significantly reduced the number of colonic tumors by 28%. Expression of *β-catenin* and *cyclin D1* as a *Wnt* target gene was reduced in tumors of the green tea group. Interestingly, *RXRα* expression was selectively downregulated early during colon carcinogenesis due to an increase in promoter methylation, whereas other retinoic acid receptors (*RARα, RARβ, RXRβ,* and *RXRγ*) were all expressed. *RXRα* silencing was independent of *β-catenin*, and could be reversed by green tea intervention [64]. This study showed that dietary levels of GTP were sufficient to reexpress silenced *RXRα* at the mRNA and protein level and to inhibit colon carcinogenesis.

13 Summary and Conclusions

As outlined above, major cellular pathways and cell functions, including drug metabolism, cell cycle regulation, potential to repair DNA damage or to induce apoptosis, response to inflammatory stimuli, cell signalling, cell growth control and differentiation, become deregulated during carcinogenesis by defects in epigenetic gene regulation. These include, among others, silencing by promoter methylation of detoxifying enzymes, tumor suppressor genes, cell cycle regulators, apoptosis-inducing and DNA repair genes, nuclear receptors, signal transducers and transcription factors, as well as modifications of histones and non-histone proteins such as *p53, NF-κB,* and *HSP90* by acetylation or methylation. Accumulating evidence indicates that dietary chemopreventive agents can prevent or reverse these alterations by affecting global DNA methylation, reexpressing tumor suppressor genes silenced by promoter methylation, and upregulating genes by altering histone and non-histone acetylation and methylation, at least in cell culture systems.

There are several challenges for future nutri-epigenetic research in cancer chemoprevention:

1. A definite link between cancer chemopreventive efficacy in animal models or human pilot studies and targeting of epigenetic mechanisms is often missing. Future investigations will have to demonstrate that chemopreventive efficacy is mediated by epigenetic gene regulation.
2. Some of the described nutri-epigenetic effects appear to be cell type or organ-specific. Underlying mechanisms for these differences have not yet been addressed.
3. Given the fact that epigenetics plays an important role in gene regulation during development, timing of dietary chemopreventive interventions might be critical to target epigenetic deregulation during tumorigenesis. Epigenetic alterations are considered as early events during cancer development. Consequently, interventions with chemopreventive agents might have to start early after birth to be most effective, and cancer preventive effects through epigenetic mechanisms might have been underestimated in studies performed so far. The question of "critical time windows" for application should be addressed in more detail in the future, both in direction of cancer prevention and with respect to potential harmful effects.
4. Frequency of application might also be a critical determinant of chemopreventive efficacy. Several studies have reported that inhibition of HDACs and consequent histone hyperacetylation is a transient effect. Although these activities have been demonstrated in rodent models and in humans, it is not yet clear whether occasional consumption of dietary HDAC inhibitors, for example from cruciferous vegetables would result in long-term epigenetic regulation of gene expression and downstream chemopreventive effects. This also applies to other epigenetic mechanisms.
5. Some interventions are apparently more effective when applied in combination, as exemplified by the combined application of ATRA with DNMT or HDAC inhibitors. This aspect has not been systematically investigated in nutri-epigenetics, but might be relevant when comparing activities of isolated compounds with complex extracts or food items.
6. Most investigations on epigenetic effects have so far only been performed in a targeted candidate gene approach. It becomes more and more clear that epigenetic gene regulation is coordinated in an intricate network and involves a crosstalk between effects on DNA methylation, histone modifications, and miRNA expression. To understand fully the potential impact of epigenetic gene regulation and to target it for chemoprevention, we need to consider the epigenome as an interactive three-dimensional system. Future investigations on DNA methylation changes and the modulation of activating and repressive histone marks at a genome-wide level will improve our understanding of mechanistic links. These analyses will also provide important clues as to whether

Appendix

Table 1 Effect of natural compounds on DNA methylation in cancer models in vitro and in vivo (for a review see [20–36])

Agent	Source	Mechanism	Organ	Target, effect	Reference
Apigenin	Celery, chamomile	DNMTi			[37]
Apple polyphenols	Apples	↓ Promoter meth ↓ DNMT expr	Colon	hMLH1, p14ARF, p16	[38]
Apple polyphenols (in vivo)	Apples	↑ Global DNA meth ↑ Promoter meth	Apc$^{Min/+}$ mice	↓ Adenoma numbers Line-1, Igf2, P2rx7	[284]
B vitamins (B$_2$, B$_6$, B$_{12}$)	Meat, nuts	Synthesis of SAM from methionine			Review in [39]
Baicalein	Scutellaria baicalensis	DNMTi			[40]
Betaine	Spinach, beets, wheat	Synthesis of SAM from methionine			Review in [39]
Betanin	Beetroot	DNMTi			[40]
Black raspberry extract	Black raspberries	↓ DNMT expr ↓ Promoter meth	Colon (phase 1 clinical trial)	SFRP2, SFRP5, WIF1, PAX6, Line-1	[41]
Butyrate		↓ Promoter meth	Colon	RARβ2	[42]
Caffeic acid	Coffee	DNMTi	Breast	RARβ	[43]
Catechins	Green tea	DNMTi, SAM: SAH			[44]
Chlorogenic acid	Coffee, apples	DNMTi	Prostate	GSTP1, MBD2	[45]
Chlorogenic acid		DNMTi	Breast	RARβ	[43]
Chlorogenic acid derivatives	Synthetic	DNMTi	Rec. DNMT3a		[46]

Cancer Chemoprevention and Nutri-Epigenetics: State of the Art and Future... 103

					Review in [39]
Choline	Egg, milk, meat	Synthesis of SAM from methionine			
Cruciferous vegetables (in vivo)		↓ DNA meth ↔ Promoter meth	Human PBMC of heavy smokers 4 weeks intervention	*Line1* *RASSF1A, ARF, CDKN2, MLH1, MTHFR*	[47]
Curcumin	Turmeric	DNMTi, ↓ 5mC ↓ Promoter meth ↓CGI meth ↑ MeCP2 binding ↓ H3K27me3 binding	Leukemia Prostate (mouse) Prostate	*Nrf-2* *Neurog-1*	[48] [49] [50]
		↓ Promoter meth ↓ Promoter meth	Cervix Leukemia	*RARβ2* p53 pathway	[51] [52]
Cyanidin	Blueberries	DNMTi			[40]
Disulfiram	Synthetic	DNMTi ↓ 5mC levels ↓ Promoter meth	Prostate	*APC, RARB*	[53]
Ellagic acid	Berries	DNMTi			[40]
Epicatechin	Green tea	DNMTi, SAM: SAH			[44]
(−)-Epigallocatechin gallate (EGCG)	Green tea	DNMTi, SAM: SAH	Breast	*RARβ*	[44]
		↓ Promoter meth	Breast	*hTERT*	[54]
		DNMTi	Prostate	*GSTP1*	[45]
		DNMTi	Esophagus, Colon, prostate	*p16, RARβ, MGMT, hMLH1*	[55]
		↔ Methylation	Bladder, colon, prostate	*p16, MAGE-A1, Alu, LINE*	[56]
		↔ 5mC level	Colon, leukemia		[57]
		↓ Promoter meth	Colon	*p16* via folate metabolism	[58]

(continued)

Table 1 (continued)

Agent	Source	Mechanism	Organ	Target, effect	Reference
		↑ mRNA expr	Esophagus	*p16, MGMT*	[37]
		↑ mRNA expr	Lung, esophagus	*RARβ, p16, DAPK*	[37]
		↓ Promoter meth	Oral cavity	*RECK*	[59]
		↓ Promoter meth	Lung	*WIF-1*	[60]
		DNMTi act/expr ↓ 5mC	Skin	*p16, p21*	[61]
proEGCG	Prodrug	DNMTi ↓ Promoter meth	Breast	*hTERT*	[62]
EGCG (in vivo)	Green tea	↓ SAM levels ↔ SAH, methionine, homocysteine	Plasma, small intestine, liver in healthy mice		[37]
Green tea polyphenols (in vivo)	Green tea	↔ DNA meth, 5mC ↔ Promoter meth	Prostate, gut, liver in TRAMP mice	*B1 repetitive elements, MAGE-a8 IRX3, CACNA1A, CDKN2A, NRX2*	[63]
		↓ Promoter meth	Colon, small intestine in AOM-treated mice	*RXRα* ↓ Colon tumors	[64]
		↓ Promoter meth ↔ Promoter meth	Gastric cancer patients	↓ *CDX2, BMP2* ↔ *p16, CACNA2D3, GATA5, ER*	[65]
Fisetin	Strawberries	DNMTi, SAM: SAH			[44]
Flavonoids (in vivo)	Green tea and soy products	↓ DNA meth ↔ Promoter meth	Human PBMC of heavy smokers 4 weeks intervention	*Line1* ↔ *RASSF1A, ARF, CDKN2, MLH1, MTHFR*	[47]
Folate	Green vegetables	Synthesis of SAM from methionine	Various	Maintenance of genomic stability, regulation of purine and pyrimidine biosynthesis ⇒ DNA biosynthesis, DNA repair, proliferation	Reviewed in [23, 39, 66–74]

Folic acid (in vivo)	Supplement	↑ CGI meth		Colorectal mucosa	ERα, SFRP1	[75]
Folate (in vivo)	Green vegetables			Liver, colon in mouse, rat; healthy individuals; patients with colonic adenoma and colon cancer		reviewed in [23, 66–68, 76]
Folate, green vegetables, multivitamins		Protection against methylation		Cohort-based study with 1,100 participants	p16, MGMT, RASSF1A, DAPK, GATA4, GATA5, PAX5α, PAX5β	[77]
Galangin	Propolis, galangal root	DNMTi				[40]
Garcinol	Mangosteen tree	DNMTi				[37]
Genistein, daidzein	Soy beans	DNMTi, ↓ promoter meth		Esophagus, prostate	p16, RARβ, MGMT	[78]
		↓ Promoter meth	Breast	GSTP1	[79]
		↓ Promoter meth	Colon	RARβ2	[42]
		↓ Promoter meth	Colon	Wnt5a	[80]
		↓ DNMT expr	Breast	hTERT	[81]
		↓ Promoter meth			
		DNMTi			
		↓ Promoter meth	Kidney	BTG	[82]
		↓ Promoter meth	Prostate	↓ GSTP1, EPHB2 ↔ RASSF1A, BRCA1	[83]
		↓ Promoter meth	Prostate	↓ BRCA1, GSTP1, EPHB2	[84]
		↑ Protein expr			
		↓ Promoter meth	Cervix	RARβ2	[51]
		↓ Promoter meth Genome wide analysis	Embryonic stem cells	Ucp1, Syt11	[85]
Genistein (in vitro and in vivo)	Soy beans	↓ Promoter meth	Endometrium	SF-1	[86]

(continued)

Table 1 (continued)

Agent	Source	Mechanism	Organ	Target, effect	Reference
Genistein (in vivo)		↑ DNA meth	Bone marrow	*Repetitive elements*	[83]
		↑ DNA meth in prostate	Brain, kidney, liver, spleen, prostate, testes of healthy mice		[87]
		↑ Methylation	tail, brain, kidney, liver in A^{vy} mice	A^{vy} *intracisternal A particle (IAP) murine retrotransposon*	[88]
		↑↓ Promoter meth	Cynomolgus monkeys	Fat tissue: *ABCG5, TBX5, HoxB1* Muscle: *HoxA5, HoxA11, NTRK3*	[89]
		↓ Promoter meth	Uterus in healthy mice, intact and ovarextomized (OVX)	*Nsbp1*	[90]
Soy isoflavones	Soy beans	↓ Promoter meth ↔ Promoter meth	prostate	*GSTP1* and *EPHB2* *BRCA1* and *RASSF1A*	[83]
		↓ Promoter meth	Pancreas, liver in healthy mice	*Acta1*	[91]
Soy isoflavones (in vivo)	Soy beans	Promoter meth	Human intervention trial	↓ *RARβ2, CCDN2* ↔ *ER, p16, RASSF1A*	[92]
Hesperetin	Citrus fruit	DNMTi			[37]
Hydroxycinnamic acid	Fruit	DNMTi			[37]
Luteolin	Parsley, celery	DNMTi			[37]
Lycopene	Tomatoes	↓ Promoter meth	Breast	*GSTP1, RARβ, HIN1*	[79]
Mahanine	Asian vegetables	DNMTi	Prostate, lung, breast, pancreas, vulva, ovaries	*RASSF1A*	[93]
Mahanine derivative	Synthetic	DNMTi	Prostate	*RASSF1A*	[94]
Mahanine derivative (in vivo)	Synthetic		Prostate xenograft	↓ Tumor volume	[94]
Methionine	Dairy products, nuts, fish	Synthesis of SAM			Review in [39]
Mithramycin A (MMA)		↓ Promoter meth ↓ DNMT1 expr	Lung	*SLIT2, TIMP3*	[95]

Myricetin	Fruit, herbs, vegetables	DNMTi; SAM: SAH			[37, 40, 44]
Naringenin	Citrus fruit	DNMTi			[37]
Nordihydroguaiaretic acid (NDGA)	Creosote bush	↓ Promoter meth	Breast	E-cadherin, p16	[96, 97]
Parthenolide	Feverfew	↔ DNA meth	Liver	LINE-1	[98]
		DNMTi; ↓ DNMT expr; ↓ 5mC	Breast	HIN-1	[99]
Parthenolide (in vivo)		↓ DNA meth; ↓ DNMT expr	Human leukemia Xenograft	↓ Tumor volume	[99]
Phenylethyl isothiocyanate (PEITC)	Watercress	↓ Promoter meth	Prostate	GSTP1	[100]
PEITC (in vivo)		↓ Promoter meth	Prostate of TRAMP and wt mice	MGMT; ↓ Tumor incidence	[101]
Phenylhexyl isothiocyanate (PHI)	Synthetic	↓ Promoter meth	Myeloma	p16	[102]
Phloretin	Apples	DNMTi			[40]
Piceatannol	Grapes	DNMTi			[40]
Protocatechuic acid	Açaí oil, olives	DNMTi			[40]
Quercetin	Ubiquitous	DNMTi; SAM: SAH			[37, 44]
Resveratrol	Grapes	↓ Promoter meth	Colon	p16	[283]
		DNMTi		BRCA1	[40]
		↓ MBD2 recruitment	Breast	BRCA1	[40, 103]
		↓ Promoter meth; ↓ DNMT expr	Breast	PTEN	[104]
		↓ Promoter meth	Breast	RARβ2	[105]

(continued)

Table 1 (continued)

Agent	Source	Mechanism	Organ	Target, effect	Reference
Retinoic acid		↓ Promoter meth	Leukemia	RARβ2	[105, 106]
		↓ Promoter meth	Leukemia	hTERT	[107]
		↔ DNA meth	Leukemia	RARβ	[108]
		↓ Promoter meth	Breast	PTEN	[104]
		↓ Promoter meth	Breast	RARβ2	[105]
		↓ Promoter meth (genome wide) ↓ DNMT1, 3B expr	Neuroblastoma	iNOS	[109]
Retinoic acid (in vivo)		↔ Promoter meth	Breast cancer patients	RARβ2	[110]
Rosmarinic acid	Rosemary	DNMTi			[40]
Selenomethionine		↔ Promoter meth	Prostate	GSTP1, RASSF1A	[111]
Sinapic acid	Rapeseed	DNMTi			[40]
Sodium selenite	Inorganic	DNMTi			[112]
		↑ Global DNA meth	Colon	p53	[113]
		↑ Global DNA meth	Colon		[114]
		↓ DNMT1 expr	Colon		[114]
		↓ DNMT1 expr	Prostate	GSTP1, APC, CSR1	[115]
		↓ Global DNA meth			[115]
		↓ Liver SAM: SAH	Intestine in DMH-treated rats	↓ Aberrant crypt foci formation	[114]
		↓ Global DNA meth	Rat intestine	↓ Aberrant crypt foci formation	[116]

Sodium selenite (in vivo)	Anorganic	↓ Global DNA meth	Liver, colon in rats	[113]
Sulforaphane	Broccoli	↓ DNMT expr ↓ Promoter meth	Breast	hTERT [31]
Syringic acid	Açaí oil	DNMTi		[40]
Thearubigins	Black tea	DNMTi	Recombinant DNMT3a	[46]
Vitamin D		↓ Weak promoter meth	Breast	RARβ2 [105]
		↓ Promoter meth ↓ DNMT expr	Breast	PTEN [104]
Vitamin E (in vivo)	Seed oils	↔ DNA meth ↔ Promoter meth	Rat liver	SDR5A1, GCLM [117]

CGI CpG island, *MeCP2* methylated CpG binding protein 2, *DNMTi* inhibition of DNMT activity, *expr* expression, *meth* methylation, *SAM:SAH* modulation of the SAM to SAH ratio through alternative mechanisms, ↓ reduction, inhibition, ↔ no effect, ↑ induction, stimulation

Table 2 Effect of natural compounds on acetylation of histones and non-histone substrates in cancer models in vitro and in vivo (for a review see [20, 22, 25–27, 29–36, 121, 122])

Agent	Source	Mechanism	Organ/cell type	Target, effect	Reference
Allylmercaptan	Garlic	HDACi ↑ ac-H3 and ac-H4	Colon	p21	[123]
Anacardic acid	Cashew nuts	HATi: p300, PCAF HATi: Tip60 HATi	Cervix, embryonic kidney Leukemia, tongue, lung, prostate	ATM, DNA PKs IκBα, ↓ p65ac, NF-κB-dependent IAP1, XIAP, Bcl-2, Bcl-xL, c-FLIP, cyclin D1, c-Myc, Cox-2, VEGF, ICAM-1, MMP-9	[124] [125] [126]
Apicidin	Fungal metabolite	HDACi ↑ ac-H4	Cervix and others	Gelsolin, p21,DNMT1	[127, 128]
Butyrate	Fermentation	HDACi ↑ ac-H3 and ac-H4 ↑ ac-H3 and ac-H4	Colon Colon	↑ p21 ↓ Cell proliferation ↓ Bcl-2	[129] [130]
		HDACi	T lymphocytes	DR5, caspases 8 and 10	[131]
		HDACi ↑ ac-histones	Leukemia	Cyclin D1, B1, c-Myc ↑ p21	[132]
Butyrate (in vivo)	Supplement	↑ ac-H3	DMH-treated mice	↑p53ac	[133]
Cambinol	Synthetic	SIRTi	Lung, lymphoma	↓ Tumor growth	[134]
Cambinol (in vivo)	Synthetic		Burkitt lymphoma xenograft		[134]
Chalcone derivative	Synthetic	SIRTi	Embryonic kidney	↑p53ac, p21	[135]
Curcumin	Turmeric	↓ HDAC1, HDAC3 expr ↓ p300 (HAT) expr	B-cell lymphoma	↓ Proliferation Notch 1	[136]
	Turmeric	HATi: p300/CBP ↓ ac-H3, H4		↓ p53ac	[137, 138]

Curcumin (in vitro and in vivo)	Turmeric	↓ HDAC4 expr	Medulloblastoma	*Tubulin* ↓ Xenograft growth	[139]
Delphinidin	Pomegranate	HATi: *p300* ↔ *PCAF, SIRT1,* *HDACs, HMTs*	Synoviocytes	↓ *p65ac,* ↑ *cytosolic IκBα,* ↓ *NF-κB-dependent Cox-2, IL-6, IL-1β, TNF-α*	[140]
Diallyldisulfide	Garlic	↑ ac-H3, ac-H4	Leukemia, colon, liver, breast, prostate	*p21*	[141]
Diallyldisulfide (in vivo)		↑ Transient ac-H4	Rat colon		[142]
Dihydrocoumarin	Sweet clover	SIRTi	Leukemia	↑ *p53ac*	[143]
Diindolylmethane (DIM)	Broccoli metabolite	↑ *HDAC1, 2, 3,* 8 degradation ↓ *HDAC1, 2, 3* expr	Colon	*p21, p27,* apoptosis	[144]
		↓ acH4	Breast	↓ *Cox2* expression	[145]
Diindolylmethane (in vivo)		↓ *HDAC1, 2* expr	Colon cancer xenografts	*p21*	[144]
(−)-Epigallocatechin gallate (EGCG)	Green tea	↑ HAT *p300* expr ↓ HDAC1 exp ↑ ac-H3, H3K9ac, ac-H4	Breast	*ERα*	[146]
		↓ HDAC activity ↑ ac-H3, ac-H4	Skin	*p16, p21*	[61]
		HATi: *p300* *PCAF, Tip60* ↔ SIRT1, HDACs, HMTs	Leukemia, B-cells	↓ *p65ac,* ↑ *cytosolic IκBα,* ↓ p65-binding to IL6 promoter, ↓ *NF-κB*-dependent *Cox-2, IL-6, NOS-2, XIAP, Bcl-2, Bcl-xL,, cyclin D1, c-Myc* ↓ EBV-mediated *IL-6, IL-12,* Cell transformation	[147]
		HATi: *p300* ↓ ac-H3 ↔ SIRT1, HDACs, HMTs	Prostate	↓ AR-mediated *PSA, NKX3.1* ↓ ac-AR, ↓ AR nuclear translocation, ↓ Proliferation	[148]

(continued)

Table 2 (continued)

Agent	Source	Mechanism	Organ/cell type	Target, effect	Reference
EGCG (in vivo)			TNFα-stimulated mice	↓ Serum *IL-6*, ↓ *p65ac* in macrophages, ↓ cytokine expression	[149]
ProEGCG	Prodrug	HDACi ↓ ac-H3, H3K9ac	Breast	*hTERT*	[62]
Gallic acid	Rose flowers	HATi: *p300 PCAF, Tip60* ↔ SIRT1, HDACs, HMTs	Lung	↑ Promoter binding *MAD1, E2F1*; ↓ binding *c-Myc* ↓ *p65ac*, ↑ cytosolic *IκBα*, ↓ p65-binding to IL6 promoter, ↓ *NF-κB*-dependent *Cox-2, IL-6, IL-1β, NOS-2, XIAP, Bcl-2, Bcl-xL*,, *cyclin D1*, *c-Myc*	[149]
Gallic acid (in vivo)			LPS-stimulated mice	↓ Serum *IL-6*, ↓ *p65ac* in macrophages, ↓ cytokine expression	[149]
Garcinol	Mangosteen tree	HATi: *p300* HATi: *p300, PCAF* ↑ ac-H4, acH2B ↓ HDAC expr	Cervix	↓ Global gene expression	[150] [151]
Genistein	Soy beans	↑ HAT expr *p300, PCAF, CREBBP, HAT1* ↑ ac-H3, ac-H4 ↑ ac-H3K9 ↓ SIRT1 expr	Prostate Prostate Prostate	↑ *HSP90ac* promotes dissociation and degradation of AR *p21, p16^INK4a* ↓ *Akt* signaling through *PTEN, CYLD, p53, FOXO3a*	[152] [153] [154]
Genistein, equol, AglyMax	Soy beans	↑ ac-histones	In vitro	ERα-mediated	[155]

Cancer Chemoprevention and Nutri-Epigenetics: State of the Art and Future...

Compound	Source	Mechanism	Tissue/Cell	Effect	Ref
β-Methylselenopyruvate α-Keto-γ-methylseleno-butyrate	Metabolites of selenium compounds	HDACi	Colon	↑ ac-H3, p21^WAF1	[156]
Parthenolide	Feverfew	↑ HDAC1 degrad ↑ ac-H3	Breast	HDAC1 degradation through ATM, ↑ cell death	[99, 157]
Phenylethyl isothiocyanate (PEITC)	Watercress	↑ ac-H3	Prostate	p27, p21, c-Myc	[158]
Phenylhexyl isothiocyanate (PHI)	Synthetic	HDACi HATa ↑ ac-histones	Prostate, leukemia, myeloma, hepatoma	p21, p27 Bcl-2	[102, 159–162]
In vivo		↑ ac-H3 and ac-H4	Bone marrow of AML patients		[163]
Resveratrol	Grapes	SIRT1a ↓ MTA1/NuRD corepressor complex	Bone Prostate	↓ p53ac ↑ p53ac, recruitment to p21 and Bax promoters ↑ Apoptosis	[164] [164, 165]
Retinoic acid (ATRA)		↑ ac-histones at CpG islands	Leukemia	↑ ac-H4 binding at HOXA1 and satellite DNA ↑RARβ, CD11b, HCK, OS-9, HOXA1, c-myc, c-myb, hTERT mRNA expr	[108]
			Leukemia, breast	↓ H3K9ac at hTERT promoter	[107, 166, 167]
Retinoic acid (in vitro and in vivo)			Breast, prostate, larynx	↑ ac at RARβ P2 promoter ↓ T47D xenograft growth	[110]
Silimarin	Milk thistle	SIRTa	Melanoma	Bax	[168]
Sulforaphane (SFN)	Broccoli	HDACi ↑ ac-H3, ac-H4	Embryonic kidney, colon, prostate	p21, Bax	[169, 170]
		HDACi	Prostate	↑ HSP90ac promotes dissociation and degradation of AR	[171]

(continued)

Table 2 (continued)

Agent	Source	Mechanism	Organ/cell type	Target, effect	Reference
		HDACi ↓ HDAC cl. I/II expr ↑ ac-H3	Prostate cancer vs. normal prostate	p21, ↑ac-tubulin	[172]
		HDACi ↑ ac-H3, H3K9ac ↔ HAT activity	Breast	hTERT MAD1, c-Myc, CTCF	[173]
		HDACi ↔ ac-H3, H4	Breast	↑ G$_2$/M phase arrest ↑ Apoptosis	[174]
		↓ HDAC3 expr	Colon	SMRT corepressor complex Pin1, 14-3-3	[175]
Sulforaphane (in vivo)	Broccoli	HDACi ↑ ac-H3, ac-H4	Colon mucosa, ileum, colon, prostate, PBMC of wt mice; ileum, colon of APC$^{Min/+}$ mice	p21, Bax	[176]
		HDACi ↑ ac-histones	Prostate cancer xenografts	Bax ↓ growth of PC3 xenograft	[177]
Broccoli sprouts (in vivo)		HDACi ↑ ac-H3, ac-H4 (both transient)	Human PBMC		[172, 178]
Ursodeoxycholic acid	Endogenous secondary bile acid	↓ ac-histones ↔ HDACi ↑ HDAC6 expr	Colon	E-cadherin, CK8, 18, 19	[179]

ac acetylation, *AglyMax* fermented soybean germs, *AR* androgen receptor, *degrade* degradation, *HDACi* HDAC inhibitor, *HATa* activator of HAT proteins, *HATi* inhibitor of HAT activity, *SIRTi* inhibitor of SIRT deacetylases, *SITRa* activator of SIRT, ↓ reduction, inhibition, ↔ no effect, ↑ induction, stimulation

Table 3 Effect of natural compounds on histone methylation in cancer models in vitro and in vivo (for a review see [29, 30, 34])

Agent	Source	Mechanism	Organ	Target, effect	Reference
Chaetocin		HMTi: *SUV39* ↓H3K9me2, ↓H3K9me3		*p21*	[180, 181]
		↓H3K9me2 and H3K9me3 at *p15* and *E-cadherin* promoter	Leukemia	*p15*^{INK4B} and *E-cadherin*	[182]
Curcumin	Turmeric	↓*EZH2* expr ↓H3K27me3	Breast	Via *MAPK* pathway	[183]
(−)-Epigallocatechin gallate (EGCG)	Green tea	↓HMT expr: *BMI-1, SUZ12, EZH2, Eed* ↓H3K27me3	Skin	*p21, p27, Bax, Bcl-xL* ↑Effect in combination with SAH hydrolase inhibitor	[184, 185]
		↓H3K9me	Skin	*p16, p21*	[61]
		↓HMT *SUV39H1* expr ↑H3K4me2, ↓H3K9me3	Breast	*ERα*	[146]
Genistein	Soy beans	↓H3K9me2	Prostate	*Akt* signaling through *PTEN, CYLD, p53, FOXO3a*	[154]
Polyamine analogs	Synthetic	HDMi: *LSD1* ↑H3K4me2, ac-H3K9	Colon	*sFRP1, sFRP4, sFRP5, GATA5*	[186]
PG11144 (*cis*), PG11150 (*trans*)	Synthetic	HDMi: *LSD1* ↑H3K4me, H3K4me2	Colon	*SFRP1, SFRP2*	[187]
n-3 Polyunsaturated fatty acid (n-3 PUFA) DHA, EPA	Fish oil	↓*EZH2* expr ↓H3K27me3 and H3K9me3	Breast	*E-cadherin, IGFBP3*	[188]

HTMi inhibitor of histone methyltransferases, *HDMi* inhibitor of histone demethylases, *expr* expression, ↓ reduction, inhibition, ↔ no effect, ↑ induction, stimulation

nutri-epigenetic effects are specific for certain pathways or selective for subsets of genes. The emergence of novel technologies such as next-generation sequencing for genome-wide assessment of DNA methylation and localization of histone marks, the expected drop in sequencing costs, and the development of bioinformatic tools to integrate systematically available information will facilitate this type of analyses in future chemoprevention studies.

References

1. Henikoff S, Matzke MA (1997) Exploring and explaining epigenetic effects. Trends Genet 13(8):293–295, doi:10.1016/S0168-9525(97)01219-5, pii: S0168952597012195
2. Qiu J (2006) Epigenetics: unfinished symphony. Nature 441(7090):143–145. doi:10.1038/441143a
3. Jaenisch R, Bird A (2003) Epigenetic regulation of gene expression: how the genome integrates intrinsic and environmental signals. Nat Genet 33(suppl):245–254. doi:10.1038/ng1089
4. Choudhuri S (2011) From Waddington's epigenetic landscape to small noncoding RNA: some important milestones in the history of epigenetics research. Toxicol Mech Methods 21(4):252–274. doi:10.3109/15376516.2011.559695
5. Berdasco M, Esteller M (2010) Aberrant epigenetic landscape in cancer: how cellular identity goes awry. Dev Cell 19(5):698–711. doi:10.1016/j.devcel.2010.10.005, pii: S1534-5807(10)00458-2
6. Payer B, Lee JT (2008) X chromosome dosage compensation: how mammals keep the balance. Annu Rev Genet 42:733–772. doi:10.1146/annurev.genet.42.110807.091711
7. Illingworth R, Kerr A, Desousa D, Jorgensen H, Ellis P, Stalker J, Jackson D, Clee C, Plumb R, Rogers J, Humphray S, Cox T, Langford C, Bird A (2008) A novel CpG island set identifies tissue-specific methylation at developmental gene loci. PLoS Biol 6(1):e22. doi:10.1371/journal.pbio.0060022, pii: 07-PLBI-RA-3186
8. Stilling RM, Fischer A (2011) The role of histone acetylation in age-associated memory impairment and Alzheimer's disease. Neurobiol Learn Mem 96(1):19–26. doi:10.1016/j.nlm.2011.04.002, pii: S1074-7427(11)00064-5
9. Suter MA, Aagaard-Tillery KM (2009) Environmental influences on epigenetic profiles. Semin Reprod Med 27(5):380–390. doi:10.1055/s-0029-1237426
10. Fraga MF, Ballestar E, Paz MF, Ropero S, Setien F, Ballestar ML, Heine-Suner D, Cigudosa JC, Urioste M, Benitez J, Boix-Chornet M, Sanchez-Aguilera A, Ling C, Carlsson E, Poulsen P, Vaag A, Stephan Z, Spector TD, Wu YZ, Plass C, Esteller M (2005) Epigenetic differences arise during the lifetime of monozygotic twins. Proc Natl Acad Sci USA 102(30):10604–10609. doi:10.1073/pnas.0500398102
11. Wong AH, Gottesman II, Petronis A (2005) Phenotypic differences in genetically identical organisms: the epigenetic perspective. Hum Mol Genet 14(Spec no. 1):R11–R18. doi:10.1093/hmg/ddi116, pii: 14/suppl_1/R11
12. Jones PA, Baylin SB (2007) The epigenomics of cancer. Cell 128(4):683–692. doi:10.1016/j.cell.2007.01.029, pii: S0092-8674(07)00127-4
13. Bruce KD, Cagampang FR (2011) Epigenetic priming of the metabolic syndrome. Toxicol Mech Methods 21(4):353–361. doi:10.3109/15376516.2011.559370
14. Beaulieu N, Morin S, Chute IC, Robert MF, Nguyen H, MacLeod AR (2002) An essential role for DNA methyltransferase DNMT3B in cancer cell survival. J Biol Chem 277(31): 28176–28181. doi:10.1074/jbc.M204734200

15. Linhart HG, Lin H, Yamada Y, Moran E, Steine EJ, Gokhale S, Lo G, Cantu E, Ehrich M, He T, Meissner A, Jaenisch R (2007) Dnmt3b promotes tumorigenesis in vivo by gene-specific de novo methylation and transcriptional silencing. Genes Dev 21(23):3110–3122. doi:10.1101/gad.1594007, pii: 21/23/3110
16. Esteller M (2007) Cancer epigenomics: DNA methylomes and histone-modification maps. Nat Rev Genet 8(4):286–298. doi:10.1038/nrg2005
17. Kopelovich L, Crowell JA, Fay JR (2003) The epigenome as a target for cancer chemoprevention. J Natl Cancer Inst 95(23):1747–1757. doi:10.1093/jnci/dig109
18. Gama-Sosa MA, Slagel VA, Trewyn RW, Oxenhandler R, Kuo KC, Gehrke CW, Ehrlich M (1983) The 5-methylcytosine content of DNA from human tumors. Nucleic Acids Res 11(19):6883–6894. doi:10.1093/nar/11.19.6883
19. Goelz SE, Vogelstein B, Hamilton SR, Feinberg AP (1985) Hypomethylation of DNA from benign and malignant human colon neoplasms. Science 228(4696):187–190. doi:10.1126/science.2579435
20. Huang J, Plass C, Gerhauser C (2011) Cancer chemoprevention by targeting the epigenome. Curr Drug Targets 12(13):1925–1956, doi:10.2174/138945011798184155
21. Verma M, Maruvada P, Srivastava S (2004) Epigenetics and cancer. Crit Rev Clin Lab Sci 41(5–6):585–607. doi:10.1080/10408360490516922
22. Hauser AT, Jung M (2008) Targeting epigenetic mechanisms: potential of natural products in cancer chemoprevention. Planta Med 74(13):1593–1601. doi:10.1055/s-2008-1081347
23. Johnson IT, Belshaw NJ (2008) Environment, diet and CpG island methylation: epigenetic signals in gastrointestinal neoplasia. Food Chem Toxicol 46(4):1346–1359. doi:10.1016/j.fct.2007.09.101, pii: S0278-6915(07)00451-6
24. Arasaradnam RP, Commane DM, Bradburn D, Mathers JC (2008) A review of dietary factors and its influence on DNA methylation in colorectal carcinogenesis. Epigenetics 3(4):193–198, doi:10.4161/epi.3.4.6508
25. Molinie B, Georgel P (2009) Genetic and epigenetic regulations of prostate cancer by genistein. Drug News Perspect 22(5):247–254. doi:10.1358/dnp.2009.22.5.1378633
26. Choi S-W, Friso S (eds) (2009) Nutrients and epigenetics. CRC, Boca Raton. doi:10.1201/9781420063561.ch5
27. Gilbert ER, Liu D (2010) Flavonoids influence epigenetic-modifying enzyme activity: structure–function relationships and the therapeutic potential for cancer. Curr Med Chem 17(17):1756–1768, pii: BSP/CMC/E-Pub/105
28. Li Y, Tollefsbol TO (2010) Impact on DNA methylation in cancer prevention and therapy by bioactive dietary components. Curr Med Chem 17(20):2141–2151, doi:10.2174/092986710791299966
29. vel Szic KS, Ndlovu MN, Haegeman G, Vanden Berghe W (2010) Nature or nurture: let food be your epigenetic medicine in chronic inflammatory disorders. Biochem Pharmacol 80(12):1816–1832. doi:10.1016/j.bcp.2010.07.029, pii:S0006-2952(10)00567-8
30. Link A, Balaguer F, Goel A (2010) Cancer chemoprevention by dietary polyphenols: promising role for epigenetics. Biochem Pharmacol 80(12):1771–1792. doi:10.1016/j.bcp.2010.06.036, pii: S0006-2952(10)00470-3
31. Meeran SM, Ahmed A, Tollefsbol TO (2010) Epigenetic targets of bioactive dietary components for cancer prevention and therapy. Clin Epigenetics 1(3–4):101–116. doi:10.1007/s13148-010-0011-5
32. Reuter S, Gupta SC, Park B, Goel A, Aggarwal BB (2011) Epigenetic changes induced by curcumin and other natural compounds. Genes Nutr 6(2):93–108. doi:10.1007/s12263-011-0222-1
33. Vanden Berghe W (2012) Epigenetic impact of dietary polyphenols in cancer chemoprevention: lifelong remodeling of our epigenomes. Pharmacol Res 65(6):565–576. doi:10.1016/j.phrs.2012.03.007, pii: S1043-6618(12)00050-3

34. Khan SI, Aumsuwan P, Khan IA, Walker LA, Dasmahapatra AK (2012) Epigenetic events associated with breast cancer and their prevention by dietary components targeting the epigenome. Chem Res Toxicol 25(1):61–73. doi:10.1021/tx200378c
35. Malireddy S, Kotha SR, Secor JD, Gurney TO, Abbott JL, Maulik G, Maddipati KR, Parinandi NL (2012) Phytochemical antioxidants modulate Mammalian cellular epigenome: implications in health and disease. Antioxid Redox Signal 17(2):327–339. doi:10.1089/ars.2012.4600
36. Stefanska B, Karlic H, Varga F, Fabianowska-Majewska K, Haslberger AG (2012) Epigenetic mechanisms in anti-cancer actions of bioactive food components – the implications in cancer prevention. Br J Pharmacol. doi:10.1111/j.1476-5381.2012.02002.x
37. Fang M, Chen D, Yang CS (2007) Dietary polyphenols may affect DNA methylation. J Nutr 137(1 Suppl):223S–228S, pii: 137/1/223S
38. Fini L, Piazzi G, Daoud Y, Selgrad M, Maegawa S, Garcia M, Fogliano V, Romano M, Graziani G, Vitaglione P, Carmack SW, Gasbarrini A, Genta RM, Issa JP, Boland CR, Ricciardiello L (2011) Chemoprevention of intestinal polyps in ApcMin/+ mice fed with western or balanced diets by drinking annurca apple polyphenol extract. Cancer Prev Res (Phila) 4(6):907-15. doi:10.1158/1940-6207.CAPR-10-0359
39. Chen J, Xu X (2010) Diet, epigenetic, and cancer prevention. Adv Genet 71:237–255. doi:10.1016/B978-0-12-380864-6.00008-0
40. Paluszczak J, Krajka-Kuzniak V, Baer-Dubowska W (2010) The effect of dietary polyphenols on the epigenetic regulation of gene expression in MCF7 breast cancer cells. Toxicol Lett 192(2):119–125. doi:10.1016/j.toxlet.2009.10.010, pii: S0378-4274(09)01451-9
41. Wang LS, Arnold M, Huang YW, Sardo C, Seguin C, Martin E, Huang TH, Riedl K, Schwartz S, Frankel W, Pearl D, Xu Y, Winston J 3rd, Yang GY, Stoner G (2011) Modulation of genetic and epigenetic biomarkers of colorectal cancer in humans by black raspberries: a phase I pilot study. Clin Cancer Res 17(3):598–610. doi:10.1158/1078-0432.CCR-10-1260
42. Spurling CC, Suhl JA, Boucher N, Nelson CE, Rosenberg DW, Giardina C (2008) The short chain fatty acid butyrate induces promoter demethylation and reactivation of RARbeta2 in colon cancer cells. Nutr Cancer 60(5):692–702. doi:10.1080/01635580802008278, pii: 902435515
43. Lee WJ, Zhu BT (2006) Inhibition of DNA methylation by caffeic acid and chlorogenic acid, two common catechol-containing coffee polyphenols. Carcinogenesis 27(2):269–277. doi:10.1093/carcin/bgi206
44. Lee WJ, Shim JY, Zhu BT (2005) Mechanisms for the inhibition of DNA methyltransferases by tea catechins and bioflavonoids. Mol Pharmacol 68(4):1018–1030. doi:10.1124/mol.104.008367
45. Pandey M, Shukla S, Gupta S (2009) Promoter demethylation and chromatin remodeling by green tea polyphenols leads to re-expression of GSTP1 in human prostate cancer cells. Int J Cancer. doi:10.1002/ijc.24988
46. Rajavelu A, Tulyasheva Z, Jaiswal R, Jeltsch A, Kuhnert N (2011) The inhibition of the mammalian DNA methyltransferase 3a (Dnmt3a) by dietary black tea and coffee polyphenols. BMC Biochem 12:16. doi:10.1186/1471-2091-12-16
47. Scoccianti C, Ricceri F, Ferrari P, Cuenin C, Sacerdote C, Polidoro S, Jenab M, Hainaut P, Vineis P, Herceg Z (2011) Methylation patterns in sentinel genes in peripheral blood cells of heavy smokers: influence of cruciferous vegetables in an intervention study. Epigenetics 6(9):1114–1119, doi:10.4161/epi.6.9.16515
48. Liu Z, Xie Z, Jones W, Pavlovicz RE, Liu S, Yu J, Li PK, Lin J, Fuchs JR, Marcucci G, Li C, Chan KK (2009) Curcumin is a potent DNA hypomethylation agent. Bioorg Med Chem Lett 19(3):706–709. doi:10.1016/j.bmcl.2008.12.041, pii: S0960-894X(08)01551-5
49. Khor TO, Huang Y, Wu TY, Shu L, Lee J, Kong AN (2011) Pharmacodynamics of curcumin as DNA hypomethylation agent in restoring the expression of Nrf2 via promoter CpGs demethylation. Biochem Pharmacol 82(9):1073–1078. doi:10.1016/j.bcp.2011.07.065, pii: S0006-2952(11)00523-5

50. Shu L, Khor TO, Lee JH, Boyanapalli SS, Huang Y, Wu TY, Saw CL, Cheung KL, Kong AN (2011) Epigenetic CpG demethylation of the promoter and reactivation of the expression of neurog1 by curcumin in prostate LNCaP cells. AAPS J. doi:10.1208/s12248-011-9300-y
51. Jha AK, Nikbakht M, Parashar G, Shrivastava A, Capalash N, Kaur J (2010) Reversal of hypermethylation and reactivation of the RARbeta2 gene by natural compounds in cervical cancer cell lines. Folia Biol (Praha) 56(5):195–200, pii: FB2010A0026
52. Vilas-Zornoza A, Agirre X, Martin-Palanco V, Martin-Subero JI, San Jose-Eneriz E, Garate L, Alvarez S, Miranda E, Rodriguez-Otero P, Rifon J, Torres A, Calasanz MJ, Cruz Cigudosa J, Roman-Gomez J, Prosper F (2011) Frequent and simultaneous epigenetic inactivation of TP53 pathway genes in acute lymphoblastic leukemia. PLoS One 6(2):e17012. doi:10.1371/journal.pone.0017012
53. Lin J, Haffner MC, Zhang Y, Lee BH, Brennen WN, Britton J, Kachhap SK, Shim JS, Liu JO, Nelson WG, Yegnasubramanian S, Carducci MA (2011) Disulfiram is a DNA demethylating agent and inhibits prostate cancer cell growth. Prostate 71(4):333–343. doi:10.1002/pros.21247
54. Berletch JB, Liu C, Love WK, Andrews LG, Katiyar SK, Tollefsbol TO (2008) Epigenetic and genetic mechanisms contribute to telomerase inhibition by EGCG. J Cell Biochem 103(2):509–519. doi:10.1002/jcb.21417
55. Fang MZ, Wang Y, Ai N, Hou Z, Sun Y, Lu H, Welsh W, Yang CS (2003) Tea polyphenol (−)-epigallocatechin-3-gallate inhibits DNA methyltransferase and reactivates methylation-silenced genes in cancer cell lines. Cancer Res 63(22):7563–7570
56. Chuang JC, Yoo CB, Kwan JM, Li TW, Liang G, Yang AS, Jones PA (2005) Comparison of biological effects of non-nucleoside DNA methylation inhibitors versus 5-aza-2′-deoxycytidine. Mol Cancer Ther 4(10):1515–1520. doi:10.1158/1535-7163.MCT-05-0172, pii: 4/10/1515
57. Stresemann C, Brueckner B, Musch T, Stopper H, Lyko F (2006) Functional diversity of DNA methyltransferase inhibitors in human cancer cell lines. Cancer Res 66(5):2794–2800. doi:10.1158/0008-5472.CAN-05-2821, pii: 66/5/2794
58. Navarro-Peran E, Cabezas-Herrera J, Campo LS, Rodriguez-Lopez JN (2007) Effects of folate cycle disruption by the green tea polyphenol epigallocatechin-3-gallate. Int J Biochem Cell Biol 39(12):2215–2225. doi:10.1016/j.biocel.2007.06.005, pii: S1357-2725(07)00183-5
59. Kato K, Long NK, Makita H, Toida M, Yamashita T, Hatakeyama D, Hara A, Mori H, Shibata T (2008) Effects of green tea polyphenol on methylation status of RECK gene and cancer cell invasion in oral squamous cell carcinoma cells. Br J Cancer 99(4):647–654. doi:10.1038/sj.bjc.6604521
60. Gao Z, Xu Z, Hung MS, Lin YC, Wang T, Gong M, Zhi X, Jablon DM, You L (2009) Promoter demethylation of WIF-1 by epigallocatechin-3-gallate in lung cancer cells. Anticancer Res 29(6):2025–2030, pii: 29/6/2025
61. Nandakumar V, Vaid M, Katiyar SK (2011) (−)-Epigallocatechin-3-gallate reactivates silenced tumor suppressor genes, Cip1/p21 and p16INK4a, by reducing DNA methylation and increasing histones acetylation in human skin cancer cells. Carcinogenesis 32(4):537–544. doi:10.1093/carcin/bgq285
62. Meeran SM, Patel SN, Chan TH, Tollefsbol TO (2011) A novel prodrug of epigallocatechin-3-gallate: differential epigenetic hTERT repression in human breast cancer cells. Cancer Prev Res (Phila) 4(8):1243–1254. doi:10.1158/1940-6207.CAPR-11-0009
63. Morey Kinney SR, Zhang W, Pascual M, Greally JM, Gillard BM, Karasik E, Foster BA, Karpf AR (2009) Lack of evidence for green tea polyphenols as DNA methylation inhibitors in murine prostate. Cancer Prev Res (Phila) 2(12):1065–1075. doi:10.1158/1940-6207.CAPR-09-0010
64. Volate SR, Muga SJ, Issa AY, Nitcheva D, Smith T, Wargovich MJ (2009) Epigenetic modulation of the retinoid X receptor alpha by green tea in the azoxymethane-Apc Min/+ mouse model of intestinal cancer. Mol Carcinog 48(10):920–933. doi:10.1002/mc.20542

65. Yuasa Y, Nagasaki H, Akiyama Y, Hashimoto Y, Takizawa T, Kojima K, Kawano T, Sugihara K, Imai K, Nakachi K (2009) DNA methylation status is inversely correlated with green tea intake and physical activity in gastric cancer patients. Int J Cancer 124(11): 2677–2682. doi:10.1002/ijc.24231
66. Duthie SJ (2010) Folate and cancer: how DNA damage, repair and methylation impact on colon carcinogenesis. J Inherit Metab Dis. doi:10.1007/s10545-010-9128-0
67. Kim YI (2004) Folate, colorectal carcinogenesis, and DNA methylation: lessons from animal studies. Environ Mol Mutagen 44(1):10–25. doi:10.1002/em.20025
68. Kim YI (2005) Nutritional epigenetics: impact of folate deficiency on DNA methylation and colon cancer susceptibility. J Nutr 135(11):2703–2709, pii: 135/11/2703
69. McKay JA, Mathers JC (2011) Diet induced epigenetic changes and their implications for health. Acta Physiol (Oxf) 202(2):103–118. doi:10.1111/j.1748-1716.2011.02278.x
70. Park LK, Friso S, Choi SW (2012) Nutritional influences on epigenetics and age-related disease. Proc Nutr Soc 71:75–83. doi:10.1017/S0029665111003302
71. Jang H, Mason JB, Choi SW (2005) Genetic and epigenetic interactions between folate and aging in carcinogenesis. J Nutr 135(12 Suppl):2967S–2971S, pii: 135/12/2967S
72. Kim KC, Friso S, Choi SW (2009) DNA methylation, an epigenetic mechanism connecting folate to healthy embryonic development and aging. J Nutr Biochem 20(12):917–926. doi:10.1016/j.jnutbio.2009.06.008, pii: S0955-2863(09)00134-X
73. Burdge GC, Lillycrop KA (2010) Nutrition, epigenetics, and developmental plasticity: implications for understanding human disease. Annu Rev Nutr 30:315–339. doi:10.1146/annurev.nutr.012809.104751
74. Duthie SJ (2011) Epigenetic modifications and human pathologies: cancer and CVD. Proc Nutr Soc 70(1):47–56. doi:10.1017/S0029665110003952
75. Wallace K, Grau MV, Levine AJ, Shen L, Hamdan R, Chen X, Gui J, Haile RW, Barry EL, Ahnen D, McKeown-Eyssen G, Baron JA, Issa JP (2010) Association between folate levels and CpG Island hypermethylation in normal colorectal mucosa. Cancer Prev Res (Phila) 3(12):1552–1564. doi:10.1158/1940-6207.CAPR-10-0047, pii: 3/12/1552
76. Lamprecht SA, Lipkin M (2003) Chemoprevention of colon cancer by calcium, vitamin D and folate: molecular mechanisms. Nat Rev Cancer 3(8):601–614. doi:10.1038/nrc1144
77. Stidley CA, Picchi MA, Leng S, Willink R, Crowell RE, Flores KG, Kang H, Byers T, Gilliland FD, Belinsky SA (2010) Multivitamins, folate, and green vegetables protect against gene promoter methylation in the aerodigestive tract of smokers. Cancer Res 70(2):568–574. doi:10.1158/0008-5472.CAN-09-3410
78. Fang MZ, Chen D, Sun Y, Jin Z, Christman JK, Yang CS (2005) Reversal of hypermethylation and reactivation of p16INK4a, RARbeta, and MGMT genes by genistein and other isoflavones from soy. Clin Cancer Res 11(19 Pt 1):7033–7041. doi:10.1158/1078-0432.CCR-05-0406, pii: 11/19/7033
79. King-Batoon A, Leszczynska JM, Klein CB (2008) Modulation of gene methylation by genistein or lycopene in breast cancer cells. Environ Mol Mutagen 49(1):36–45. doi:10.1002/em.20363
80. Wang Z, Chen H (2010) Genistein increases gene expression by demethylation of WNT5a promoter in colon cancer cell line SW1116. Anticancer Res 30(11):4537–4545, pii: 30/11/4537
81. Li Y, Liu L, Andrews LG, Tollefsbol TO (2009) Genistein depletes telomerase activity through cross-talk between genetic and epigenetic mechanisms. Int J Cancer 125(2): 286–296. doi:10.1002/ijc.24398
82. Majid S, Dar AA, Ahmad AE, Hirata H, Kawakami K, Shahryari V, Saini S, Tanaka Y, Dahiya AV, Khatri G, Dahiya R (2009) BTG3 tumor suppressor gene promoter demethylation, histone modification and cell cycle arrest by genistein in renal cancer. Carcinogenesis 30(4):662–670. doi:10.1093/carcin/bgp042

83. Vardi A, Bosviel R, Rabiau N, Adjakly M, Satih S, Dechelotte P, Boiteux JP, Fontana L, Bignon YJ, Guy L, Bernard-Gallon DJ (2010) Soy phytoestrogens modify DNA methylation of GSTP1, RASSF1A, EPH2 and BRCA1 promoter in prostate cancer cells. In Vivo 24(4):393–400, pii: 24/4/393
84. Adjakly M, Bosviel R, Rabiau N, Boiteux JP, Bignon YJ, Guy L, Bernard-Gallon D (2011) DNA methylation and soy phytoestrogens: quantitative study in DU-145 and PC-3 human prostate cancer cell lines. Epigenomics 3(6):795–803. doi:10.2217/epi.11.103
85. Sato N, Yamakawa N, Masuda M, Sudo K, Hatada I, Muramatsu M (2011) Genome-wide DNA methylation analysis reveals phytoestrogen modification of promoter methylation patterns during embryonic stem cell differentiation. PLoS One 6(4):e19278. doi:10.1371/journal.pone.0019278, pii: PONE-D-10-06293
86. Matsukura H, Aisaki K, Igarashi K, Matsushima Y, Kanno J, Muramatsu M, Sudo K, Sato N (2011) Genistein promotes DNA demethylation of the steroidogenic factor 1 (SF-1) promoter in endometrial stromal cells. Biochem Biophys Res Commun 412(2):366–372. doi:10.1016/j.bbrc.2011.07.104, pii: S0006-291X(11)01339-8
87. Day JK, Bauer AM, DesBordes C, Zhuang Y, Kim BE, Newton LG, Nehra V, Forsee KM, MacDonald RS, Besch-Williford C, Huang TH, Lubahn DB (2002) Genistein alters methylation patterns in mice. J Nutr 132(8 Suppl):2419S–2423S
88. Dolinoy DC, Weidman JR, Waterland RA, Jirtle RL (2006) Maternal genistein alters coat color and protects Avy mouse offspring from obesity by modifying the fetal epigenome. Environ Health Perspect 114(4):567–572. doi:10.1289/ehp.8700
89. Howard TD, Ho SM, Zhang L, Chen J, Cui W, Slager R, Gray S, Hawkins GA, Medvedovic M, Wagner JD (2011) Epigenetic changes with dietary soy in cynomolgus monkeys. PLoS One 6(10):e26791. doi:10.1371/journal.pone.0026791, pii: PONE-D-11-12849
90. Tang WY, Newbold R, Mardilovich K, Jefferson W, Cheng RY, Medvedovic M, Ho SM (2008) Persistent hypomethylation in the promoter of nucleosomal binding protein 1 (Nsbp1) correlates with overexpression of Nsbp1 in mouse uteri neonatally exposed to diethylstilbestrol or genistein. Endocrinology 149(12):5922–5931. doi:10.1210/en.2008-0682
91. Guerrero-Bosagna CM, Sabat P, Valdovinos FS, Valladares LE, Clark SJ (2008) Epigenetic and phenotypic changes result from a continuous pre and post natal dietary exposure to phytoestrogens in an experimental population of mice. BMC Physiol 8:17. doi:10.1186/1472-6793-8-17
92. Qin W, Zhu W, Shi H, Hewett JE, Ruhlen RL, MacDonald RS, Rottinghaus GE, Chen YC, Sauter ER (2009) Soy isoflavones have an antiestrogenic effect and alter mammary promoter hypermethylation in healthy premenopausal women. Nutr Cancer 61(2):238–244. doi:10.1080/01635580802404196, pii: 908919410
93. Jagadeesh S, Sinha S, Pal BC, Bhattacharya S, Banerjee PP (2007) Mahanine reverses an epigenetically silenced tumor suppressor gene RASSF1A in human prostate cancer cells. Biochem Biophys Res Commun 362(1):212–217. doi:10.1016/j.bbrc.2007.08.005, pii: S0006-291X(07)01709-3
94. Sheikh KD, Banerjee PP, Jagadeesh S, Grindrod SC, Zhang L, Paige M, Brown ML (2010) Fluorescent epigenetic small molecule induces expression of the tumor suppressor ras-association domain family 1A and inhibits human prostate xenograft. J Med Chem 53(6):2376–2382. doi:10.1021/jm9011615
95. Lin RK, Hsu CH, Wang YC (2007) Mithramycin A inhibits DNA methyltransferase and metastasis potential of lung cancer cells. Anticancer Drugs 18(10):1157–1164. doi:10.1097/CAD.0b013e3282a215e9, pii: 00001813-200711000-00006
96. Cui Y, Lu C, Liu L, Sun D, Yao N, Tan S, Bai S, Ma X (2008) Reactivation of methylation-silenced tumor suppressor gene p16INK4a by nordihydroguaiaretic acid and its implication in G1 cell cycle arrest. Life Sci 82(5–6):247–255. doi:10.1016/j.lfs.2007.11.013, pii: S0024-3205(07)00843-0

97. Cui Y, Lu C, Kang A, Liu L, Tan S, Sun D, Hu J, Ma X (2008) Nordihydroguaiaretic acid restores expression of silenced E-cadherin gene in human breast cancer cell lines and xenografts. Anticancer Drugs 19(5):487–494. doi:10.1097/CAD.0b013e3282fd5310
98. Byun HM, Choi SH, Laird PW, Trinh B, Siddiqui MA, Marquez VE, Yang AS (2008) 2'-Deoxy-N4-[2-(4-nitrophenyl)ethoxycarbonyl]-5-azacytidine: a novel inhibitor of DNA methyltransferase that requires activation by human carboxylesterase 1. Cancer Lett 266(2):238–248. doi:10.1016/j.canlet.2008.02.069, pii: S0304-3835(08)00171-7
99. Liu Z, Liu S, Xie Z, Pavlovicz RE, Wu J, Chen P, Aimiuwu J, Pang J, Bhasin D, Neviani P, Fuchs JR, Plass C, Li PK, Li C, Huang TH, Wu LC, Rush L, Wang H, Perrotti D, Marcucci G, Chan KK (2009) Modulation of DNA methylation by a sesquiterpene lactone parthenolide. J Pharmacol Exp Ther 329(2):505–514. doi:10.1124/jpet.108.147934
100. Wang LG, Beklemisheva A, Liu XM, Ferrari AC, Feng J, Chiao JW (2007) Dual action on promoter demethylation and chromatin by an isothiocyanate restored GSTP1 silenced in prostate cancer. Mol Carcinog 46(1):24–31. doi:10.1002/mc.20258
101. Wang LG, Chiao JW (2010) Prostate cancer chemopreventive activity of phenethyl isothiocyanate through epigenetic regulation (Review). Int J Oncol 37(3):533–539. doi:10.3892/ijo_00000702
102. Lu Q, Lin X, Feng J, Zhao X, Gallagher R, Lee MY, Chiao JW, Liu D (2008) Phenylhexyl isothiocyanate has dual function as histone deacetylase inhibitor and hypomethylating agent and can inhibit myeloma cell growth by targeting critical pathways. J Hematol Oncol 1:6. doi:10.1186/1756-8722-1-6
103. Papoutsis AJ, Lamore SD, Wondrak GT, Selmin OI, Romagnolo DF (2010) Resveratrol prevents epigenetic silencing of BRCA-1 by the aromatic hydrocarbon receptor in human breast cancer cells. J Nutr 140(9):1607–1614. doi:10.3945/jn.110.123422
104. Stefanska B, Salame P, Bednarek A, Fabianowska-Majewska K (2012) Comparative effects of retinoic acid, vitamin D and resveratrol alone and in combination with adenosine analogues on methylation and expression of phosphatase and tensin homologue tumour suppressor gene in breast cancer cells. Br J Nutr 107(6):781–790. doi:10.1017/S0007114511003631
105. Stefanska B, Rudnicka K, Bednarek A, Fabianowska-Majewska K (2010) Hypomethylation and induction of retinoic acid receptor beta 2 by concurrent action of adenosine analogues and natural compounds in breast cancer cells. Eur J Pharmacol 638(1–3):47–53. doi:10.1016/j.ejphar.2010.04.032, pii: S0014-2999(10)00363-8
106. Di Croce L, Raker VA, Corsaro M, Fazi F, Fanelli M, Faretta M, Fuks F, Lo Coco F, Kouzarides T, Nervi C, Minucci S, Pelicci PG (2002) Methyltransferase recruitment and DNA hypermethylation of target promoters by an oncogenic transcription factor. Science 295(5557):1079–1082. doi:10.1126/science.1065173, pii: 295/5557/1079
107. Liu L, Saldanha SN, Pate MS, Andrews LG, Tollefsbol TO (2004) Epigenetic regulation of human telomerase reverse transcriptase promoter activity during cellular differentiation. Genes Chromosomes Cancer 41(1):26–37. doi:10.1002/gcc.20058
108. Nouzova M, Holtan N, Oshiro MM, Isett RB, Munoz-Rodriguez JL, List AF, Narro ML, Miller SJ, Merchant NC, Futscher BW (2004) Epigenomic changes during leukemia cell differentiation: analysis of histone acetylation and cytosine methylation using CpG island microarrays. J Pharmacol Exp Ther 311(3):968–981. doi:10.1124/jpet.104.072488
109. Das S, Foley N, Bryan K, Watters KM, Bray I, Murphy DM, Buckley PG, Stallings RL (2010) MicroRNA mediates DNA demethylation events triggered by retinoic acid during neuroblastoma cell differentiation. Cancer Res 70(20):7874–7881. doi:10.1158/0008-5472.CAN-10-1534
110. Sirchia SM, Ren M, Pili R, Sironi E, Somenzi G, Ghidoni R, Toma S, Nicolo G, Sacchi N (2002) Endogenous reactivation of the RARbeta2 tumor suppressor gene epigenetically silenced in breast cancer. Cancer Res 62(9):2455–2461

111. Ramachandran K, Navarro L, Gordian E, Das PM, Singal R (2007) Methylation-mediated silencing of genes is not altered by selenium treatment of prostate cancer cells. Anticancer Res 27(2):921–925
112. Fiala ES, Staretz ME, Pandya GA, El-Bayoumy K, Hamilton SR (1998) Inhibition of DNA cytosine methyltransferase by chemopreventive selenium compounds, determined by an improved assay for DNA cytosine methyltransferase and DNA cytosine methylation. Carcinogenesis 19(4):597–604. doi:10.1093/carcin/19.4.597
113. Davis CD, Uthus EO, Finley JW (2000) Dietary selenium and arsenic affect DNA methylation in vitro in Caco-2 cells and in vivo in rat liver and colon. J Nutr 130(12):2903–2909
114. Davis CD, Uthus EO (2002) Dietary selenite and azadeoxycytidine treatments affect dimethylhydrazine-induced aberrant crypt formation in rat colon and DNA methylation in HT-29 cells. J Nutr 132(2):292–297
115. Xiang N, Zhao R, Song G, Zhong W (2008) Selenite reactivates silenced genes by modifying DNA methylation and histones in prostate cancer cells. Carcinogenesis 29(11):2175–2181. doi:10.1093/carcin/bgn179
116. Davis CD, Uthus EO (2003) Dietary folate and selenium affect dimethylhydrazine-induced aberrant crypt formation, global DNA methylation and one-carbon metabolism in rats. J Nutr 133(9):2907–2914
117. Fischer A, Gaedicke S, Frank J, Doring F, Rimbach G (2010) Dietary vitamin E deficiency does not affect global and specific DNA methylation patterns in rat liver. Br J Nutr 104:935–940. doi:10.1017/S0007114510001649
118. Kouzarides T (2007) Chromatin modifications and their function. Cell 128(4):693–705. doi:10.1016/j.cell.2007.02.005, pii: S0092-8674(07)00184-5
119. Bannister AJ, Kouzarides T (2011) Regulation of chromatin by histone modifications. Cell Res 21(3):381–395. doi:10.1038/cr.2011.22
120. Fullgrabe J, Kavanagh E, Joseph B (2011) Histone onco-modifications. Oncogene 30(31): 3391–3403. doi:10.1038/onc.2011.121
121. Nian H, Delage B, Ho E, Dashwood RH (2009) Modulation of histone deacetylase activity by dietary isothiocyanates and allyl sulfides: studies with sulforaphane and garlic organosulfur compounds. Environ Mol Mutagen 50(3):213–221. doi:10.1002/em.20454
122. Druesne-Pecollo N, Latino-Martel P (2011) Modulation of histone acetylation by garlic sulfur compounds. Anticancer Agents Med Chem 11(3):254–259, doi:10.2174/187152011795347540
123. Nian H, Delage B, Pinto JT, Dashwood RH (2008) Allyl mercaptan, a garlic-derived organosulfur compound, inhibits histone deacetylase and enhances Sp3 binding on the P21WAF1 promoter. Carcinogenesis 29(9):1816–1824. doi:10.1093/carcin/bgn165
124. Balasubramanyam K, Swaminathan V, Ranganathan A, Kundu TK (2003) Small molecule modulators of histone acetyltransferase p300. J Biol Chem 278(21):19134–19140. doi:10.1074/jbc.M301580200
125. Sun Y, Jiang X, Chen S, Price BD (2006) Inhibition of histone acetyltransferase activity by anacardic acid sensitizes tumor cells to ionizing radiation. FEBS Lett 580(18):4353–4356. doi:10.1016/j.febslet.2006.06.092, pii: S0014-5793(06)00816-7
126. Sung B, Pandey MK, Ahn KS, Yi T, Chaturvedi MM, Liu M, Aggarwal BB (2008) Anacardic acid (6-nonadecyl salicylic acid), an inhibitor of histone acetyltransferase, suppresses expression of nuclear factor-kappaB-regulated gene products involved in cell survival, proliferation, invasion, and inflammation through inhibition of the inhibitory subunit of nuclear factor-kappaBalpha kinase, leading to potentiation of apoptosis. Blood 111(10):4880–4891. doi:10.1182/blood-2007-10-117994
127. Han JW, Ahn SH, Park SH, Wang SY, Bae GU, Seo DW, Kwon HK, Hong S, Lee HY, Lee YW, Lee HW (2000) Apicidin, a histone deacetylase inhibitor, inhibits proliferation of tumor cells via induction of p21WAF1/Cip1 and gelsolin. Cancer Res 60(21):6068–6074

128. You JS, Kang JK, Lee EK, Lee JC, Lee SH, Jeon YJ, Koh DH, Ahn SH, Seo DW, Lee HY, Cho EJ, Han JW (2008) Histone deacetylase inhibitor apicidin downregulates DNA methyltransferase 1 expression and induces repressive histone modifications via recruitment of corepressor complex to promoter region in human cervix cancer cells. Oncogene 27(10): 1376–1386. doi:10.1038/sj.onc.1210776
129. Wu JT, Archer SY, Hinnebusch B, Meng S, Hodin RA (2001) Transient vs. prolonged histone hyperacetylation: effects on colon cancer cell growth, differentiation, and apoptosis. Am J Physiol Gastrointest Liver Physiol 280(3):G482–G490
130. Archer SY, Meng S, Shei A, Hodin RA (1998) p21(WAF1) is required for butyrate-mediated growth inhibition of human colon cancer cells. Proc Natl Acad Sci USA 95(12):6791–6796
131. Nakata S, Yoshida T, Horinaka M, Shiraishi T, Wakada M, Sakai T (2004) Histone deacetylase inhibitors upregulate death receptor 5/TRAIL-R2 and sensitize apoptosis induced by TRAIL/APO2-L in human malignant tumor cells. Oncogene 23(37):6261–6271. doi:10.1038/sj.onc.1207830
132. Myzak MC, Dashwood RH (2006) Histone deacetylases as targets for dietary cancer preventive agents: lessons learned with butyrate, diallyl disulfide, and sulforaphane. Curr Drug Targets 7(4):443–452
133. Lu R, Wang X, Sun DF, Tian XQ, Zhao SL, Chen YX, Fang JY (2008) Folic acid and sodium butyrate prevent tumorigenesis in a mouse model of colorectal cancer. Epigenetics 3(6): 330–335, doi:10.4161/epi.3.6.7125
134. Heltweg B, Gatbonton T, Schuler AD, Posakony J, Li H, Goehle S, Kollipara R, Depinho RA, Gu Y, Simon JA, Bedalov A (2006) Antitumor activity of a small-molecule inhibitor of human silent information regulator 2 enzymes. Cancer Res 66(8):4368–4377. doi:10.1158/0008-5472.CAN-05-3617, pii: 66/8/4368
135. Kahyo T, Ichikawa S, Hatanaka T, Yamada MK, Setou M (2008) A novel chalcone polyphenol inhibits the deacetylase activity of SIRT1 and cell growth in HEK293T cells. J Pharmacol Sci 108(3):364–371, doi:10.1254/jphs.08203FP, pii: JST.JSTAGE/jphs/08203FP
136. Chen Y, Shu W, Chen W, Wu Q, Liu H, Cui G (2007) Curcumin, both histone deacetylase and p300/CBP-specific inhibitor, represses the activity of nuclear factor kappa B and Notch 1 in Raji cells. Basic Clin Pharmacol Toxicol 101(6):427–433. doi:10.1111/j.1742-7843.2007.00142.x, pii: PTO142
137. Balasubramanyam K, Varier RA, Altaf M, Swaminathan V, Siddappa NB, Ranga U, Kundu TK (2004) Curcumin, a novel p300/CREB-binding protein-specific inhibitor of acetyltransferase, represses the acetylation of histone/nonhistone proteins and histone acetyltransferase-dependent chromatin transcription. J Biol Chem 279(49):51163–51171. doi:10.1074/jbc.M409024200
138. Kang J, Chen J, Shi Y, Jia J, Zhang Y (2005) Curcumin-induced histone hypoacetylation: the role of reactive oxygen species. Biochem Pharmacol 69(8):1205–1213. doi:10.1016/j.bcp.2005.01.014, pii: S0006-2952(05)00068-7
139. Lee SJ, Krauthauser C, Maduskuie V, Fawcett PT, Olson JM, Rajasekaran SA (2011) Curcumin-induced HDAC inhibition and attenuation of medulloblastoma growth in vitro and in vivo. BMC Cancer 11:144. doi:10.1186/1471-2407-11-144
140. Seong AR, Yoo JY, Choi K, Lee MH, Lee YH, Lee J, Jun W, Kim S, Yoon HG (2011) Delphinidin, a specific inhibitor of histone acetyltransferase, suppresses inflammatory signaling via prevention of NF-kappaB acetylation in fibroblast-like synoviocyte MH7A cells. Biochem Biophys Res Commun 410(3):581–586. doi:10.1016/j.bbrc.2011.06.029, pii: S0006-291X(11)00987-9
141. Druesne-Pecollo N, Chaumontet C, Latino-Martel P (2008) Diallyl disulfide increases histone acetylation in colon cells in vitro and in vivo. Nutr Rev 66(suppl 1):S39–S41. doi:10.1111/j.1753-4887.2008.00066.x, pii: NURE066

142. Druesne-Pecollo N, Chaumontet C, Pagniez A, Vaugelade P, Bruneau A, Thomas M, Cherbuy C, Duee PH, Martel P (2007) In vivo treatment by diallyl disulfide increases histone acetylation in rat colonocytes. Biochem Biophys Res Commun 354(1):140–147. doi:10.1016/j.bbrc.2006.12.158, pii: S0006-291X(06)02841-5
143. Olaharski AJ, Rine J, Marshall BL, Babiarz J, Zhang L, Verdin E, Smith MT (2005) The flavoring agent dihydrocoumarin reverses epigenetic silencing and inhibits sirtuin deacetylases. PLoS Genet 1(6):e77. doi:10.1371/journal.pgen.0010077
144. Li Y, Li X, Guo B (2010) Chemopreventive agent 3,3′-diindolylmethane selectively induces proteasomal degradation of class I histone deacetylases. Cancer Res 70(2):646–654. doi:10.1158/0008-5472.CAN-09-1924
145. Degner SC, Papoutsis AJ, Selmin O, Romagnolo DF (2009) Targeting of aryl hydrocarbon receptor-mediated activation of cyclooxygenase-2 expression by the indole-3-carbinol metabolite 3,3′-diindolylmethane in breast cancer cells. J Nutr 139(1):26–32. doi:10.3945/jn.108.099259
146. Li Y, Yuan YY, Meeran SM, Tollefsbol TO (2010) Synergistic epigenetic reactivation of estrogen receptor-alpha (ERalpha) by combined green tea polyphenol and histone deacetylase inhibitor in ERalpha-negative breast cancer cells. Mol Cancer 9:274. doi:10.1186/1476-4598-9-274
147. Choi KC, Jung MG, Lee YH, Yoon JC, Kwon SH, Kang HB, Kim MJ, Cha JH, Kim YJ, Jun WJ, Lee JM, Yoon HG (2009) Epigallocatechin-3-gallate, a histone acetyltransferase inhibitor, inhibits EBV-induced B lymphocyte transformation via suppression of RelA acetylation. Cancer Res 69(2):583–592. doi:10.1158/0008-5472.CAN-08-2442, pii: 69/2/583
148. Lee YH, Kwak J, Choi HK, Choi KC, Kim S, Lee J, Jun W, Park HJ, Yoon HG (2012) EGCG suppresses prostate cancer cell growth modulating acetylation of androgen receptor by anti-histone acetyltransferase activity. Int J Mol Med 30(1):69–74. doi:10.3892/ijmm.2012.966
149. Choi KC, Lee YH, Jung MG, Kwon SH, Kim MJ, Jun WJ, Lee J, Lee JM, Yoon HG (2009) Gallic acid suppresses lipopolysaccharide-induced nuclear factor-kappaB signaling by preventing RelA acetylation in A549 lung cancer cells. Mol Cancer Res 7(12):2011–2021. doi:10.1158/1541-7786.MCR-09-0239
150. Arif M, Pradhan SK, Thanuja GR, Vedamurthy BM, Agrawal S, Dasgupta D, Kundu TK (2009) Mechanism of p300 specific histone acetyltransferase inhibition by small molecules. J Med Chem 52(2):267–277. doi:10.1021/jm800657z
151. Balasubramanyam K, Altaf M, Varier RA, Swaminathan V, Ravindran A, Sadhale PP, Kundu TK (2004) Polyisoprenylated benzophenone, garcinol, a natural histone acetyltransferase inhibitor, represses chromatin transcription and alters global gene expression. J Biol Chem 279(32):33716–33726. doi:10.1074/jbc.M402839200
152. Basak S, Pookot D, Noonan EJ, Dahiya R (2008) Genistein down-regulates androgen receptor by modulating HDAC6-Hsp90 chaperone function. Mol Cancer Ther 7(10):3195–3202. doi:10.1158/1535-7163.MCT-08-0617, pii: 7/10/3195
153. Majid S, Kikuno N, Nelles J, Noonan E, Tanaka Y, Kawamoto K, Hirata H, Li LC, Zhao H, Okino ST, Place RF, Pookot D, Dahiya R (2008) Genistein induces the p21WAF1/CIP1 and p16INK4a tumor suppressor genes in prostate cancer cells by epigenetic mechanisms involving active chromatin modification. Cancer Res 68(8):2736–2744. doi:10.1158/0008-5472.CAN-07-2290, pii: 68/8/2736
154. Kikuno N, Shiina H, Urakami S, Kawamoto K, Hirata H, Tanaka Y, Majid S, Igawa M, Dahiya R (2008) Genistein mediated histone acetylation and demethylation activates tumor suppressor genes in prostate cancer cells. Int J Cancer 123(3):552–560. doi:10.1002/ijc.23590
155. Hong T, Nakagawa T, Pan W, Kim MY, Kraus WL, Ikehara T, Yasui K, Aihara H, Takebe M, Muramatsu M, Ito T (2004) Isoflavones stimulate estrogen receptor-mediated core histone acetylation. Biochem Biophys Res Commun 317(1):259–264. doi:10.1016/j.bbrc.2004.03.041, pii: S0006291X04005029
156. Nian H, Bisson WH, Dashwood WM, Pinto JT, Dashwood RH (2009) Alpha-keto acid metabolites of organoselenium compounds inhibit histone deacetylase activity in human colon cancer cells. Carcinogenesis 30(8):1416–1423. doi:10.1093/carcin/bgp147

157. Gopal YN, Arora TS, Van Dyke MW (2007) Parthenolide specifically depletes histone deacetylase 1 protein and induces cell death through ataxia telangiectasia mutated. Chem Biol 14(7):813–823. doi:10.1016/j.chembiol.2007.06.007, pii: S1074-5521(07)00212-8
158. Wang LG, Liu XM, Fang Y, Dai W, Chiao FB, Puccio GM, Feng J, Liu D, Chiao JW (2008) De-repression of the p21 promoter in prostate cancer cells by an isothiocyanate via inhibition of HDACs and c-Myc. Int J Oncol 33(2):375–380. doi:10.3892/ijo_00000018
159. Beklemisheva AA, Fang Y, Feng J, Ma X, Dai W, Chiao JW (2006) Epigenetic mechanism of growth inhibition induced by phenylhexyl isothiocyanate in prostate cancer cells. Anticancer Res 26(2A):1225–1230
160. Ma X, Fang Y, Beklemisheva A, Dai W, Feng J, Ahmed T, Liu D, Chiao JW (2006) Phenylhexyl isothiocyanate inhibits histone deacetylases and remodels chromatins to induce growth arrest in human leukemia cells. Int J Oncol 28(5):1287–1293
161. Huang YQ, Ma XD, Zhen RJ, Chiao JW, Liu DL (2007) Experiment study of PHI on histone methylation and acetylation in Molt-4 cells. Zhonghua Xue Ye Xue Za Zhi 28(9):612–615
162. Huang YQ, Ma XD, Lai YD, Wang XZ, Chiao JW, Liu DL (2010) Phenylhexyl isothiocyanate(PHI) regulates histone methylation and acetylation and induces apoptosis in SMMC-7721 cells. Zhonghua Gan Zang Bing Za Zhi 18(3):209–212
163. Xiao L, Huang Y, Zhen R, Chiao JW, Liu D, Ma X (2010) Deficient histone acetylation in acute leukemia and the correction by an isothiocyanate. Acta Haematol 123(2):71–76. doi:10.1159/000264628
164. Howitz KT, Bitterman KJ, Cohen HY, Lamming DW, Lavu S, Wood JG, Zipkin RE, Chung P, Kisielewski A, Zhang LL, Scherer B, Sinclair DA (2003) Small molecule activators of sirtuins extend *Saccharomyces cerevisiae* lifespan. Nature 425(6954):191–196. doi:10.1038/nature01960
165. Kai L, Samuel SK, Levenson AS (2010) Resveratrol enhances p53 acetylation and apoptosis in prostate cancer by inhibiting MTA1/NuRD complex. Int J Cancer 126(7):1538–1548. doi:10.1002/ijc.24928
166. Phipps SM, Love WK, White T, Andrews LG, Tollefsbol TO (2009) Retinoid-induced histone deacetylation inhibits telomerase activity in estrogen receptor-negative breast cancer cells. Anticancer Res 29(12):4959–4964, pii: 29/12/4959
167. Love WK, Berletch JB, Andrews LG, Tollefsbol TO (2008) Epigenetic regulation of telomerase in retinoid-induced differentiation of human leukemia cells. Int J Oncol 32(3):625–631
168. Li L-H, Wu L-J, Tashiro S-I, Onodera S, Uchiuni F (2007) Activation of SIRT1 pathway and modulation of the cell cycle were involved in silimarin's protection against UV-induced A375-S2 cell apoptosis. J Asian Nat Prod Res 9:245–252. doi:10.1080/10286020600604260
169. Myzak MC, Karplus PA, Chung FL, Dashwood RH (2004) A novel mechanism of chemoprotection by sulforaphane: inhibition of histone deacetylase. Cancer Res 64(16):5767–5774. doi:10.1158/0008-5472.CAN-04-1326, pii: 64/16/5767
170. Myzak MC, Hardin K, Wang R, Dashwood RH, Ho E (2006) Sulforaphane inhibits histone deacetylase activity in BPH-1, LnCaP and PC-3 prostate epithelial cells. Carcinogenesis 27(4):811–819. doi:10.1093/carcin/bgi265
171. Gibbs A, Schwartzman J, Deng V, Alumkal J (2009) Sulforaphane destabilizes the androgen receptor in prostate cancer cells by inactivating histone deacetylase 6. Proc Natl Acad Sci USA 106(39):16663–16668. doi:10.1073/pnas.0908908106
172. Clarke JD, Hsu A, Yu Z, Dashwood RH, Ho E (2011) Differential effects of sulforaphane on histone deacetylases, cell cycle arrest and apoptosis in normal prostate cells versus hyperplastic and cancerous prostate cells. Mol Nutr Food Res 55(7):999–1009. doi:10.1002/mnfr.201000547
173. Meeran SM, Patel SN, Tollefsbol TO (2010) Sulforaphane causes epigenetic repression of hTERT expression in human breast cancer cell lines. PLoS One 5(7):e11457. doi:10.1371/journal.pone.0011457

174. Pledgie-Tracy A, Sobolewski MD, Davidson NE (2007) Sulforaphane induces cell type-specific apoptosis in human breast cancer cell lines. Mol Cancer Ther 6(3):1013–1021. doi:10.1158/1535-7163.MCT-06-0494
175. Rajendran P, Delage B, Dashwood WM, Yu TW, Wuth B, Williams DE, Ho E, Dashwood RH (2011) Histone deacetylase turnover and recovery in sulforaphane-treated colon cancer cells: competing actions of 14-3-3 and Pin1 in HDAC3/SMRT corepressor complex dissociation/reassembly. Mol Cancer 10:68. doi:10.1186/1476-4598-10-68
176. Myzak MC, Dashwood WM, Orner GA, Ho E, Dashwood RH (2006) Sulforaphane inhibits histone deacetylase in vivo and suppresses tumorigenesis in Apc-minus mice. FASEB J 20(3):506–508. doi:10.1096/fj.05-4785fje
177. Myzak MC, Tong P, Dashwood WM, Dashwood RH, Ho E (2007) Sulforaphane retards the growth of human PC-3 xenografts and inhibits HDAC activity in human subjects. Exp Biol Med (Maywood) 232(2):227–234, pii: 232/2/227
178. Dashwood RH, Ho E (2007) Dietary histone deacetylase inhibitors: from cells to mice to man. Semin Cancer Biol 17(5):363–369. doi:10.1016/j.semcancer.2007.04.001, pii: S1044-579X(07)00024-7
179. Akare S, Jean-Louis S, Chen W, Wood DJ, Powell AA, Martinez JD (2006) Ursodeoxycholic acid modulates histone acetylation and induces differentiation and senescence. Int J Cancer 119(12):2958–2969. doi:10.1002/ijc.22231
180. Cherrier T, Suzanne S, Redel L, Calao M, Marban C, Samah B, Mukerjee R, Schwartz C, Gras G, Sawaya BE, Zeichner SL, Aunis D, Van Lint C, Rohr O (2009) p21(WAF1) gene promoter is epigenetically silenced by CTIP2 and SUV39H1. Oncogene 28(38):3380–3389. doi:10.1038/onc.2009.193
181. Greiner D, Bonaldi T, Eskeland R, Roemer E, Imhof A (2005) Identification of a specific inhibitor of the histone methyltransferase SU(VAR)3-9. Nat Chem Biol 1(3):143–145. doi:10.1038/nchembio721
182. Lakshmikuttyamma A, Scott SA, DeCoteau JF, Geyer CR (2010) Reexpression of epigenetically silenced AML tumor suppressor genes by SUV39H1 inhibition. Oncogene 29(4):576–588. doi:10.1038/onc.2009.361
183. Hua WF, Fu YS, Liao YJ, Xia WJ, Chen YC, Zeng YX, Kung HF, Xie D (2010) Curcumin induces down-regulation of EZH2 expression through the MAPK pathway in MDA-MB-435 human breast cancer cells. Eur J Pharmacol 637(1–3):16–21. doi:10.1016/j.ejphar.2010.03.051, pii: S0014-2999(10)00278-5
184. Balasubramanian S, Adhikary G, Eckert RL (2010) The Bmi-1 polycomb protein antagonizes the (−)-epigallocatechin-3-gallate-dependent suppression of skin cancer cell survival. Carcinogenesis 31(3):496–503. doi:10.1093/carcin/bgp314
185. Choudhury SR, Balasubramanian S, Chew YC, Han B, Marquez VE, Eckert RL (2011) (−)-Epigallocatechin-3-gallate and DZNep reduce polycomb protein level via a proteasome-dependent mechanism in skin cancer cells. Carcinogenesis 32(10):1525–1532. doi:10.1093/carcin/bgr171
186. Huang Y, Greene E, Murray Stewart T, Goodwin AC, Baylin SB, Woster PM, Casero RA Jr (2007) Inhibition of lysine-specific demethylase 1 by polyamine analogues results in reexpression of aberrantly silenced genes. Proc Natl Acad Sci USA 104(19):8023–8028. doi:10.1073/pnas.0700720104
187. Huang Y, Stewart TM, Wu Y, Baylin SB, Marton LJ, Perkins B, Jones RJ, Woster PM, Casero RA Jr (2009) Novel oligoamine analogues inhibit lysine-specific demethylase 1 and induce reexpression of epigenetically silenced genes. Clin Cancer Res 15(23):7217–7228. doi:10.1158/1078-0432.CCR-09-1293
188. Dimri M, Bommi P, Sahasrabuddhe AA, Khandekar JD, Dimri GP (2010) Dietary omega-3 polyunsaturated fatty acids suppress expression of EZH2 in breast cancer cells. Carcinogenesis 31(3):489–495. doi:10.1093/carcin/bgp305
189. Fu S, Kurzrock R (2010) Development of curcumin as an epigenetic agent. Cancer 116(20):4670–4676. doi:10.1002/cncr.25414

190. Suzuki T, Miyata N (2006) Epigenetic control using natural products and synthetic molecules. Curr Med Chem 13(8):935–958
191. Sauve AA, Wolberger C, Schramm VL, Boeke JD (2006) The biochemistry of sirtuins. Annu Rev Biochem 75:435–465. doi:10.1146/annurev.biochem.74.082803.133500
192. Mottet D, Castronovo V (2008) Histone deacetylases: target enzymes for cancer therapy. Clin Exp Metastasis 25(2):183–189. doi:10.1007/s10585-007-9131-5
193. Brooks CL, Gu W (2009) How does SIRT1 affect metabolism, senescence and cancer? Nat Rev Cancer 9(2):123–128. doi:10.1038/nrc2562
194. Smeenk L, Lohrum M (2010) Behind the scenes: unravelling the molecular mechanisms of p53 target gene selectivity (Review). Int J Oncol 37(5):1061–1070. doi:10.3892/ijo_00000757
195. Spange S, Wagner T, Heinzel T, Kramer OH (2009) Acetylation of non-histone proteins modulates cellular signalling at multiple levels. Int J Biochem Cell Biol 41(1):185–198. doi:10.1016/j.biocel.2008.08.027, pii: S1357-2725(08)00347-6
196. Brait M, Sidransky D (2011) Cancer epigenetics: above and beyond. Toxicol Mech Methods 21(4):275–288. doi:10.3109/15376516.2011.562671
197. Upadhyay AK, Cheng X (2011) Dynamics of histone lysine methylation: structures of methyl writers and erasers. Prog Drug Res 67:107–124
198. Yang XD, Lamb A, Chen LF (2009) Methylation, a new epigenetic mark for protein stability. Epigenetics 4(7):429–433, doi:10.4161/epi.4.7.9787
199. Munro S, Khaire N, Inche A, Carr S, La Thangue NB (2010) Lysine methylation regulates the pRb tumour suppressor protein. Oncogene 29(16):2357–2367. doi:10.1038/onc.2009.511
200. West LE, Gozani O (2011) Regulation of p53 function by lysine methylation. Epigenomics 3(3):361–369. doi:10.2217/EPI.11.21
201. Calin GA, Croce CM (2006) MicroRNA signatures in human cancers. Nat Rev Cancer 6(11):857–866. doi:10.1038/nrc1997
202. Winter J, Jung S, Keller S, Gregory RI, Diederichs S (2009) Many roads to maturity: microRNA biogenesis pathways and their regulation. Nat Cell Biol 11(3):228–234. doi:10.1038/ncb0309-228
203. Davis CD, Ross SA (2008) Evidence for dietary regulation of microRNA expression in cancer cells. Nutr Rev 66(8):477–482. doi:10.1111/j.1753-4887.2008.00080.x, pii: NURE080
204. Li Y, Kong D, Wang Z, Sarkar FH (2010) Regulation of microRNAs by natural agents: an emerging field in chemoprevention and chemotherapy research. Pharm Res 27(6):1027–1041. doi:10.1007/s11095-010-0105-y
205. Huang S (2002) Histone methyltransferases, diet nutrients and tumour suppressors. Nat Rev Cancer 2(6):469–476. doi:10.1038/nrc819
206. Henderson CJ, Wolf CR (2011) Knockout and transgenic mice in glutathione transferase research. Drug Metab Rev 43(2):152–164. doi:10.3109/03602532.2011.562900
207. Giudice A, Arra C, Turco MC (2010) Review of molecular mechanisms involved in the activation of the Nrf2-ARE signaling pathway by chemopreventive agents. Methods Mol Biol 647:37–74. doi:10.1007/978-1-60761-738-9_3
208. Tew KD, Townsend DM (2011) Regulatory functions of glutathione S-transferase P1-1 unrelated to detoxification. Drug Metab Rev 43(2):179–193. doi:10.3109/03602532.2011.552912
209. Nakayama M, Gonzalgo ML, Yegnasubramanian S, Lin X, De Marzo AM, Nelson WG (2004) GSTP1 CpG island hypermethylation as a molecular biomarker for prostate cancer. J Cell Biochem 91(3):540–552. doi:10.1002/jcb.10740
210. Yu S, Khor TO, Cheung KL, Li W, Wu TY, Huang Y, Foster BA, Kan YW, Kong AN (2010) Nrf2 expression is regulated by epigenetic mechanisms in prostate cancer of TRAMP mice. PLoS One 5(1):e8579. doi:10.1371/journal.pone.0008579
211. Kunnumakkara AB, Anand P, Aggarwal BB (2008) Curcumin inhibits proliferation, invasion, angiogenesis and metastasis of different cancers through interaction with multiple cell signaling proteins. Cancer Lett 269(2):199–225. doi:10.1016/j.canlet.2008.03.009, pii: S0304-3835(08)00193-6

212. Hanahan D, Weinberg RA (2011) Hallmarks of cancer: the next generation. Cell 144 (5):646–674. doi:10.1016/j.cell.2011.02.013, pii: S0092-8674(11)00127-9
213. Pan MH, Ho CT (2008) Chemopreventive effects of natural dietary compounds on cancer development. Chem Soc Rev 37(11):2558–2574. doi:10.1039/b801558a
214. Li J, Poi MJ, Tsai MD (2011) Regulatory mechanisms of tumor suppressor P16(INK4A) and their relevance to cancer. Biochemistry 50(25):5566–5582. doi:10.1021/bi200642e
215. Raish M, Dhillon VS, Ahmad A, Ansari MA, Mudassar S, Shahid M, Batra V, Gupta P, Das BC, Shukla N, Husain SA (2009) Promoter hypermethylation in tumor suppressing genes p16 and FHIT and their relationship with estrogen receptor and progesterone receptor status in breast cancer patients from Northern India. Transl Oncol 2(4):264–270. doi:10.1593/tlo.09148
216. Belinsky SA, Nikula KJ, Palmisano WA, Michels R, Saccomanno G, Gabrielson E, Baylin SB, Herman JG (1998) Aberrant methylation of p16(INK4a) is an early event in lung cancer and a potential biomarker for early diagnosis. Proc Natl Acad Sci USA 95(20):11891–11896. doi:10.1073/pnas.95.20.11891
217. Shima K, Nosho K, Baba Y, Cantor M, Meyerhardt JA, Giovannucci EL, Fuchs CS, Ogino S (2011) Prognostic significance of CDKN2A (p16) promoter methylation and loss of expression in 902 colorectal cancers: cohort study and literature review. Int J Cancer 128(5): 1080–1094. doi:10.1002/ijc.25432
218. Boultwood J, Wainscoat JS (2007) Gene silencing by DNA methylation in haematological malignancies. Br J Haematol 138(1):3–11. doi:10.1111/j.1365-2141.2007.06604.x, pii: BJH6604
219. Esteller M, Corn PG, Baylin SB, Herman JG (2001) A gene hypermethylation profile of human cancer. Cancer Res 61(8):3225–3229
220. Kim YK, Seo DW, Kang DW, Lee HY, Han JW, Kim SN (2006) Involvement of HDAC1 and the PI3K/PKC signaling pathways in NF-kappaB activation by the HDAC inhibitor apicidin. Biochem Biophys Res Commun 347(4):1088–1093. doi:10.1016/j.bbrc.2006.06.196, pii: S0006-291X(06)01550-6
221. el-Deiry WS, Tokino T, Velculescu VE, Levy DB, Parsons R, Trent JM, Lin D, Mercer WE, Kinzler KW, Vogelstein B (1993) WAF1, a potential mediator of p53 tumor suppression. Cell 75(4):817–825. doi:10.1016/0092-8674(93)90500-P
222. Harper JW, Adami GR, Wei N, Keyomarsi K, Elledge SJ (1993) The p21 Cdk-interacting protein Cip1 is a potent inhibitor of G1 cyclin-dependent kinases. Cell 75(4):805–816. doi:10.1016/0092-8674(93)90499-G
223. Abbas T, Dutta A (2009) p21 in cancer: intricate networks and multiple activities. Nat Rev Cancer 9(6):400–414. doi:10.1038/nrc2657
224. Ocker M, Schneider-Stock R (2007) Histone deacetylase inhibitors: signalling towards p21cip1/waf1. Int J Biochem Cell Biol 39(7–8):1367–1374. doi:10.1016/j.biocel.2007.03.001, pii: S1357-2725(07)00079-9
225. Gartel AL, Radhakrishnan SK (2005) Lost in transcription: p21 repression, mechanisms, and consequences. Cancer Res 65(10):3980–3985. doi:10.1158/0008-5472.CAN-04-3995, pii: 65/10/3980
226. Claus R, Lubbert M (2003) Epigenetic targets in hematopoietic malignancies. Oncogene 22(42):6489–6496. doi:10.1038/sj.onc.1206814
227. Fang JY, Lu YY (2002) Effects of histone acetylation and DNA methylation on p21(WAF1) regulation. World J Gastroenterol 8(3):400–405
228. Riggs MG, Whittaker RG, Neumann JR, Ingram VM (1977) n-Butyrate causes histone modification in HeLa and Friend erythroleukaemia cells. Nature 268(5619):462–464. doi:10.1038/268462a0
229. Candido EP, Reeves R, Davie JR (1978) Sodium butyrate inhibits histone deacetylation in cultured cells. Cell 14(1):105–113. doi:10.1016/0092-8674(78)90305-7
230. Sealy L, Chalkley R (1978) The effect of sodium butyrate on histone modification. Cell 14(1):115–121. doi:10.1016/0092-8674(78)90306-9

231. Hamer HM, Jonkers D, Venema K, Vanhoutvin S, Troost FJ, Brummer RJ (2008) Review article: the role of butyrate on colonic function. Aliment Pharmacol Ther 27(2):104–119. doi:10.1111/j.1365-2036.2007.03562.x, pii: APT3562
232. Jin Z, El-Deiry WS (2005) Overview of cell death signaling pathways. Cancer Biol Ther 4(2):139–163, doi:10.4161/cbt.4.2.1508, pii: 1508
233. Strathmann J, Gerhauser C (2012) Anti-proliferative and apoptosis-inducing properties of Xanthohumol, a prenylated chalcone from hops (*Humulus lupulus* L.). In: Diederich M (ed) Natural compounds as inducers of cell death. Springer, Heidelberg
234. Verkerk R, Schreiner M, Krumbein A, Ciska E, Holst B, Rowland I, De Schrijver R, Hansen M, Gerhauser C, Mithen R, Dekker M (2009) Glucosinolates in Brassica vegetables: the influence of the food supply chain on intake, bioavailability and human health. Mol Nutr Food Res 53(suppl 2):S219. doi:10.1002/mnfr.200800065
235. Cheung KL, Kong AN (2010) Molecular targets of dietary phenethyl isothiocyanate and sulforaphane for cancer chemoprevention. AAPS J 12(1):87–97. doi:10.1208/s12248-009-9162-8
236. Lea MA, Rasheed M, Randolph VM, Khan F, Shareef A, desBordes C (2002) Induction of histone acetylation and inhibition of growth of mouse erythroleukemia cells by S-allylmercaptocysteine. Nutr Cancer 43(1):90–102. doi:10.1207/S15327914NC431_11
237. Higdon JV, Delage B, Williams DE, Dashwood RH (2007) Cruciferous vegetables and human cancer risk: epidemiologic evidence and mechanistic basis. Pharmacol Res 55(3):224–236. doi:10.1016/j.phrs.2007.01.009, pii: S1043-6618(07)00032-1
238. Ellis L, Atadja PW, Johnstone RW (2009) Epigenetics in cancer: targeting chromatin modifications. Mol Cancer Ther 8(6):1409–1420. doi:10.1158/1535-7163.MCT-08-0860
239. Dammann R, Schagdarsurengin U, Seidel C, Strunnikova M, Rastetter M, Baier K, Pfeifer GP (2005) The tumor suppressor RASSF1A in human carcinogenesis: an update. Histol Histopathol 20(2):645–663
240. Agathanggelou A, Cooper WN, Latif F (2005) Role of the Ras-association domain family 1 tumor suppressor gene in human cancers. Cancer Res 65(9):3497–3508. doi:10.1158/0008-5472.CAN-04-4088, pii: 65/9/3497
241. Tommasi S, Dammann R, Zhang Z, Wang Y, Liu L, Tsark WM, Wilczynski SP, Li J, You M, Pfeifer GP (2005) Tumor susceptibility of Rassf1a knockout mice. Cancer Res 65(1):92–98, pii: 65/1/92
242. Shivakumar L, Minna J, Sakamaki T, Pestell R, White MA (2002) The RASSF1A tumor suppressor blocks cell cycle progression and inhibits cyclin D1 accumulation. Mol Cell Biol 22(12):4309–4318. doi:10.1128/MCB.22.12.4309-4318.2002
243. Hanahan D, Weinberg RA (2000) The hallmarks of cancer. Cell 100(1):57–70. doi:10.1016/S0092-8674(00)81683-9
244. Herranz D, Serrano M (2010) SIRT1: recent lessons from mouse models. Nat Rev Cancer 10(12):819–823. doi:10.1038/nrc2962
245. Vivanco I, Sawyers CL (2002) The phosphatidylinositol 3-kinase AKT pathway in human cancer. Nat Rev Cancer 2(7):489–501. doi:10.1038/nrc839
246. Hollander MC, Blumenthal GM, Dennis PA (2011) PTEN loss in the continuum of common cancers, rare syndromes and mouse models. Nat Rev Cancer 11(4):289–301. doi:10.1038/nrc3037
247. Michie AM, McCaig AM, Nakagawa R, Vukovic M (2010) Death-associated protein kinase (DAPK) and signal transduction: regulation in cancer. FEBS J 277(1):74–80. doi:10.1111/j.1742-4658.2009.07414.x, pii: EJB7414
248. Jacinto FV, Esteller M (2007) Mutator pathways unleashed by epigenetic silencing in human cancer. Mutagenesis 22(4):247–253. doi:10.1093/mutage/gem009
249. Esteller M (2000) Epigenetic lesions causing genetic lesions in human cancer: promoter hypermethylation of DNA repair genes. Eur J Cancer 36(18):2294–2300, doi:10.1016/S0959-8049(00)00303-8, pii: S0959804900003038
250. Sawan C, Vaissiere T, Murr R, Herceg Z (2008) Epigenetic drivers and genetic passengers on the road to cancer. Mutat Res 642(1–2):1–13. doi:10.1016/j.mrfmmm.2008.03.002, pii: S0027-5107(08)00061-4

251. Jacinto FV, Esteller M (2007) MGMT hypermethylation: a prognostic foe, a predictive friend. DNA Repair (Amst) 6(8):1155–1160. doi:10.1016/j.dnarep.2007.03.013, pii: S1568-7864(07)00131-0
252. Fang MZ, Jin Z, Wang Y, Liao J, Yang GY, Wang LD, Yang CS (2005) Promoter hypermethylation and inactivation of O(6)-methylguanine-DNA methyltransferase in esophageal squamous cell carcinomas and its reactivation in cell lines. Int J Oncol 26(3):615–622
253. Gerhauser C (2008) Cancer chemopreventive potential of apples, apple juice, and apple components. Planta Med 74(13):1608–1624. doi:10.1055/s-0028-1088300
254. McCabe MT, Low JA, Daignault S, Imperiale MJ, Wojno KJ, Day ML (2006) Inhibition of DNA methyltransferase activity prevents tumorigenesis in a mouse model of prostate cancer. Cancer Res 66(1):385–392. doi:10.1158/0008-5472.CAN-05-2020, pii: 66/1/385
255. Karin M (2006) Nuclear factor-kappaB in cancer development and progression. Nature 441(7092):431–436. doi:10.1038/nature04870
256. Mantovani A, Allavena P, Sica A, Balkwill F (2008) Cancer-related inflammation. Nature 454(7203):436–444. doi:10.1038/nature07205
257. Coussens LM, Werb Z (2002) Inflammation and cancer. Nature 420(6917):860–867. doi:10.1038/nature01322
258. Chaturvedi MM, Sung B, Yadav VR, Kannappan R, Aggarwal BB (2011) NF-kappaB addiction and its role in cancer: 'one size does not fit all'. Oncogene 30(14):1615–1630. doi:10.1038/onc.2010.566
259. Perkins ND (2012) The diverse and complex roles of NF-kappaB subunits in cancer. Nat Rev Cancer 12(2):121–132. doi:10.1038/nrc3204
260. Chen L, Fischle W, Verdin E, Greene WC (2001) Duration of nuclear NF-kappaB action regulated by reversible acetylation. Science 293(5535):1653–1657. doi:10.1126/science.1062374, pii: 293/5535/1653
261. Chen LF, Greene WC (2003) Regulation of distinct biological activities of the NF-kappaB transcription factor complex by acetylation. J Mol Med (Berl) 81(9):549–557. doi:10.1007/s00109-003-0469-0
262. Ghizzoni M, Haisma HJ, Maarsingh H, Dekker FJ (2011) Histone acetyltransferases are crucial regulators in NF-kappaB mediated inflammation. Drug Discov Today 16(11–12): 504–511. doi:10.1016/j.drudis.2011.03.009, pii: S1359-6446(11)00103-6
263. Aggarwal BB, Shishodia S (2006) Molecular targets of dietary agents for prevention and therapy of cancer. Biochem Pharmacol 71(10):1397–1421. doi:10.1016/j.bcp.2006.02.009, pii: S0006-2952(06)00095-5
264. Padhye S, Ahmad A, Oswal N, Sarkar FH (2009) Emerging role of Garcinol, the antioxidant chalcone from *Garcinia indica* Choisy and its synthetic analogs. J Hematol Oncol 2:38. doi:10.1186/1756-8722-2-38
265. Prasad S, Ravindran J, Sung B, Pandey MK, Aggarwal BB (2010) Garcinol potentiates TRAIL-induced apoptosis through modulation of death receptors and antiapoptotic proteins. Mol Cancer Ther 9(4):856–868. doi:10.1158/1535-7163.MCT-09-1113
266. Singh S, Aggarwal BB (1995) Activation of transcription factor NF-kappa B is suppressed by curcumin (diferuloylmethane) [corrected]. J Biol Chem 270(42):24995–25000. doi:10.1074/jbc.270.18.10631
267. Feldman BJ, Feldman D (2001) The development of androgen-independent prostate cancer. Nat Rev Cancer 1(1):34–45. doi:10.1038/35094009
268. DeMarzo AM, Nelson WG, Isaacs WB, Epstein JI (2003) Pathological and molecular aspects of prostate cancer. Lancet 361(9361):955–964. doi:10.1016/S0140-6736(03)12779-1
269. Wang D, Tindall DJ (2011) Androgen action during prostate carcinogenesis. Methods Mol Biol 776:25–44. doi:10.1007/978-1-61779-243-4_2
270. Zhou J, Geng G, Wu JH (2009) Synthesis and in vitro characterization of ionone-based chalcones as novel antiandrogens effective against multiple clinically relevant androgen receptor mutants. Invest New Drugs. doi:10.1007/s10637-009-9251-7

271. Shah S, Small E (2010) Emerging biological observations in prostate cancer. Expert Rev Anticancer Ther 10(1):89–101. doi:10.1586/era.09.161
272. Walsh PC (2010) Chemoprevention of prostate cancer. N Engl J Med 362(13):1237–1238. doi:10.1056/NEJMe1001045, pii: 362/13/1237
273. Barker N, Clevers H (2006) Mining the Wnt pathway for cancer therapeutics. Nat Rev Drug Discov 5(12):997–1014. doi:10.1038/nrd2154
274. Klaus A, Birchmeier W (2008) Wnt signalling and its impact on development and cancer. Nat Rev Cancer 8(5):387–398. doi:10.1038/nrc2389
275. Martinez P, Blasco MA (2011) Telomeric and extra-telomeric roles for telomerase and the telomere-binding proteins. Nat Rev Cancer 11(3):161–176. doi:10.1038/nrc3025
276. Phipps SM, Love WK, Mott TE, Andrews LG, Tollefsbol TO (2009) Differential expression of epigenetic modulators during human embryonic stem cell differentiation. Mol Biotechnol 41(3):201–207. doi:10.1007/s12033-008-9118-8
277. Gronemeyer H, Gustafsson JA, Laudet V (2004) Principles for modulation of the nuclear receptor superfamily. Nat Rev Drug Discov 3(11):950–964. doi:10.1038/nrd1551
278. Delage B, Dashwood RH (2008) Dietary manipulation of histone structure and function. Annu Rev Nutr 28:347–366. doi:10.1146/annurev.nutr.28.061807.155354
279. Esteller M (2007) Epigenetic gene silencing in cancer: the DNA hypermethylome. Hum Mol Genet 16(Spec no. 1):R50–R59. doi:10.1093/hmg/ddm018, pii: 16/R1/R50
280. Niles RM (2007) Biomarker and animal models for assessment of retinoid efficacy in cancer chemoprevention. Acta Pharmacol Sin 28(9):1383–1391. doi:10.1111/j.1745-7254.2007.00685.x
281. Tang XH, Albert M, Scognamiglio T, Gudas LJ (2009) A DNA methyltransferase inhibitor and all-trans retinoic acid reduce oral cavity carcinogenesis induced by the carcinogen 4-nitroquinoline 1-oxide. Cancer Prev Res (Phila) 2(12):1100–1110. doi:10.1158/1940-6207.CAPR-09-0136
282. Chen D, Cui QC, Yang H, Dou QP (2006) Disulfiram, a clinically used anti-alcoholism drug and copper-binding agent, induces apoptotic cell death in breast cancer cultures and xenografts via inhibition of the proteasome activity. Cancer Res 66(21):10425–10433. doi:10.1158/0008-5472.CAN-06-2126, pii: 66/21/10425
283. Tan S, Wang C, Lu C, Zhao B, Cui Y, Shi X, Ma X (2009) Quercetin is able to demethylate the p16INK4a gene promoter. Chemotherapy 55(1):6–10. doi:10.1159/000166383
284. Fini L, Piazzi G, Daoud Y, Selgrad M, Maegawa S, Garcia M, Fogliano V, Romano M, Graziani G, Vitaglione P, Carmack SW, Gasbarrini A, Genta RM, Issa JP, Boland CR, Ricciardiello L (2011) Chemoprevention of intestinal polyps in ApcMin/+ mice fed with western or balanced diets by drinking annurca apple polyphenol extract. Cancer Prev Res (Phila) 4(6):907–15. doi:10.1158/1940-6207.CAPR-10-0359

A Perspective on Dietary Phytochemicals and Cancer Chemoprevention: Oxidative Stress, Nrf2, and Epigenomics

Zheng-Yuan Su, Limin Shu, Tin Oo Khor, Jong Hun Lee, Francisco Fuentes, and Ah-Ng Tony Kong

Abstract Oxidative stress is caused by an imbalance of reactive oxygen species (ROS)/reactive nitrogen species (RNS) and the antioxidative stress defense systems in cells. ROS/RNS or carcinogen metabolites can attack intracellular proteins, lipids, and nucleic acids, which can result in genetic mutations, carcinogenesis, and other diseases. Nrf2 plays a critical role in the regulation of many antioxidative stress/antioxidant and detoxification enzyme genes, such as glutathione S-transferases (GSTs), NAD(P)H:quinone oxidoreductase 1 (NQO1), UDP-glucuronyl transferases (UGTs), and heme oxygenase-1 (HO-1), directly via the antioxidant response element (ARE). Recently, many studies have shown that dietary phytochemicals possess cancer chemopreventive potential through the induction of Nrf2-mediated antioxidant/detoxification enzymes and anti-inflammatory signaling pathways to protect organisms against cellular damage caused by oxidative stress. In addition, carcinogenesis can be caused by epigenetic alterations such as DNA methylation and histone modifications in tumor–suppressor genes and oncogenes. Interestingly, recent studies have shown that several naturally occurring dietary phytochemicals can epigenetically modify the chromatin, including reactivating Nrf2 via demethylation of CpG islands and the inhibition of histone deacetylases (HDACs) and/or histone acetyltransferases (HATs). The advancement and development of dietary phytochemicals in cancer chemoprevention research requires the integration of the

---

Z.-Y. Su, L. Shu, T.O. Khor, J.H. Lee, and A.-N.T. Kong (✉)
Department of Pharmaceutics, Center for Cancer Prevention Research, Ernest-Mario School of Pharmacy, Rutgers, the State University of New Jersey, 160 Frelinghuysen Road, Piscataway, NJ 08854, USA
e-mail: KongT@pharmacy.rutgers.edu

F. Fuentes
Department of Pharmaceutics, Center for Cancer Prevention Research, Ernest-Mario School of Pharmacy, Rutgers, the State University of New Jersey, 160 Frelinghuysen Road, Piscataway, NJ 08854, USA

Departamento de Agricultura del Desierto y Biotecnología, Universidad Arturo Prat, Casilla 121, Iquique, Chile

known, and as-yet-unknown, compounds with the Nrf2-mediated antioxidant, detoxification, and anti-inflammatory systems and their in vitro and in vivo epigenetic mechanisms; human clinical efficacy studies must also be performed.

Keywords Antioxidant response · Inflammation · Keap1 · Nrf2

Contents

1 Introduction 134
2 Oxidative Stress and Cancer 135
   2.1 Oxidative Stress 135
   2.2 Oxidative Stress and Cancer 135
   2.3 The Antioxidant Defense System in Carcinogenesis 136
   2.4 Antioxidant Gene Regulation and the Antioxidant Response Element 138
   2.5 The Regulation of Nrf2 Activation 138
   2.6 Cancer Chemoprevention by Dietary Phytochemicals 139
3 Nrf2-Mediated Antioxidant and Detoxification Systems
   and Anti-inflammation and Cancer Prevention 139
   3.1 Nrf2 and the Antioxidant and Detoxification Systems 140
   3.2 Nrf2 and Anti-inflammation 141
4 Cancer Prevention by Dietary Phytochemicals Via the Nrf2 Pathway 142
5 Epigenetic Alterations in Cancer 147
   5.1 DNA Methylation 148
   5.2 Histone Modifications 148
6 The Epigenomic Reactivation of Nrf2 by Dietary Phytochemicals 149
   6.1 Curcumin 150
   6.2 The Isothiocyanates Sulforaphane and Phenethyl Isothiocyanate 152
   6.3 Tea Polyphenols 153
   6.4 Genistein 154
7 Conclusions 155
References 156

1 Introduction

Cancer chemoprevention is a major cancer preventive strategy that utilizes naturally occurring dietary phytochemicals or therapeutic drugs with relatively low toxicity. Phytochemicals, along with physical activity and mental relaxation, can inhibit, retard, or reverse carcinogenesis. With the advent of modern technology and instrumentation, many studies on dietary phytochemicals have been performed, including studies on their chemistry, biological activities, and mechanisms of action at the cellular level, in in vivo animal model systems, and in clinical trials. Carcinogenic species, such as environmental pollutants, dietary mutagens and radiation, may result in the production of reactive oxygen species (ROS) and/or reactive nitrogen species (RNS), which further react with cellular molecules such as proteins, lipids, and DNA to induce carcinogenesis. Dietary phytochemicals not only directly scavenge ROS/RNS but also indirectly remove carcinogenic reactive intermediates via the transcription factor Nrf2 [nuclear factor erythroid 2 p45 (NF-E2)-related factor 2] antioxidant and detoxification system. When Nrf2 is released from Kelch-like ECH associated protein 1 (Keap1) and translocates to the nucleus, Nrf2 binds to antioxidant responsive elements (AREs) in the promoter/enhancer region of phase II detoxification and antioxidant enzyme genes with the Maf protein. Recent research has also shown that the reactivation of Nrf2 might be regulated by dietary phytochemicals through epigenetic modifications such as DNA methylation and histone modification. In this review we will summarize the correlations among oxidative stress, Nrf2 and cancer. The cancer chemopreventive effects of dietary phytochemicals on the activation of Nrf2-mediated antioxidant, detoxification and anti-inflammatory systems through Nrf2–Keap 1 and epigenetic pathways will also be discussed with regard to their roles in blocking the initiation of carcinogenesis.

2 Oxidative Stress and Cancer

2.1 Oxidative Stress

Free radicals are molecules or molecular fragments containing one or more unpaired electrons. The human body is under attack from free radicals, including superoxide ($O_2^- \bullet$), nitric oxide (NO) and hydroxyl ions (OH$\bullet$) [1]. Hydrogen peroxide, superoxide, and hydroxyl radicals are more generally known as ROS generated as byproducts of the metabolism of oxygen, whereas nitrite, nitrate, and peroxynitrite, referred to as RNS, are generated as the products of NO metabolism [2]. ROS/RNS are generated through various processes, including mitochondria-catalyzed electron transport reactions, UV irradiation, X-rays and gamma rays, inflammatory processes, lipid peroxidation (LPO), and environmental pollutants [3].

Oxidative stress is an imbalance between the generation of ROS/RNS and the antioxidative stress defense systems [4, 5]. Cumulatively produced ROS/RNS in the body induce a cellular redox imbalance and subsequent biomolecular damage. Oxidative stress is a common pathogenic mechanism in aging and the development of various types of cancers and neurodegenerative diseases, such as Alzheimer's disease (AD), Parkinson's disease (PD), and amyotrophic lateral sclerosis (ALS) [6, 7].

2.2 Oxidative Stress and Cancer

Reactive species are well recognized for playing a dual role as both deleterious and beneficial species. ROS/RNS are important intracellular signaling molecules that play key roles in various physiological processes, including apoptosis [8]. ROS/RNS can regulate Bcl-2 expression levels, thereby impacting the function of Bcl-2 to induce cell death through the necrotic or apoptotic pathway [8]. Apoptotic regulation involves receptor activation, a change in the expression levels of the Bcl-2 family of proteins, caspase activation, and mitochondrial dysfunction [9]. C-Jun N-terminal kinase (JNK), or stress-activated protein kinase (SAPK), members of the mitogen-activated protein kinase superfamily (MAPK), are also involved in ROS/RNS-mediated cell death [10]. When at low to moderate concentrations, ROS may induce cellular senescence and apoptosis and play a beneficial physiological role as antitumorigenic species [11, 12]. However, ROS act as second messengers in signal-transduction pathways [13] and are considered to be important mediators of damage to cell structures, including lipids and membranes, proteins, and DNA [14].

Increased levels of reactive species are associated with oncogenic stimulation, and oxidative stress can be considered an important class of carcinogen [11]. Chronic inflammation is associated with an increased risk of various types of human cancers, and inflammation is associated with the induction of oxidative/nitrosative stress and LPO, which generate excess ROS/RNS and DNA-reactive aldehydes [15]. Cancer development is characterized by the cumulative action of multiple events in a single cell with initiation, promotion, and progression stages; the ROS are involved in all stages [16].

The initiation stage involves a non-lethal mutation in DNA [17]. Both ROS and RNS have been shown to be involved in DNA damage [18, 19]. The DNA mutations caused by reactive species include point mutations, deletions, insertions, chromosomal translocations, crosslinks, and other modifications. An early study demonstrated that DNA alterations by oxidative stress through 8-hydroxyguanine (8-OH-G) mutations, which may arise from the formation of 8-OH-dG, involve the GC $\rightarrow$ TA transversion [17]. This type of modified DNA is relatively easily formed, is mutagenic and carcinogenic, and can be used as a potential biomarker of carcinogenesis [20]. Direct DNA damage or genomic instability coupled with altered gene expression and changes in protein conformation occur simultaneously in cancer development [12].

The promotion stage is characterized by the clonal expansion of initiated cells by the induction of cell proliferation and the failure to induce cell death. A high level of oxidative stress is cytotoxic and induces cell apoptosis or necrosis. However, if the oxidative stress is present continuously at a relatively low level, cell division and subsequent tumor growth is stimulated [21]. Progression is an irreversible stage of the carcinogenic process. Further genetic damage and the disruption of chromosome integrity occur at this stage, corresponding to a cell transition from benign to malignant [21, 22].

2.3 *The Antioxidant Defense System in Carcinogenesis*

Antioxidants may be characterized as acting either through the inhibition of ROS generation or through the direct scavenging of free radicals [12, 23]. In living organisms the effects of ROS/RNS are balanced by the antioxidant action, which is composed of both enzymatic and nonenzymatic antioxidants. Antioxidants directly remove free radicals and maintain the intracellular redox status [24].

The nonenzymatic antioxidants include vitamin C (L-ascorbate), vitamin E, carotenoids, selenium, flavonoids, and thiol antioxidants such as glutathione, thioredoxin (Txn), and lipoic acid. [11, 17, 25]. Vitamin C is a water-soluble antioxidant and an enzyme cofactor present in plants and some animals. Humans must obtain vitamin C through the diet because of the inability to synthesize this nutrient endogenously. There are two chemical forms of vitamin C: the reduced form (ascorbic acid, AA) and the oxidized form (dehydroascorbic acid, DHA). Reduced AA is the more predominant chemical structure in the human body, and it is a potent antioxidant that efficiently quenches damaging free radicals. Many in vivo studies have shown a beneficial role of vitamin C in cancer prevention and treatment [26]. However, at high concentrations, vitamin C also serves as a prooxidant promoting ROS levels [26]. Vitamin C can also cooperate with vitamin E to regenerate alpha-tocopherol radicals in membranes and lipoproteins [27]. Vitamin E is a fat-soluble vitamin that exists in eight different forms, and this vitamin also serves as both an anti- and a pro-oxidant via different mechanisms [26].

The enzymatic antioxidants include superoxide dismutases (SODs), catalase, and glutathione peroxidases (GPxs) [27]. SODs are the major antioxidant defense systems against $O_2^- \bullet$ and consist of three isoforms in mammals: SOD1 (the cytoplasmic Cu/ZnSOD), SOD2 (the mitochondrial MnSOD), and SOD3 (the extracellular Cu/ZnSOD). All of the SOD isoforms require a catalytic metal (Cu or Mn) for activation [28]. Catalase is an enzyme that degrades hydrogen peroxide, reducing H_2O_2 to water and oxidizing it to molecular oxygen [29]. Glutathione S-transferases (GSTs) and GPxs are important in the defense against free-radical-induced oxidative damage [30, 31].

The thiol-containing small molecules, such as glutathione (GSH), are major intracellular antioxidants. γ-Glutamyl cysteine synthase (γGCS), including the glutamate cysteine ligase (Gcl), catalytic (Gclc), and modifier (Gclm) subunits, is essential for the biosynthesis of GSH. Some small thiol-containing compounds, such as Txn, glutaredoxins, and periredoxins, undergo rapid oxidization and regeneration and serve as substrates for antioxidant enzymes [24]. In addition to the above-described antioxidant enzymes (SODs, catalase, and GPxs), which inactivate ROS/RNS directly, the antioxidant system also includes enzymes such as glutathione reductase (GSR), NAD(P)H:quinone oxidoreductase 1 (NQO1), UDP-glucuronyl transferases (UGTs), and thioredoxin reductase (Txnd), sulfiredoxin (Srx), and GSTs, which recycle thiols or facilitate the excretion of oxidized and reactive secondary metabolites (e.g., quinones, epoxides, aldehydes, and peroxides) through reduction/conjugation reactions. In antioxidant systems there are other stress response proteins, such as heme oxygenase-1 (HO-1) and -2 (HO-2), metallothionines, and heat shock proteins that also provide cellular protection against various oxidant or pro-oxidant insults [24].

2.4 Antioxidant Gene Regulation and the Antioxidant Response Element

Most of the antioxidant genes listed above contain *cis*-acting antioxidant response elements (AREs) with a functional consensus sequence of 5′-RTGAYnnnGCR-3′ (where R = A or G and Y = C or T) [32]. The AREs have been widely used to screen for potential inducers of antioxidant enzymes [12, 32]. At the transcription level, the antioxidant enzymes are largely regulated by the binding of a particular transcription factor known as nuclear factor erythroid 2p45 (NF-E2)-related factor 2 (Nrf2) to the ARE [33, 34]. Nrf2 was first isolated in 1994 from a hemin-induced K562 erythroid cell line belonging to the basic leucine zipper nuclear transcription factor family, which share regions of homology with that of the *Drosophila* cap "n" collar (CNC) protein [35, 36]. The human Nrf2 showed a high sequence homology to the known p45 subunit of nuclear factor erythroid 2 (NF-E2) [35, 36]. The importance of Nrf2 was demonstrated with Nrf2-knockout mice, which were found to contain lower levels of detoxifying enzymes than wild-type mice and were susceptible to xenobiotics and environmental poisons [37, 38].

Nrf2 activity is mainly regulated by Kelch-like ECH-associated protein 1 (Keap1), a homolog of the *Drosophila* actin-binding protein Kelch, which binds to the actin cytoskeleton. Under homeostatic conditions, Nrf2 is mainly retained in the cytosol by the Keap1 protein [39]. Upon a challenge by oxidative or chemical stress, Nrf2 can be released from the Keap 1 sequestration and translocates to the nucleus [39, 40]. In the nucleus, Nrf2 selectively heterodimerizes with Maf, activation transcription factor (ATF), and/or members of the AP-1 family of leucine zipper proteins to trigger the transcription of its target genes [41, 42].

2.5 The Regulation of Nrf2 Activation

The MAPKs include extracellular signal-regulated kinases (ERKs), JNK, and protein 38 (p38). The MAPK cascade, protein kinase C (PKC), and phosphatidylinositol 3-kinase (PI3K) are involved in the activation of Nrf2–Keap1 with significant cross talk. Numerous studies have revealed that ERK and JNK have a positive effect on ARE-mediated activities [12, 43, 44] and that the phosphorylation of Nrf2 by p38 may inhibit Nrf2 activation by increasing Keap1/Nrf2 binding [45]. Nrf2 can be directly phosphorylated by PKC at serine 40 [46–49], and PI3K signaling facilitates Nrf2 nuclear translocation [50–53]. The direct phosphorylation of Nrf2 by MAPKs, however, has only a slight effect on Nrf2 translocation and activity [54]. However, recent evidence suggests that oxidative stress-mediated posttranscriptional control of Nrf2 activation may also play a role in the regulation of Nrf2 activation [23, 55].

2.6 Cancer Chemoprevention by Dietary Phytochemicals

Phytochemicals from dietary plants and medicinal herbs are becoming increasingly important factors in cancer chemoprevention or adjuvant chemotherapy because many of these plants exhibit effects on cell death and intracellular redox status modulation [40]. Many flavonoids and polyphenolic antioxidants, such as catechins, epigallocatechin gallate (EGCG), and curcumin, exert their anti-inflammatory and antioxidative effects through phase II detoxification/antioxidant enzymes that are mediated by integrated Nrf2 [12, 25, 56, 57]. One phytochemical compound may act on multiple pathways. For example, curcumin has an anti-inflammatory effect by inhibiting NF-κB by blocking IκB degradation. Curcumin has also been shown to regulate the antioxidant response by inhibiting the phosphorylation of Akt and ERK [58, 59]. In addition, curcumin regulates cell death by decreasing the expression levels of tumor necrosis factor-α and endogenous Bcl-2 and Bcl-xL [60, 61]. EGCG has been shown to have multiple effects on the cell cycle and on anti-inflammatory and anticancer regulation through the modulation of NF-κB, COX-2, DNA methyl transferase 1 (DNMT1), ERK-1/2, p38, and matrix metalloproteinase-2 (MMP2) [62–64].

3 Nrf2-Mediated Antioxidant and Detoxification Systems and Anti-inflammation and Cancer Prevention

Oxidative stress results in various pathological conditions and diseases such as inflammation and cancer because oxidative stress causes biochemical alterations in cellular components such as proteins, nucleic acids, and lipids [14]. Oxidative stress is caused by the imbalance between ROS formation and cellular antioxidant capacity. The antioxidant system in cells mitigates the toxic attack and ROS potential. Thiol-containing small molecules, such as GSH and Txn, which belong to the nonenzymatic antioxidant system, can eliminate ROS directly [65]. Enzymes such as catalase, GPx, and peroxiredoxins (Prdx) can remove ROS via catalytic reactions accompanied by GSH or Txn [66, 67].

Xenobiotics come from various drugs, carcinogens and environmental chemicals, and they are typically converted into intermediate molecules that may contain nucleophilic or electrophilic groups through the catalytic action of phase I enzymes such as cytochrome P450 enzymes [68, 69]. Some xenobiotic metabolites may possess toxic or carcinogenesis potentials, and the induction of oxidative stress may be one of the inducible phenomena. However, most if not all hydrophobic xenobiotic metabolites are eliminated after conjugation with hydrophilic molecules such as GSH and glucuronic acid by phase II detoxification and antioxidant enzymes [70].

Nrf2 is a crucial regulator in the induction of the phase II antioxidant and detoxification enzyme genes, which protect cells from damage resulting from oxidative and electrophilic attack [71, 72]. Therefore, dietary phytochemicals will be indirect antioxidants that improve cellular antioxidant capacity by enhancing the gene expression of phase II antioxidant and detoxification enzymes via the Nrf2 pathway.

3.1 Nrf2 and the Antioxidant and Detoxification Systems

The principal phase II antioxidant and detoxification enzymes include the classical conjugating enzymes such as GSTs and UGTs, reduction enzymes such as NQOs, and stress response enzymes such as HO-1 [67, 73]. Many phase II antioxidant and detoxification genes are regulated through the ARE in the promoter [74]. Nrf2 has been demonstrated in extensive studies to be an essential transcription factor for the regulation of the ARE [42, 75–77]. Nrf2 that has translocated from the cytoplasm to the nucleus interacts with other bZIP transcription factor partners, such as small Maf proteins (Maf F, Maf G, and Maf K) and ATF4, and transactivates AREs [78–81]. Many chemicals induce the expression of ARE-driven genes through the translocation of Nrf2, including phenolic antioxidants, such as BHA and *tert*-butyl hydroxyquinone (tBHQ); isothiocyanates, such as sulforaphane (SFN) and PEITC; and synthetic triterpenoids, such as oleanane [82–88].

GSTs have seven distinct classes based on amino-acid sequences, the physical structure of the genes and immunological cross-reactivity; these classes include alpha (α), mu (μ), omega (ω), pi (π), sigma (σ), theta (θ), and zeta (ζ) [89]. GSTs scavenge endogenous and exogenous electrophiles, such as epoxides, aldehydes, and peroxides, in cells [89]. A number of studies have demonstrated that Nrf2 plays a crucial role in the regulation of GSTs. Nrf2 induces significant changes in the mRNA expression levels of many subtypes of mouse hepatic GSTs [75]. GST mRNA and protein expression levels are decreased in Nrf2-KO mice compared with wild-type mice, and elevated Nrf2 activation in the liver resulted in a marked increase of GST mRNA expression in Keap1-knockdown mice [75, 90]. Chemopreventive synthetic antioxidants, such as butylated hydroxyanisole (BHA) and ethoxyquin, increased the expression of GSTs in the mouse liver through Nrf2 induction [91]. In addition, lithocholic acid, the most toxic bile acid, has been shown to increase hepatic glutathione and GST activity in wild-type mice compared with Nrf2-KO mice [92].

UGTs are important enzymes for the excretion of water-soluble glucuronides transformed from toxic exogenous (such as drugs, pesticides, and carcinogens) and endogenous (such as bilirubin, steroids, and hormones) compounds through a conjugation reaction [93]. UGTs play a critical protective role against environmental chemicals and carcinogens. For example, UGT-deficient cultured rat skin fibroblast is more susceptible to B[*a*]P carcinogenesis [94]. The reduction of DMBA–DNA adduct formation was found in breast cancer cells with elevated UGT1A1 [95]. It has also been found that tBHQ induces the UGT1A1 mRNA level and enzyme activity in the liver and intestine in UGT1A transgenic mice [96].

Lower basal mRNA expression levels of UGTs such as UGT1A6, UGT1A9, UGT2B34, UGT2B35, and UGT2B36 were observed in Nrf2-knockout mice compared with wild-type mice [86, 97, 98]. It has been demonstrated that Nrf2 up-regulates UGT activity and promotes a conjugation reaction of 4-aminobiphenyl (ABP) from tobacco smoke with glucuronic acid in the liver, which might protect the liver against ABP [99]. The GST activity was reduced in the liver and small intestine of Nrf2 KO mice, and oltipraz, a chemopreventive agent, does not affect the expression levels of these enzymes in Nrf2-KO mice compared with wild-type mice [100].

NQO1 is a cytosolic flavoprotein and facilitates the detoxification and excretion of endogenous and exogenous chemicals through a reduction reaction from quinones to hydroquinones [101, 102]. It has been reported that the disruption of NQO1 contributed to a higher susceptibility to B[a]P-induced skin carcinogenesis in mice [103]. Lower Nqo1 expression and activity were found in the liver, small intestine, and forestomach of Nrf2-KO mice [75, 99, 100]. Early carcinogenesis induced by cyclophosphamide, which causes oxidative stress in the rat liver, can be effectively inhibited by the powerful antioxidant astaxanthin accompanied by an increase in NQO-1 and HO-1 as mediated through the Nrf2-ARE pathway [104]. The lycopene metabolite apo-8′-lycopenal induced the accumulation of nuclear Nrf2, which resulted in an increase in HO-1 and NQO-1 expression levels in human hepatoma HepG2 cells [105]. In addition, NQO1 mRNA and protein expression levels can be increased by curcumin as mediated by restoring Nrf2 expression through DNA demethylation on Nrf2 promoter CpG islands [106].

HO-1 exhibits both antioxidative and anti-inflammatory capacities. HO-1 catalyzes the catabolism of the pro-oxidant heme to produce bilirubin and carbon monoxide, which have antioxidative and anti-inflammatory effects, respectively [107–109]. HO-1 mRNA and protein expression levels are induced when cells are exposed to oxidative stress that results in cellular injury [110], and Nrf2 is a critical transcription factor that regulates the induction of the HO-1 gene [111]. The administration of toxic paraquat and cadmium chloride induced the expression of HO-1 mRNA and protein in peritoneal macrophages of wild-type mice but not in Nrf2-KO mice [112]. Nordihydroguaiaretic acid (NDGA), a cancer chemopreventive agent, induced the protein expression of Nrf2 and HO-1 in kidney-derived LLC-PK1, in HEK293T cells, and in wild-type MEFs, but not in Nrf2-KO MEFs [113]. Berberine is an important active compound in the Chinese herb *Rhizoma coptidis*. Berberine promoted HO-1 mRNA and protein expression levels mediated by Nrf2 activation through the PI 3-kinase/AKT pathway in rat brain astrocytes [114].

3.2 Nrf2 and Anti-inflammation

In addition to oxidative stress, Nrf2 also participates in the protection against inflammation in cells [115–120]. It has been shown that lipopolysaccharide (LPS) increased NADPH oxidase-dependent ROS generation and the levels of TNF-alpha, IL-6 and chemokines (Mip2 and Mcp-1) in the peritoneal neutrophils from Nrf2-KO mice compared with wild-type mice [121]. Nrf2 is a crucial regulator that has been shown to modulate the innate immune response and survival during experimental sepsis using Nrf2-deficient mice and Nrf2-deficient mouse embryonic fibroblasts [122]. Some findings have suggested that there is cross-talk between Nrf2 and inflammation [123]. The Nrf2/ARE signaling pathway may be negatively regulated by proinflammatory signaling [124]. It was hypothesized that NF-κB/p65 could result in the inactivation of Nrf2 through the selective deprivation of the CREB binding protein (CBP) from Nrf2 [124]. NF-κB/p65 also promotes the interaction of HDAC3 with either CBP or MafK, which results in the repression of ARE [124].

It has been reported that Nrf2 mitigates chemical-induced pulmonary injury and inflammation [125, 126]. The genetic ablation of Nrf2 resulted in severe tobacco-smoke-induced emphysema, airway inflammation, and asthma in mice [127, 128]. The major reason for the expression of these phenotypes is that a disruption of Nrf2 caused lower antioxidant gene expression levels, enhanced the expression levels of the T helper type 2 cytokines interleukin (IL)-4 and IL-13 in bronchoalveolar lavage fluid and in splenocytes, and increased alveolar cell apoptosis after allergen challenge [127, 128]. The Nrf2-KO mice are also more susceptible to DSS-induced colitis. More severe colonic colitis was observed in Nrf2-KO mice, including the loss of colonic crypts, the massive infiltration of inflammatory cells, and anal bleeding, than in wild-type mice [117]. A lower induction of phase II antioxidant and detoxification enzymes, such as HO-1, NQO1, UGT1A1, and GSTM1, and a higher induction of proinflammatory biomarkers, such as interleukin (IL)-1β, IL-6, TNF-α, nitric oxide synthetase (iNOS), and cyclooxygenase 2 (COX2), were observed in Nrf2-KO mice [117]. It has also been shown that indirect antioxidants protected animals from inflammatory damage via Nrf2 activation, which may be a cancer-preventive mechanism [121, 129], and that Nrf2 is required for sulforaphane (SFN)-mediated anti-inflammatory response [130].

4 Cancer Prevention by Dietary Phytochemicals Via the Nrf2 Pathway

Chemoprevention involves the use of dietary compounds or synthetic chemicals to inhibit the development of invasive cancer. Chemoprevention can involve preventing carcinogens from reaching the target sites, from undergoing metabolic activation, or from subsequently interacting with crucial cellular macromolecules such as DNA, RNA, and proteins at the initiation stage. In addition, chemoprevention can inhibit the malignant transformation of initiated cells at either the promotion or the progression stage [71, 131, 132].

In this context, the induction of phase II detoxification and antioxidant enzymes is assumed to be one of the most effective ways to prevent carcinogenesis by both endogenous and exogenous carcinogens [133]. Thus, several dietary compounds that exhibit antioxidant activity and function as inducers and/or cell signals have been reported to increase phase II detoxification enzymes, and these compounds may act as chemopreventive agents [134, 135]. Most of these phase II detoxification enzymes are known to be induced by promoting the nuclear translocation of Nrf2 and its subsequent binding to the ARE sequence in those enzyme genes, leading to transcriptional activation [136]. Thus, Nrf2 is considered the major regulatory pathway of cytoprotective gene expression against oxidative and/or electrophilic stress [137].

Several studies have used in vitro and in vivo approaches involving natural dietary compounds to show that Nrf2 controls the expression of ARE-mediated gene expression and to demonstrate the role of Nrf2 in cancer chemoprevention [138, 139]. Some examples of Nrf2 inducers include curcumin from turmeric [106]; indole-3-carbinol (I3C), 3,3'-diindolylmethane (DIM), phenethyl isothiocyanate (PEITC), and sulforaphane (SFN) from cruciferous vegetables [56, 140]; epigallocatechin-3-gallate (EGCG) from green tea [141]; resveratrol from grapes [142], gamma-tocopherol-enriched mixed tocopherols from soybeans and corn oil [143]; and other compounds described in Table 1. To date, the Nrf2 downstream genes identified can be grouped into the following categories: intracellular redox-balancing proteins, which reduce the levels of ROS with enzymes such as glutamate cysteine ligase (GCL), GPx, Txn, Txnd, peroxiredoxin (Prx), and HO-1; phase II detoxifying enzymes, which metabolize xenobiotics into less toxic forms and/or catalyze conjugation reactions to increase the solubility of xenobiotics, thereby facilitating their elimination [133] with enzymes like HO-1, NQO1, GSTs, GSR, glutamate–cysteine ligase (the catalytic subunit, GCLC and the modifier subunit, GCLM), microsomal epoxide hydrolase 1 (mEH), and the UGT1 family polypeptide A6 (UGT1A6) [150]; and transporters, which control the uptake and efflux of endogenous substances and xenobiotics such as the multidrug resistance-associated protein (MRP) [112, 133]. Thus, this complicated crosstalk among various molecular targets and signaling pathways constitutes an elaborate network that responds coordinately to various xenobiotics, including carcinogens, drugs, and dietary bioactive compounds [134].

Interestingly, the Nrf2 pathway has also been connected to the inflammatory response by studies using the TRAMP mouse model of prostate carcinogenesis [154]. Mice lacking the Nrf2 pathway have proven to be more susceptible to experimentally induced colitis; as expected, these mice express low levels of phase II detoxification and antioxidant enzymes (i.e., HO1, NQO-1, UGST1A1, GST) and exhibit an increased expression of proinflammatory cytokines/mediators [i.e., cyclooxygenase-2 (COX-2), inducible nitric oxide synthase (iNOS), interleukin 1β (IL-1β), interleukin 6 (IL-6), and tumor necrosis factor α (TNF-α)] [117]. In contrast, extracts from *Chrysanthemum zawadskii* (CZ) and licorice *Glycyrrhiza uralensis* (LE) have been shown (using in vitro and in vivo approaches) to possess a strong inhibitory effect against NF-κB-mediated inflammation and to have a strong activation of the Nrf2-ARE-antioxidative stress-signaling pathways [155].

Table 1 Dietary cancer chemopreventive compounds that activate Nrf2

Compound(s)	Plant	Model	The molecular targets of induction or suppression	The concentration required for Nrf2 induction	Reference
Avicins	*Acacia victoriae*	HepG2, human hepatoma cell line	↑NQO1, HO-1, ferritin, bilirubin, GSH, GST, TRX$_{red}$; ↓NF-κB	0.25–2 μg/mL	[144]
		SKH-1, albino hairless (hr/hr) female mice	↑NQO1, HO-1	0.4–0.8 mg (avicin extract)	
Cafestol:Kahweol (1:1)	Coffee (*Coffea arabica*)	C57BL/6, Nrf2 knockout mice (liver, small and large intestine)	↑NQO1, GSTA1, UGT1A6, GCLC	3–6 wt%	[145]
		Embryonic fibroblasts, Nrf2 knockout mice	↓NQO1	20 μg/mL C + K	[145]
Carnosol	Rosemary (*Rosmarinus officinalis*)	HepG2, human hepatoma cell line	↑GSH, GCLM, GCLC; ↓NF-κB	5 μmol/L	[146]
Chlorogenic acid (CGA)	Coffee (*Coffea arabica*)	JB6, mouse epidermal cell line	↑NQO1, and GST; ↓AP-1, NF-κB, JNKs, p38, ERKs, and MAPK	5–40 μM	[147]
Curcumin	Turmeric (*Curcuma longa*)	TRAMP C1, mouse prostate tumor cell line	↑NQO-1; ↓DNMTs	2.5–5 μM	[106]
Epigallocatechin-3-gallate (EGCG)	Green tea (*Camellia sinensis*)	C57BL/6J, Nrf2 knockout mice (liver and small intestine)	↑ and ↓ regulation of several genes related to detoxification, transport, cell growth and apoptosis, cell adhesion, kinase, and transcription	200 mg/kg	[141]
γ-Tocopherol-enriched mixed tocopherol	Soybean (*Glycine max*)	[TRAMP 3 C57BL/6] F1 or as [TRAMP 3 C57BL/6] F2 offspring	↑HO-1, GPx, catalase, SOD	0.10%	[143]
Garlic organosulfur compounds	Garlic (*Allium sativum*)	HepG2-C8, transfected human hepatoblastoma pARE-TI-luciferase cell line	↑NQO1, HO-1, JNKs, p38, ERKs	10–500 μM	[148]

Glyceollins	Soybean (*Glycine max*) exposed to *Aspergillus sojae*	Hepa1c1c7 and its mutant (BPRc1), mouse hepatoma cell lines and HepG2-C8, transfected human hepatoblastoma pARE-TI-luciferase cell line	↑NQO1, HO1, γ-GCS, GR; activation of the PI3K pathway	0.187–3 µg/mL	[148]
Indole-3-carbinol (I3C)	Cruciferous vegetables	HepG2-C8, transfected human hepatoblastoma pARE-TI-luciferase cell line	↑HO-1, NQO1, SOD1, UGT1A1, GSTm2	25–75 µM	[56]
		TRAMP C1, mouse prostate tumor cell line	↑GCLC, NQO-1, HO-1	25–75 µM	[149]
		[TRAMP × C57BL/6] F1 or [TRAMP × C57BL/6] F2 offspring	↑Cleaved caspases-3 and −7 and p21; ↓cleaved PARP, cyclin D1	1 wt%	[149]
Lycopene metabolites	Tomato (*Solanum lycopersicum*)	BEAS-2B, human bronchial epithelial cell line	↑HO-1, NQO1, GST	1–10 µM	[150]
Phenethyl isothiocyanate (PEITC)	Cruciferous vegetables	HepG2-C8, transfected human hepatoma pARE-TI-luciferase cell line	↑HO-1, NQO1, SOD1, UGT1A1, GSTm2	1–10 µM	[56]
Sulforaphane (SFN)	Cruciferous vegetables	HepG2-C8, transfected human hepatoma pARE-TI-luciferase cell line	↑HO-1, NQO1, SOD1, UGT1A1, GSTm2	1–10 µM	[56]
Resveratrol	Grape (*Vitis vinifera*)	MCF10A, human mammary epithelial cell line	↑BRCA1, GCLC, UGT1A1	1–5 µM	[142]
trans-Cinnamic aldehyde		MDA-MB-231, human breast carcinoma cell line	↑NQO1, HO-1	1–25 µM	[151]

(continued)

Table 1 (continued)

Compound(s)	Plant	Model	The molecular targets of induction or suppression	The concentration required for Nrf2 induction	Reference
	Cinnamon (*Cinnamomum verum*)				
Zerumbone	Tropical ginger (*Zingiber zerumbet*)	JB6 Cl41, mouse epidermal cell line	↑HO-1	10 μM	[152]
		HR-1 hairless, Nrf2 knockout mice	↑HO-1	10 mmol/200 mL acetone (topical)	[152]
3,3′-Diindolylmethane (DIM)	Cruciferous vegetables	HepG2-C8, transfected human hepatoma pARE-TI-luciferase cell line	↑HO-1, NQO1, SOD1, UGT1A1, GSTm2	25–75 μM	[56]
6-(Methylsulfinyl)hexyl isothiocyanate (6-MSITC)	Wasabi (*Wasabia japonica*)	HepG2, human hepatoma cell line	↑NQO1, ↓Keap1	5–20 μM	[153]

AP-1 activator protein 1, *BRCA1* breast cancer 1, *DNMT* DNA methyltransferases, *ERK* extracellular signal-regulated protein kinase, *GCLC* glutamate–cysteine ligases, catalytic heavy subunit, *GCLM* glutamate–cysteine ligases, modulatory light subunit, *GPx* glutathione peroxidase, *GR* glutathione reductase, *GSH* glutathione, *GST* glutathione S-transferases, *GSTA1* glutathione S-transferase class Alpha 1, *HO-1* heme oxygenase 1, *JNK* c-Jun N-terminal kinase, *Keap1* Kelch-like ECH-associated protein 1, *MAPKs* mitogen-activated protein kinases, *NF-κB* nuclear factor kappa-B, *NQO1* NAD(P)H:quinone oxidoreductase 1, *PARP* poly (ADP-ribose) polymerase, *PI3K* phosphoinositide 3-kinase, *SOD* superoxide dismutase, *TRXred* thioredoxin reductase, *UGT1A6* UDP-glucuronosyl transferase 1A6, *γ-GCS* gamma glutamyl cysteine synthase

Other studies have suggested Nrf2 involvement with MAPK pathways, including the ERK, JNK, and p38 pathways, in chemical-induced detoxifying enzyme regulation [148, 156]. For example, it has been demonstrated that blocking the ERK pathway attenuates the induction of ARE-mediated gene expression by tBHQ and SFN in human hepatoma HepG2 cells and in the murine hepatoma Hepa1c1c7 cells, whereas inhibition of the p38 pathway shows an opposite effect, implying the involvement of MAPKs in the modulation of ARE-mediated gene expression [157, 158]. These MAPKs, such as ERK, JNK, and p38, have also been activated by treatment with diallyl trisulfide (DATS), one of the three major organosulfur compounds of garlic. However, the inhibition of MAPKs did not affect DATS-induced ARE activity in HepG2-ARE-C8 cells (human hepatoma cells transfected with pARE-TI-luciferase) [148].

5 Epigenetic Alterations in Cancer

Cancer is caused by a series of genetic changes in tumor suppressor genes and oncogenes. However, a large amount of evidence has shown that epigenetic alterations such as DNA methylation and histone modifications can also contribute to carcinogenesis [159]. The term "epigenetics" was first defined as "the causal interactions between genes and their products, which bring the phenotype into being" by the developmental biologist Conrad H. Waddington in 1942 [160]. The concept of epigenetics has evolved as well. As Wolffe defined it, epigenetics became "the study of heritable changes in gene expression that occur without a change in DNA sequence" [161].

In cancer, hypermethylation of the promoter regions of certain tumor suppressor genes is thought to be the most relevant epigenetic change associated with malignant transformation. These heritable changes occur through the methylation of cytosine bases in the DNA and by post-transcriptional modifications of histones [162]. For example, hypermethylation of the CpG island located in the promoter region of tumor suppressor genes such as $p16^{ink4a}$ and BRCA1 results in gene silencing [163, 164]. Histones also play a pivotal role in epigenetic modification. Histone modification is known to regulate gene expression and chromatin structure, which are closely associated with DNA methylation [165].

Unlike genetic changes, epigenetic alterations are potentially reversible. Epigenetically modified genes can be restored, whereas genetic mutations are permanent. Transcriptionally repressed genes that are silenced by epigenetic alteration can be reactivated by epigenetic modification because these silenced genes are still intact. The removal of the methyl groups from the silenced tumor suppressor genes reverses the expression of these genes, leading to the recovery of function [166]. Therefore, the study of epigenetic targets and the mechanism of inhibition can be a novel approach to halt or delay carcinogenesis. The application of drugs to target epigenetic alterations represents a new and fascinating approach in the field of cancer prevention and therapy. With their relatively low toxicity levels and promising effects, dietary chemopreventive phytochemicals may provide a plausible avenue for epigenetic chemoprevention.

We present two important epigenetic mechanisms, DNA methylation and histone modification, that are of interest for cancer chemoprevention. Specific inhibitors of these epigenetic alterations and the dietary chemopreventive phytochemicals that have potential as epigenetic modifiers are also presented in this review.

5.1 DNA Methylation

DNA methylation is the most extensively studied epigenetic event. In mammalian cells, DNA methylation is the addition of a methyl group to the 5' position of cytosine bases in CpG dinucleotides by DNA methyltransferases (DNMT) [167, 168]. The CpG dinucleotides are not distributed evenly throughout the genome but instead tend to group in regions known as CpG islands [168]. Approximately 60% of the human genome promoters are linked to CpG islands. Most CpG sites throughout the genome are known to be methylated. In contrast, the majority of CpG islands usually remain unmethylated in undifferentiated normal cells [168, 169]. These unmethylated CpG islands have an open structure and accord closely with the adjacent transcriptional promoter, leading the genes to remain transcriptionally active [170]. However, in cancer cells, the hypermethylation of CpG islands is known to cause gene silencing by preventing the recruitment of transcriptional protein from DNA [171]. In addition, DNA methylation can interact with various methyl-CpG binding domain proteins (MBDs), such as MBD1–MBD4 and methyl CpG binding protein 2 (MeCP2), by providing the binding site [172, 173]. These binding proteins can interact with a co-repressor complex, including histone deacetylases (HDACs), resulting in transcriptional repression [174, 175].

The primary goal of DNA methylation studies is to find DNMT inhibitors. However, other molecules are also involved in epigenetic mechanisms. Among the DNMT inhibitors, 5-azacytidine and 5-aza-2-deoxycytidine are the most widely studied epigenetic modifiers [176, 177]. However, there are many studies showing that DNA methylation is an essential function in normal mammalian cells [169]. In a mutant-DNMT mouse model, homozygous mouse embryos exhibited delayed development and did not survive past mid-gestation [178]. DNMT 3a and 3b are essential for de novo DNA methylation and mouse development. The inactivation of both genes by gene targeting blocks de novo methylation in embryonic stem cells and arrests embryonic development [179]. Thus, the genetic disruption of DNMTs in a mouse model shows that a balanced DNMT activity is important to maintaining cellular homeostasis. Accumulating evidence demonstrates that the DNA methylation of genes in most human cancers, similar to mutations and deletions, causes the transcriptional silencing of tumor suppressor genes [180].

5.2 Histone Modifications

Together with DNA methylation, histone modification plays an important role in gene expression and tumorigenesis by influencing chromatin structure [159, 181].

Chromatin is present in eukaryotic cells and is a densely packed macromolecular complex that is composed of DNA, histones, and non-histone proteins. The functional roles of chromatin are to package DNA into a small volume to fit within the nucleus and to influence gene expression and DNA replication. The nucleosome, the basic subunit of chromatin, is composed of a histone octamer that consists of an H3/H4 tetramer and two H2A/H2B dimers, and 146 bp of DNA is wrapped around this octamer. Higher-order structuring of nucleosomes results in a compact 30-nm fiber, which is then condensed to form chromosomes. The stability of these more highly folded structures is maintained by the addition of histones. The chromatin structure, which is closely involved in gene expression, is regulated by post-translational modifications of histones [182–184]. There are two different forms of chromatin structure: heterochromatin (condensed) and euchromatin (extended) [185]. In general, heterochromatin is a tightly packed structure, and it is difficult for transcription factors to access heterochromatin, which represses gene transcription. In contrast, euchromatin is loosely packed and more accessible to transcription factors, which enables active gene expression [186]. Histone proteins contain a globular C-terminal domain and an unsaturated N-terminal tail, which are amino-terminal residues protruding from nucleosomes [182]. Most histone modifications occur at the lysine, arginine, and serine residues of the N-terminal tails extending from the histone core by post-transcriptional modifications such as acetylation, methylation, phosphorylation, ubiquitination, and sumoylation [182, 187, 188]. The chromatin structure can be regulated through these modifications, which provide different levels of accessibility to transcription factors [189]. Various histone modifications are potentially reversible through the addition and removal of covalent alterations at the histone tail [181].

Interestingly, methylation on a lysine residue at histone H3 appears to induce two opposite structures, transcriptionally active chromatin or inactive chromatin, depending on which residue is methylated. Methylation at lysine 4 (Lys4) at the histone H3 tail is known to be associated with transcriptionally active chromatin, whereas methylation at lysine 9 (Lys9) in the same histone tail is reported to be related to transcriptionally repressed chromatin [185, 190, 191]. Moreover, important findings suggest that the methylation of H3 Lys9 might be required for DNA methylation [192, 193]. DNMT inhibitors, such as 5-azacytidine and 5-aza-2-deoxycytidine, trichostatin A and suberoylanilide hydroxamic acid (SAHA), are widely used as HDAC inhibitors in many studies [177, 194].

6 The Epigenomic Reactivation of Nrf2 by Dietary Phytochemicals

Epigenetic modification plays a prominent role in the development and differentiation of various cells in an organism. Defects in the epigenome have been implicated in many diseases and are known to be influenced, in whole or at least in part, by environmental factors. It is apparent that environmental factors, diet, and lifestyle have an impact on the development of various cancers in humans. Hence, minimizing exposure to environmental carcinogens, maintaining a healthier lifestyle, and consuming a healthy diet are thought to be reasonable approaches for cancer prevention. In addition to genetic mutations, epigenetic alterations play an important role in cancer development. It is believed that epigenetic changes arise before genetic alterations. The potential of dietary phytochemicals as cancer chemopreventive/anticancer agents through epigenetic modification has been demonstrated in many studies. In this chapter we will provide an overview of cancer epigenetics and discuss the potential for (and challenges of) using dietary phytochemicals as epigenetic modifiers for cancer chemoprevention.

The inclusion of epigenetics in the National Institutes of Health (NIH) research portfolio and roadmap in 2008 has indicated the urgent need for research in epigenetic mechanisms of diseases, including cancer. Unlike genetic mutations, changes in gene expression due to epigenetic regulation during carcinogenesis can be reversed or prevented by chemicals. Therefore, the pharmacological targeting of epigenetic events has emerged as a promising approach to treating or preventing cancers.

Several HDAC and DNMT inhibitors have been approved for the treatment of hematological malignancies and are currently at different phases of clinical trials [195, 196]. Similarly, the DNA-hypomethylating agents 5-azacitdine and 5-aza-2′-deoxycytidine (decitabine) have been tested in myelodysplastic syndrome (MDS), acute myelogenous leukemia (AML), and chronic myelogenous leukemia (CML) patients with some encouraging outcomes [197–200]. HDAC inhibitors, such as vorinostat (suberoylanilide hydroxamic acid or SAHA), belinostat, romidepsin, and panobinostat, have been used to treat hematological malignancies and solid tumors [201, 202]. The development of HDAC or DNMT inhibitors as anticancer drugs has been hindered by their adverse side effects [203]. Accumulating evidence suggests that some dietary phytochemicals may exert their cancer chemopreventive/anticancer effects via epigenetic modifications [204–206]. In this chapter, we focus on a few of the most widely studied dietary compounds as epigenetic modifiers.

6.1 Curcumin

Hailed as "Indian solid gold," curcumin is a polyphenolic compound derived from the *Curcuma longa* plant. Despite its poor bioavailability, curcumin has been shown to be a strong anticancer agent against different types of cancers in animals and with in vitro cell culture systems [207]. At least 33 proteins have been identified as being targeted by curcumin. The potential of curcumin in targeting epigenetic modifications has recently been revealed [207].

6.1.1 Curcumin as a DNA Hypomethylation Agent

DNA methylation is a heritable epigenetic modification that modulates the transcriptional plasticity of the genome. The hypermethylation of promoter CpG islands, particularly at tumor suppressor genes, plays a causative role in carcinogenesis. In fact, recent findings suggest that epigenetic alterations may precede genetic mutations [159]. DNA methylation is regulated by DNA methyltransferases (DNMT1, DNMT3a, and DNMT3b) to transfer a methyl group from the methyl donor S-adenosyl-L-methionine (SAM) to cytosine residues at the C-5 position [208]. There are contradicting reports on the potential for curcumin as a DNMT inhibitor. Using a molecular docking approach, curcumin has been shown to bind covalently to the catalytic thiolate of C1226 of DNMT1, leading to its inhibitory effect [209]. In contrast, Medina-Franco et al. [210] found that curcumin has little or no pharmacologically relevant DNMT inhibitory activity. However, we have recently reported that curcumin can restore the expression of the Nrf2 and Neurog1 genes through DNA demethylation [57, 106]. Similarly, Jha et al. demonstrated that curcumin can reverse CpG hypermethylation, leading to the activation of the RARβ2 gene in cervical cancer cell lines [211]. However, in another report, demethoxycurcumin and bisdemethoxycurcumin, but not curcumin, were found to be able to demethylate the WIF-1 promoter region in A549 cells [212]. Further research is necessary to explain these discrepancies.

6.1.2 The Effect of Curcumin on Histone Modification

Post-translational histone modifications, including acetylation, methylation, phosphorylation, and ubiquitination, are important epigenetic events that regulate gene expression. Histone acetylation catalyzed by histone acetyltransferases (HATs) and HDACs is one of the most studied histone modifications. An accumulating body of evidence suggests that alterations in HAT and HDAC activity occur in cancer [213]. Curcumin has been reported to be a strong inhibitor of both HDACs and HATs. Curcumin is a specific inhibitor of the p300/CREB-binding protein (CBP) HAT activity but not of p300/CBP-associated factor, as demonstrated by Balasubramanyam et al. [214]. In agreement with this finding, Morimoto et al. found that the inhibition of p300 HAT activity by curcumin prevented heart failure in rats; Li et al. reported that curcumin possesses a protective effect against cardiac hypertrophy, inflammation, and fibrosis through the suppression of p300-HAT activity [215, 216]. The p300 and CBP proteins are transcriptional coactivators that function partially through their intrinsic HAT activities [217]. In addition to histones, p300 and CBP acetylate several non-histone proteins, including p53 [218]. Interestingly, curcumin was found to be able to inhibit p300-mediated acetylation of p53 in vivo [214]. In addition, Kang et al. reported that curcumin induces histone hypoacetylation in brain cancer cells, leading to the induction of apoptosis through a (PARP)- and caspase 3-mediated manner [219]. Mechanistically, Marcu et al.

proposed that curcumin is a selective HAT inhibitor. The covalent binding of curcumin with p300 leads to a conformational change, resulting in a decreased binding efficiency of histones H3, H4, and acetyl CoA [220]. In addition to HAT, curcumin was found to be a strong inhibitor of HDACs. Chen et al. reported that curcumin significantly suppresses the expression of p300, HDAC1, and HDAC3 in Raji cells [221]. Similarly, Liu et al. reported the inhibitory effect of [222]. In a study by Bora-Tatar et al., curcumin was found to be the strongest HDAC inhibitor among 33 carboxylic acid derivatives tested [223]. Curcumin-induced HDAC4 inhibition in medulloblastoma was also recently reported [224].

6.2 The Isothiocyanates Sulforaphane and Phenethyl Isothiocyanate

Isothiocyanates (ITCs) are biologically active hydrolysis products of glucosinolates from cruciferous vegetables such as broccoli, brussels sprouts, cabbage, cauliflower, Chinese cabbage, and watercress. Studies have shown that PEITC and SFN, two examples of ITCs, are strong anticancer/cancer chemopreventive agents [225]. The induction of apoptosis, cell-cycle arrest, autophagy, phase II detoxifying/antioxidant genes and the inhibition of inflammation by blocking NFKb signaling pathways are reported to be possible mechanisms by which isothiocyanates exert their anticancer/cancer chemopreventive effect [225]. The role of isothiocyanates in modulating epigenetic changes has been recently reported.

6.2.1 The Effects of SFN/PEITC on DNA Methylation

The effects of SFN on DNA methylation were first reported by Meeran et al. These researchers found that SFN treatment exhibited a dose- and time-dependent [226] suppression of DNMT1 and DNMT3a. The suppression of DNMTs by SFN is associated with the site-specific CpG demethylation of the first exon of the hTERT gene. A subsequent ChIP assay revealed that SFN increased the level of the active chromatin markers acetyl-H3, acetyl-H3K9, and acetyl-H4 but suppressed the levels of the inactive chromatin markers trimethyl-H3K9 and trimethyl-H3K27. Wang et al. reported that PEITC demethylates the promoter and restores the expression of GSTP1 in both androgen-dependent and androgen-independent LNCaP cancer cells [227]. Interestingly, PEITC was found to be more effective than 5′-aza-2′-deoxycytidine in DNA methylation.

6.2.2 The Effects of SFN/PEITC on Histone Modification

SFN is known to be a dietary HDAC inhibitor, as demonstrated in in vitro and in vivo studies [228–230]. SFN was found to suppress HDAC activity without altering protein expression levels in the human embryonic kidney 293 cells and the human colorectal cancer cell HCT116 [228]. SFN and its glutathione conjugate (SFN–GSH) were found to be less effective than the two major metabolites of SFN, SFN-cysteine and SFN-*N*-acetylcysteine, as HDAC inhibitors in vitro. A similar HDAC inhibitory effect of SFN was also observed in BPH-1, LnCaP, and PC-3 prostate epithelial cells [231]. In addition, SFN as an HDAC inhibitor is being investigated in vivo in mice and in human subjects. HDAC activity was significantly inhibited as early as 6 h after a single oral dose of 10 μmol SFN with a concomitant increase in acetylated histones H3 and H4 in the colonic mucosa [232]. More importantly, SFN was found to suppress intestinal carcinogenesis in Apc (min) mice through histone modification, as demonstrated by an increase in acetylated histones in the polyps. SFN can also suppress the growth of PC-3 xenografts by inhibiting HDAC activity [233]. In humans, a single dose of SFN-rich broccoli sprouts is sufficient to inhibit significantly HDAC activity in peripheral blood mononuclear cells (PBMCs) 3 and 6 h after consumption [233]. Like SFN, PEITC inhibits HDAC. PEITC was reported to inhibit HDAC activity and expression levels in LNCaP cells, leading to the re-expression of GSTP1 [227]. Furthermore, PEITC increases the methylation of lysine 4 of histone H3 but decreases the level of trimethylated lysine 9 of H3. Similarly, PEITC restored p21 expression through HDAC inhibition in LNCaP cells [234].

6.3 Tea Polyphenols

There is a large body of evidence indicating that bioactive polyphenolic compounds in tea (*Camellia sinensis*, Theaceae) may reduce the risk of chronic diseases, including cancers. Catechins, which include (−)-epicatechin (EC), (−)-epicatechin-3-gallate (ECG), (−)-epigallocatechin (EGC), and (−)-epigallocatechin-3-gallate (EGCG), are the most abundant compounds present in tea [235]. Among these catechins, EGCG has been identified as one of the most effective compounds. Antioxidative stress, detoxification, antiproliferation, antiinflammation, antiangiogenesis, and the induction of apoptosis have been proposed to be the mechanisms by which EGCG exerts its cancer chemopreventive effects [236]. The role of EGCG as an epigenetic modifier for cancer treatment and chemoprevention has received recent attention [205, 237].

6.3.1 The Effects of EGCG on DNA Methylation

One of the earliest reports to demonstrate the effect of EGCG on DNA methylation was the study by Fang et al. in 2003 [238]. EGCG inhibited DNMT activity, leading to a concentration-dependent and time-dependent reversal of the hypermethylation of p16 (INK4a), retinoic acid receptor beta (RARbeta), $O(6)$-methylguanine methyltransferase (MGMT), and human mutL homolog 1 (hMLH1) genes in human esophageal KYSE 510 cells. Similarly, Kato et al. found that treatment of oral cancer cells with EGCG partially reversed the hypermethylation status of the RECK gene and significantly enhanced the expression levels of RECK mRNA [239]. A dose-dependent inhibition of DNMT activity was observed in LNCaP cells after a 7-day exposure of cells to different doses of EGCG, leading to the re-expression of the GSTP1 gene [240]. In another study, EGCG treatment was found to decrease the global DNA methylation levels in A431 human skin cancer cells in a dose-dependent manner. EGCG decreased the levels of 5-methylcytosine, DNMT activity, and the mRNA and protein levels of DNMT1, DNMT3a, and DNMT3b [241]. In addition to the direct inhibitory effect on DNMT, EGCG was also found to inhibit indirectly DNMT activity by decreasing the availability of SAM [205, 242]. In contrast to the findings from in vitro studies, the in vivo hypomethylation effect of EGCG has been controversial. The oral administration of 0.3% green tea polyphenols (GTPs) to wild-type and transgenic adenocarcinomas of mouse prostate (TRAMP) mice showed decreased levels of 5-methyl-deoxycytidine (5mdC) in the liver at 12 weeks but did not alter the levels of 5mdC in the prostate, gut, and liver from WT mice at either 12 or 24 weeks of age [243]. However, EGCG treatment resulted in a significant inhibition of the UVB-induced global DNA hypomethylation pattern in the SKH-1 hairless mouse [244].

6.3.2 The Effects of EGCG on Histone Modification

In addition to its DNMT inhibitory effect, EGCG modulates gene expression via histone modification. EGCG was found to abrogate p300-induced p65 acetylation in vitro and in vivo, increase the level of cytosolic IkappaB alpha, and suppress tumor necrosis factor alpha (TNFα)-induced NF-κB activation. Despite a strong specificity for the majority of HAT enzymes, EGCG did not demonstrate activity toward HDAC, SIRT1, or HMTase [245]. However, EGCG was found to decrease HDAC activity and increase levels of acetylated lysine 9 and 14 on histone H3 (H3-Lys 9 and 14) and acetylated lysine 5, 12, and 16 on histone H4, but EGCG decreased levels of methylated H3-Lys 9 in A431 human skin cancer cells [241]. EGCG was also reported to inhibit HDAC1-3 expression and increase the levels of acetylated histone H3 (LysH9/18) and H4 levels in LNCaP cells [240]. The in vivo effect of EGCG on histone modification remains to be determined.

6.4 Genistein

Genistein is a natural isoflavone and phytoestrogen found in soy products. The antitumor properties of genistein have been extensively studied using cell culture systems and preclinical models. Epidemiological studies suggest that dietary intake of genistein is linked with a decreased risk of breast and prostate cancer [246, 247].

It has been reported that genistein can regulate gene transcription through the modulation of DNA methylation and histone modification.

6.4.1 The Effects of Genistein on DNA Methylation

The DNA hypomethylation effect of genistein on different cell lines has been previously reported. Genistein and 5aza-C treatment significantly decreased the promoter methylation of B-cell translocation gene 3 (BTG3), leading to its re-expression [248] in prostate cancer cell lines. Similarly, treatment of a squamous cervical cancer cell line, SiHa, with genistein resulted in promoter demethylation and the reactivation of the RARβ2 gene [211]. A similar promoter demethylation effect of genistein on different target genes was also observed in renal and breast cancer cell lines [248, 249]. It is believed that genistein modulates promoter demethylation through the direct inhibition of DNMTs and the methyl-CpG-binding domain 2.

6.4.2 The Effects of Genistein on Histone Modification

In addition to DNA methylation, genistein modulates gene expression through histone modification. Genistein was reported to increase acetylated histones 3, 4, and H3/K4 at the p21 and p16 transcription start sites, leading to the reactivation of the genes in human prostate cancer cells [250]. Genistein was also found to activate tumor suppressor genes, such as PTEN and CYLD, via the demethylation and acetylation of H3-K9 of the promoter region of the genes [251]. Interestingly, the suppression effect of genistein on SIRT-1 led to the acetylation of H3-K9 at the p53 and FOXO3a promoters [251].

7 Conclusions

Various toxins, such as carcinogens, environmental pollutants, solar radiation, and dietary mutagens, cause oxidative stress and inflammation and are the major drivers of cancer. Dietary phytochemicals and/or relatively nontoxic therapeutic drugs, such as cancer chemopreventive agents, are administered to inhibit, retard, or reverse the initiation and progression stages of carcinogenesis over time. The induction of the Nrf2-related antioxidant, detoxification, and anti-inflammation systems play an important role in blocking carcinogenesis. In addition to the Nrf2–Keap1 signaling pathway, epigenetic modifications are key mechanisms for the regulation of Nrf2-mediated antioxidant and detoxification genes. Therefore, a promising approach to cancer chemoprevention is the use of dietary phytochemicals to increase the expression of Nrf2 and Nrf2 downstream antioxidant and detoxification enzymes. The results from research investigating this approach may provide clinical benefits to human health.

Acknowledgments This work is supported in part by Institutional Funds and by RO1-CA073674, RO1-CA094828, R01-CA118947, and R01-CA152826 awarded to Dr. Ah-Ng Tony Kong from the National Institutes of Health (NIH).

References

1. Brüne B, Zhou J, von Knethen A (2003) Kidney Int Suppl:S22
2. Darley-Usmar V, Halliwell B (1996) Pharm Res 13:649
3. Yu BP (1994) Physiol Rev 74:139
4. Beckman KB, Ames BN (1998) Physiol Rev 78:547
5. Halliwell B, Gutteridge J (1999) Free radicals in biology and medicine. Oxford University Press, Oxford
6. Emerit J, Edeas M, Bricaire F (2004) Biomed Pharmacother 58:39
7. Cataldi A (2010) Curr Pharm Des 16:1387
8. Azad N, Iyer A, Vallyathan V, Wang L, Castranova V, Stehlik C, Rojanasakul Y (2010) Ann N Y Acad Sci 1203:1
9. Ryter SW, Kim HP, Hoetzel A, Park JW, Nakahira K, Wang X, Choi AM (2007) Antioxid Redox Signal 9:49
10. Shen HM, Liu ZG (2006) Free Radic Biol Med 40:928
11. Valko M, Rhodes CJ, Moncol J, Izakovic M, Mazur M (2006) Chem Biol Interact 160:1
12. Tan AC, Konczak I, Sze DM, Ramzan I (2011) Nutr Cancer 63:495
13. Zamocky M, Furtmuller PG, Obinger C (2008) Antioxid Redox Signal 10:1527
14. Valko M, Leibfritz D, Moncol J, Cronin MT, Mazur M, Telser J (2007) Int J Biochem Cell Biol 39:44
15. Bartsch H, Nair J (2006) Langenbecks Arch Surg 391:499
16. Klaunig JE, Kamendulis LM (2004) Annu Rev Pharmacol Toxicol 44:239
17. Valko M, Izakovic M, Mazur M, Rhodes CJ, Telser J (2004) Mol Cell Biochem 266:37
18. Dizdaroglu M, Jaruga P, Birincioglu M, Rodriguez H (2002) Free Radic Biol Med 32:1102
19. Brown GC, Borutaite V (2001) IUBMB Life 52:189
20. Shigenaga MK, Gimeno CJ, Ames BN (1989) Proc Natl Acad Sci USA 86:9697
21. Dreher D, Junod AF (1996) Eur J Cancer 32A:30
22. Carmeliet P (2000) Nat Med 6:389
23. He X, Ma Q (2009) Mol Pharmacol 76:1265
24. Reddy SP (2008) Curr Mol Med 8:376
25. Jomova K, Vondrakova D, Lawson M, Valko M (2010) Mol Cell Biochem 345:91
26. Mamede AC, Tavares SD, Abrantes AM, Trindade J, Maia JM, Botelho MF (2011) Nutr Cancer 63:479
27. Kojo S (2004) Curr Med Chem 11:1041
28. Fukai T, Ushio-Fukai M (2011) Antioxid Redox Signal 15:1583
29. Nishikawa M, Hashida M, Takakura Y (2009) Adv Drug Deliv Rev 61:319
30. Limon-Pacheco JH, Gonsebatt ME (2010) Cent Nerv Syst Agents Med Chem 10:287
31. Jung KA, Kwak MK (2010) Molecules 15:7266
32. Rushmore TH, Morton MR, Pickett CB (1991) J Biol Chem 266:11632
33. Shen G, Kong AN (2009) Biopharm Drug Dispos 30:345
34. Boutten A, Goven D, Artaud-Macari E, Boczkowski J, Bonay M (2011) Trends Mol Med 17:363
35. Moi P, Chan K, Asunis I, Cao A, Kan YW (1994) Proc Natl Acad Sci USA 91:9926
36. Chan JY, Han XL, Kan YW (1993) Proc Natl Acad Sci USA 90:11366
37. Chan K, Kan YW (1999) Proc Natl Acad Sci USA 96:12731
38. Slocum SL, Kensler TW (2011) Arch Toxicol 85:273

39. Itoh K, Wakabayashi N, Katoh Y, Ishii T, Igarashi K, Engel JD, Yamamoto M (1999) Genes Dev 13:76
40. Shu L, Cheung KL, Khor TO, Chen C, Kong AN (2010) Cancer Metastasis Rev 29:483
41. Chinenov Y, Kerppola TK (2001) Oncogene 20:2438
42. Kobayashi M, Yamamoto M (2005) Antioxid Redox Signal 7:385
43. Kong AN, Yu R, Hebbar V, Chen C, Owuor E, Hu R, Ee R, Mandlekar S (2001) Mutat Res 480–481:231
44. Hu R, Shen G, Yerramilli UR, Lin W, Xu C, Nair S, Kong AN (2006) Arch Pharm Res 29:911
45. Keum YS, Yu S, Chang PP, Yuan X, Kim JH, Xu C, Han J, Agarwal A, Kong AN (2006) Cancer Res 66:8804
46. Huang HC, Nguyen T, Pickett CB (2000) Proc Natl Acad Sci USA 97:12475
47. Huang HC, Nguyen T, Pickett CB (2002) J Biol Chem 277:42769
48. Numazawa S, Ishikawa M, Yoshida A, Tanaka S, Yoshida T (2003) Am J Physiol Cell Physiol 285:C334
49. Bloom DA, Jaiswal AK (2003) J Biol Chem 278:44675
50. Kang KW, Lee SJ, Park JW, Kim SG (2002) Mol Pharmacol 62:1001
51. Lee JM, Hanson JM, Chu WA, Johnson JA (2001) J Biol Chem 276:20011
52. Kang KW, Choi SH, Kim SG (2002) Nitric Oxide 7:244
53. Nakaso K, Yano H, Fukuhara Y, Takeshima T, Wada-Isoe K, Nakashima K (2003) FEBS Lett 546:181
54. Sun Z, Huang Z, Zhang DD (2009) PLoS One 4:e6588
55. Li W, Thakor N, Xu EY, Huang Y, Chen C, Yu R, Holcik M, Kong AN (2010) Nucleic Acids Res 38:778
56. Saw CL, Cintron M, Wu TY, Guo Y, Huang Y, Jeong WS, Kong AN (2011) Biopharm Drug Dispos 32:289
57. Shu L, Khor TO, Lee JH, Boyanapalli SS, Huang Y, Wu TY, Saw CL, Cheung KL, Kong AN (2011) AAPS J 13:606
58. Chun KS, Keum YS, Han SS, Song YS, Kim SH, Surh YJ (2003) Carcinogenesis 24:1515
59. Reuter S, Eifes S, Dicato M, Aggarwal BB, Diederich M (2008) Biochem Pharmacol 76:1340
60. Aggarwal BB, Kumar A, Bharti AC (2003) Anticancer Res 23:363
61. Deeb D, Xu YX, Jiang H, Gao X, Janakiraman N, Chapman RA, Gautam SC (2003) Mol Cancer Ther 2:95
62. Pandey M, Gupta S (2009) Front Biosci (Elite Ed) 1:13
63. Jeong WS, Kim IW, Hu R, Kong AN (2004) Pharm Res 21:649
64. Jagtap S, Meganathan K, Wagh V, Winkler J, Hescheler J, Sachinidis A (2009) Curr Med Chem 16:1451
65. Trachootham D, Lu W, Ogasawara MA, Nilsa RD, Huang P (2008) Antioxid Redox Signal 10:1343
66. Rahman I, Biswas SK, Jimenez LA, Torres M, Forman HJ (2005) Antioxid Redox Signal 7:42
67. Dinkova-Kostova AT, Talalay P (2008) Mol Nutr Food Res 52(Suppl 1):S128
68. Conney AH (2003) Annu Rev Pharmacol Toxicol 43:1
69. Jakoby WB, Ziegler DM (1990) J Biol Chem 265:20715
70. Jana S, Mandlekar S (2009) Curr Drug Metab 10:595
71. Chen C, Kong AN (2004) Free Radic Biol Med 36:1505
72. Kundu JK, Surh YJ (2010) Pharm Res 27:999
73. Hayes JD, McMahon M, Chowdhry S, Dinkova-Kostova AT (2010) Antioxid Redox Signal 13:1713
74. Wasserman WW, Fahl WE (1997) Proc Natl Acad Sci USA 94:5361
75. Itoh K, Chiba T, Takahashi S, Ishii T, Igarashi K, Katoh Y, Oyake T, Hayashi N, Satoh K, Hatayama I, Yamamoto M, Nabeshima Y (1997) Biochem Biophys Res Commun 236:313
76. Kwak MK, Egner PA, Dolan PM, Ramos-Gomez M, Groopman JD, Itoh K, Yamamoto M, Kensler TW (2001) Mutat Res 480–481:305

77. Venugopal R, Jaiswal AK (1996) Proc Natl Acad Sci USA 93:14960
78. He CH, Gong P, Hu B, Stewart D, Choi ME, Choi AM, Alam J (2001) J Biol Chem 276:20858
79. Katsuoka F, Motohashi H, Ishii T, Aburatani H, Engel JD, Yamamoto M (2005) Mol Cell Biol 25:8044
80. Igarashi K, Kataoka K, Itoh K, Hayashi N, Nishizawa M, Yamamoto M (1994) Nature 367:568
81. Motohashi H, O'Connor T, Katsuoka F, Engel JD, Yamamoto M (2002) Gene 294:1
82. Hu R, Xu C, Shen G, Jain MR, Khor TO, Gopalkrishnan A, Lin W, Reddy B, Chan JY, Kong AN (2006) Cancer Lett 243:170
83. Hu R, Xu C, Shen G, Jain MR, Khor TO, Gopalkrishnan A, Lin W, Reddy B, Chan JY, Kong AN (2006) Life Sci 79:1944
84. Kwak MK, Kensler TW (2010) Toxicol Appl Pharmacol 244:66
85. Nair S, Xu C, Shen G, Hebbar V, Gopalakrishnan A, Hu R, Jain MR, Lin W, Keum YS, Liew C, Chan JY, Kong AN (2006) Pharm Res 23:2621
86. Thimmulappa RK, Mai KH, Srisuma S, Kensler TW, Yamamoto M, Biswal S (2002) Cancer Res 62:5196
87. Yates MS, Tran QT, Dolan PM, Osburn WO, Shin S, McCulloch CC, Silkworth JB, Taguchi K, Yamamoto M, Williams CR, Liby KT, Sporn MB, Sutter TR, Kensler TW (2009) Carcinogenesis 30:1024
88. Yates MS, Kwak MK, Egner PA, Groopman JD, Bodreddigari S, Sutter TR, Baumgartner KJ, Roebuck BD, Liby KT, Yore MM, Honda T, Gribble GW, Sporn MB, Kensler TW (2006) Cancer Res 66:2488
89. Hayes JD, Flanagan JU, Jowsey IR (2005) Annu Rev Pharmacol Toxicol 45:51
90. Reisman SA, Yeager RL, Yamamoto M, Klaassen CD (2009) Toxicol Sci 108:35
91. Hayes JD, Chanas SA, Henderson CJ, McMahon M, Sun C, Moffat GJ, Wolf CR, Yamamoto M (2000) Biochem Soc Trans 28:33
92. Tan KP, Wood GA, Yang M, Ito S (2010) Br J Pharmacol 161:1111
93. Tukey RH, Strassburg CP (2000) Annu Rev Pharmacol Toxicol 40:581
94. Vienneau DS, DeBoni U, Wells PG (1995) Cancer Res 55:1045
95. Leung HY, Wang Y, Leung LK (2007) Toxicology 242:153
96. Yueh MF, Tukey RH (2007) J Biol Chem 282:8749
97. Yeager RL, Reisman SA, Aleksunes LM, Klaassen CD (2009) Toxicol Sci 111:238
98. Buckley DB, Klaassen CD (2009) Drug Metab Dispos 37:847
99. Paonessa JD, Ding Y, Randall KL, Munday R, Argoti D, Vouros P, Zhang Y (2011) Cancer Res 71:3904
100. Ramos-Gomez M, Kwak MK, Dolan PM, Itoh K, Yamamoto M, Talalay P, Kensler TW (2001) Proc Natl Acad Sci USA 98:3410
101. Nioi P, Hayes JD (2004) Mutat Res 555:149
102. Ross D (2004) Drug Metab Rev 36:639
103. Long DJ 2nd, Waikel RL, Wang XJ, Perlaky L, Roop DR, Jaiswal AK (2000) Cancer Res 60:5913
104. Tripathi DN, Jena GB (2010) Mutat Res 696:69
105. Yang CM, Huang SM, Liu CL, Hu ML (2012) J Agric Food Chem 60:1576
106. Khor TO, Huang Y, Wu TY, Shu L, Lee J, Kong AN (2011) Biochem Pharmacol 82:1073
107. Loboda A, Jazwa A, Grochot-Przeczek A, Rutkowski AJ, Cisowski J, Agarwal A, Jozkowicz A, Dulak J (2008) Antioxid Redox Signal 10:1767
108. Prestera T, Talalay P, Alam J, Ahn YI, Lee PJ, Choi AM (1995) Mol Med 1:827
109. Wunder C, Potter RF (2003) Curr Drug Targets 3:199
110. Guo X, Shin VY, Cho CH (2001) Life Sci 69:3113
111. Alam J, Stewart D, Touchard C, Boinapally S, Choi AM, Cook JL (1999) J Biol Chem 274:26071
112. Ishii T, Itoh K, Takahashi S, Sato H, Yanagawa T, Katoh Y, Bannai S, Yamamoto M (2000) J Biol Chem 275:16023

113. Rojo AI, Medina-Campos ON, Rada P, Zuniga-Toala A, Lopez-Gazcon A, Espada S, Pedraza-Chaverri J, Cuadrado A (2012) Free Radic Biol Med 52:473
114. Chen JH, Huang SM, Tan TW, Lin HY, Chen PY, Yeh WL, Chou SC, Tsai CF, Wei IH, Lu DY (2012) Int Immunopharmacol 12:94
115. Chen XL, Dodd G, Thomas S, Zhang X, Wasserman MA, Rovin BH, Kunsch C (2006) Am J Physiol 290:H1862
116. Chen XL, Kunsch C (2004) Curr Pharm Des 10:879
117. Khor TO, Huang MT, Kwon KH, Chan JY, Reddy BS, Kong AN (2006) Cancer Res 66:11580
118. Li N, Nel AE (2006) Antioxid Redox Signal 8:88
119. Rahman I, Biswas SK, Kirkham PA (2006) Biochem Pharmacol 72:1439
120. Yates MS, Kensler TW (2007) Drug News Perspect 20:109
121. Thimmulappa RK, Scollick C, Traore K, Yates M, Trush MA, Liby KT, Sporn MB, Yamamoto M, Kensler TW, Biswal S (2006) Biochem Biophys Res Commun 351:883
122. Thimmulappa RK, Lee H, Rangasamy T, Reddy SP, Yamamoto M, Kensler TW, Biswal S (2006) J Clin Invest 116:984
123. Yang H, Magilnick N, Ou X, Lu SC (2005) Biochem J 391:399
124. Liu GH, Qu J, Shen X (2008) Biochim Biophys Acta 1783:713
125. Cho HY, Reddy SP, Yamamoto M, Kleeberger SR (2004) FASEB J 18:1258
126. Ishii Y, Itoh K, Morishima Y, Kimura T, Kiwamoto T, Iizuka T, Hegab AE, Hosoya T, Nomura A, Sakamoto T, Yamamoto M, Sekizawa K (2005) J Immunol 175:6968
127. Rangasamy T, Cho CY, Thimmulappa RK, Zhen L, Srisuma SS, Kensler TW, Yamamoto M, Petrache I, Tuder RM, Biswal S (2004) J Clin Invest 114:1248
128. Rangasamy T, Guo J, Mitzner WA, Roman J, Singh A, Fryer AD, Yamamoto M, Kensler TW, Tuder RM, Georas SN, Biswal S (2005) J Exp Med 202:47
129. Khor TO, Yu S, Kong AN (2008) Planta Med 74:1540
130. Lin W, Wu RT, Wu T, Khor TO, Wang H, Kong AN (2008) Biochem Pharmacol 76:967
131. Wattenberg LW (1985) Cancer Res 45:1
132. Hu R, Saw CL, Yu R, Kong AN (2010) Antioxid Redox Signal 13:1679
133. Yu S, Kong AN (2007) Curr Cancer Drug Targets 7:416
134. Lau A, Villeneuve NF, Sun Z, Wong PK, Zhang DD (2008) Pharmacol Res 58:262
135. Finley JW, Kong AN, Hintze KJ, Jeffery EH, Ji LL, Lei XG (2011) J Agric Food Chem 59:6837
136. Kim HJ, di Luccio E, Kong AN, Kim JS (2011) Biotechnol J 6:525
137. Kobayashi A, Kang MI, Watai Y, Tong KI, Shibata T, Uchida K, Yamamoto M (2006) Mol Cell Biol 26:221
138. Surh YJ, Kundu JK, Na HK, Lee JS (2005) J Nutr 135:2993S
139. Saw CL, Wu Q, Kong AN (2010) Chinese Med 5:37
140. Xu C, Huang MT, Shen G, Yuan X, Lin W, Khor TO, Conney AH, Kong AN (2006) Cancer Res 66:8293
141. Shen G, Xu C, Hu R, Jain MR, Nair S, Lin W, Yang CS, Chan JY, Kong AN (2005) Pharm Res 22:1805
142. Kang HJ, Hong YB, Kim HJ, Wang A, Bae I (2011) Toxicol Lett 209:154
143. Barve A, Khor TO, Nair S, Reuhl K, Suh N, Reddy B, Newmark H, Kong AN (2009) Int J Cancer 124:1693
144. Haridas V, Hanausek M, Nishimura G, Soehnge H, Gaikwad A, Narog M, Spears E, Zoltaszek R, Walaszek Z, Gutterman JU (2004) J Clin Invest 113:65
145. Higgins LG, Cavin C, Itoh K, Yamamoto M, Hayes JD (2008) Toxicol Appl Pharmacol 226:328
146. Chen CC, Chen HL, Hsieh CW, Yang YL, Wung BS (2011) Acta Pharmacol Sin 32:62
147. Feng R, Lu Y, Bowman LL, Qian Y, Castranova V, Ding M (2005) J Biol Chem 280:27888
148. Chen C, Pung D, Leong V, Hebbar V, Shen G, Nair S, Li W, Kong AN (2004) Free Radic Biol Med 37:1578

149. Wu TY, Saw CL, Khor TO, Pung D, Boyanapalli SS, Kong AN (2011) Mol Carcinog
150. Lian F, Wang XD (2008) Int J Cancer 123:1262
151. Wondrak GT, Cabello CM, Villeneuve NF, Zhang S, Ley S, Li Y, Sun Z, Zhang DD (2008) Free Radic Biol Med 45:385
152. Shin JW, Ohnishi K, Murakami A, Lee JS, Kundu JK, Na HK, Ohigashi H, Surh YJ (2011) Cancer Prev Res 4:860
153. Hou DX, Korenori Y, Tanigawa S, Yamada-Kato T, Nagai M, He X, He J (2011) J Agric Food Chem 59:11975
154. Swanson HI, Njar VC, Yu Z, Castro DJ, Gonzalez FJ, Williams DE, Huang Y, Kong AN, Doloff JC, Ma J, Waxman DJ, Scott EE (2010) Drug Metab Dispos 38:539
155. Wu TY, Khor TO, Saw CL, Loh SC, Chen AI, Lim SS, Park JH, Cai L, Kong AN (2011) AAPS J 13:1
156. Yu R, Chen C, Mo YY, Hebbar V, Owuor ED, Tan TH, Kong AN (2000) J Biol Chem 275:39907
157. Yu R, Lei W, Mandlekar S, Weber MJ, Der CJ, Wu J, Kong AN (1999) J Biol Chem 274:27545
158. Yu R, Mandlekar S, Lei W, Fahl WE, Tan TH, Kong AN (2000) J Biol Chem 275:2322
159. Esteller M (2008) N Engl J Med 358:1148
160. Waddington CH (1942) Endeavour 1:18
161. Wolffe AP, Matzke MA (1999) Science 286:481
162. Baylin SB, Herman JG (2000) Trends Genet 16:168
163. Esteller M (2005) Annu Rev Pharmacol Toxicol 45:629
164. Jones PA, Baylin SB (2002) Nat Rev Genet 3:415
165. Fraga MF, Ballestar E, Villar-Garea A, Boix-Chornet M, Espada J, Schotta G, Bonaldi T, Haydon C, Ropero S, Petrie K, Iyer NG, Perez-Rosado A, Calvo E, Lopez JA, Cano A, Calasanz MJ, Colomer D, Piris MA, Ahn N, Imhof A, Caldas C, Jenuwein T, Esteller M (2005) Nat Genet 37:391
166. Issa JP (2002) J Nutr 132:2388S
167. Esteller M (2002) Oncogene 21:5427
168. Bird A (2002) Genes Dev 16:6
169. Suzuki MM, Bird A (2008) Nat Rev Genet 9:465
170. Esteller M (2007) Nat Rev Genet 8:286
171. Prendergast GC, Ziff EB (1991) Science 251:186
172. Hendrich B, Bird A (1998) Mol Cell Biol 18:6538
173. Lewis JD, Meehan RR, Henzel WJ, Maurer-Fogy I, Jeppesen P, Klein F, Bird A (1992) Cell 69:905
174. Nan X, Ng HH, Johnson CA, Laherty CD, Turner BM, Eisenman RN, Bird A (1998) Nature 393:386
175. Feng Q, Zhang Y (2001) Genes Dev 15:827
176. Lee BH, Yegnasubramanian S, Lin X, Nelson WG (2005) J Biol Chem 280:40749
177. Yu SW, Khor TO, Cheung KL, Li WG, Wu TY, Huang Y, Foster BA, Kan YW, Kong AN (2010) PLoS One 5:e8579
178. Li E, Bestor TH, Jaenisch R (1992) Cell 69:915
179. Okano M, Bell DW, Haber DA, Li E (1999) Cell 99:247
180. Herman JG, Baylin SB (2003) N Engl J Med 349:2042
181. Ellis L, Atadja PW, Johnstone RW (2009) Mol Cancer Ther 8:1409
182. Luger K, Mader AW, Richmond RK, Sargent DF, Richmond TJ (1997) Nature 389:251
183. Tremethick DJ (2007) Cell 128:651
184. Berlowitz L, Pallotta D (1972) Exp Cell Res 71:45
185. Jenuwein T, Allis CD (2001) Science 293:1074
186. Lund AH, van Lohuizen M (2004) Genes Dev 18:2315
187. Berger SL (2007) Nature 447:407
188. Kouzarides T (2007) Cell 128:693

189. Rodenhiser D, Mann M (2006) CMAJ 174:341
190. Bannister AJ, Zegerman P, Partridge JF, Miska EA, Thomas JO, Allshire RC, Kouzarides T (2001) Nature 410:120
191. Nakayama J, Rice JC, Strahl BD, Allis CD, Grewal SI (2001) Science 292:110
192. Jackson JP, Lindroth AM, Cao X, Jacobsen SE (2002) Nature 416:556
193. Tamaru H, Selker EU (2001) Nature 414:277
194. Kim SH, Kang HJ, Na H, Lee MO (2010) Breast Cancer Res 12:R22
195. Sigalotti L, Fratta E, Coral S, Cortini E, Covre A, Nicolay HJ, Anzalone L, Pezzani L, Di Giacomo AM, Fonsatti E, Colizzi F, Altomonte M, Calabro L, Maio M (2007) J Cell Physiol 212:330
196. Mai A, Altucci L (2009) Int J Biochem Cell Biol 41:199
197. Abdulhaq H, Rossetti JM (2007) Expert Opin Investig Drugs 16:1967
198. Fenaux P, Ades L (2009) Leuk Res 33(Suppl 2):S7
199. Griffiths EA, Gore SD (2008) Semin Hematol 45:23
200. Issa JP (2007) Clin Cancer Res 13:1634
201. Rasheed WK, Johnstone RW, Prince HM (2007) Expert Opin Investig Drugs 16:659
202. Prince HM, Bishton MJ, Harrison SJ (2009) Clin Cancer Res 15:3958
203. Garcia-Manero G, Assouline S, Cortes J, Estrov Z, Kantarjian H, Yang H, Newsome WM, Miller WH Jr, Rousseau C, Kalita A, Bonfils C, Dubay M, Patterson TA, Li Z, Besterman JM, Reid G, Laille E, Martell RE, Minden M (2008) Blood 112:981
204. Davis CD, Uthus EO (2004) Exp Biol Med (Maywood) 229:988
205. Fang M, Chen D, Yang CS (2007) J Nutr 137:223S
206. Yang CS, Fang M, Lambert JD, Yan P, Huang TH (2008) Nutr Rev 66(Suppl 1):S18
207. Ravindran J, Prasad S, Aggarwal BB (2009) AAPS J 11:495
208. Kanai Y (2008) Pathol Int 58:544
209. Liu Z, Xie Z, Jones W, Pavlovicz RE, Liu S, Yu J, Li PK, Lin J, Fuchs JR, Marcucci G, Li C, Chan KK (2009) Bioorg Med Chem Lett 19:706
210. Medina-Franco JL, Lopez-Vallejo F, Kuck D, Lyko F (2011) Mol Divers 15:293
211. Jha AK, Nikbakht M, Parashar G, Shrivastava A, Capalash N, Kaur J (2010) Folia Biol (Praha) 56:195
212. Liu YL, Yang HP, Gong L, Tang CL, Wang HJ (2011) Mol Med Rep 4:675
213. Timmermann S, Lehrmann H, Polesskaya A, Harel-Bellan A (2001) Cell Mol Life Sci 58:728
214. Balasubramanyam K, Varier RA, Altaf M, Swaminathan V, Siddappa NB, Ranga U, Kundu TK (2004) J Biol Chem 279:51163
215. Morimoto T, Sunagawa Y, Kawamura T, Takaya T, Wada H, Nagasawa A, Komeda M, Fujita M, Shimatsu A, Kita T, Hasegawa K (2008) J Clin Invest 118:868
216. Li HL, Liu C, de Couto G, Ouzounian M, Sun M, Wang AB, Huang Y, He CW, Shi Y, Chen X, Nghiem MP, Liu Y, Chen M, Dawood F, Fukuoka M, Maekawa Y, Zhang L, Leask A, Ghosh AK, Kirshenbaum LA, Liu PP (2008) J Clin Invest 118:879
217. Kundu TK, Palhan VB, Wang Z, An W, Cole PA, Roeder RG (2000) Mol Cell 6:551
218. Brooks CL, Gu W (2003) Curr Opin Cell Biol 15:164
219. Kang SK, Cha SH, Jeon HG (2006) Stem Cells Dev 15:165
220. Marcu MG, Jung YJ, Lee S, Chung EJ, Lee MJ, Trepel J, Neckers L (2006) Med Chem 2:169
221. Chen Y, Shu W, Chen W, Wu Q, Liu H, Cui G (2007) Basic Clin Pharmacol Toxicol 101:427
222. Liu HL, Chen Y, Cui GH, Zhou JF (2005) Acta Pharmacol Sin 26:603
223. Bora-Tatar G, Dayangac-Erden D, Demir AS, Dalkara S, Yelekci K, Erdem-Yurter H (2009) Bioorg Med Chem 17:5219
224. Lee SJ, Krauthauser C, Maduskuie V, Fawcett PT, Olson JM, Rajasekaran SA (2011) BMC Cancer 11:144
225. Cheung KL, Kong AN (2010) AAPS J 12:87
226. Meeran SM, Patel SN, Tollefsbol TO (2010) PLoS One 5:e11457
227. Wang LG, Beklemisheva A, Liu XM, Ferrari AC, Feng J, Chiao JW (2007) Mol Carcinog 46:24

228. Myzak MC, Karplus PA, Chung FL, Dashwood RH (2004) Cancer Res 64:5767
229. Clarke JD, Hsu A, Yu Z, Dashwood RH, Ho E (2011) Mol Nutr Food Res 55:999
230. Rajendran P, Delage B, Dashwood WM, Yu TW, Wuth B, Williams DE, Ho E, Dashwood RH (2011) Mol Cancer 10:68
231. Myzak MC, Hardin K, Wang R, Dashwood RH, Ho E (2006) Carcinogenesis 27:811
232. Myzak MC, Dashwood WM, Orner GA, Ho E, Dashwood RH (2006) FASEB J 20:506
233. Myzak MC, Tong P, Dashwood WM, Dashwood RH, Ho E (2007) Exp Biol Med (Maywood) 232:227
234. Wang LG, Liu XM, Fang Y, Dai W, Chiao FB, Puccio GM, Feng J, Liu D, Chiao JW (2008) Int J Oncol 33:375
235. Graham HN (1992) Prev Med 21:334
236. Yang CS, Wang H (2011) Mol Nutr Food Res 55:819
237. Li Y, Tollefsbol TO (2010) Curr Med Chem 17:2141
238. Fang MZ, Wang Y, Ai N, Hou Z, Sun Y, Lu H, Welsh W, Yang CS (2003) Cancer Res 63:7563
239. Kato K, Long NK, Makita H, Toida M, Yamashita T, Hatakeyama D, Hara A, Mori H, Shibata T (2008) Br J Cancer 99:647
240. Pandey M, Shukla S, Gupta S (2010) Int J Cancer 126:2520
241. Nandakumar V, Vaid M, Katiyar SK (2011) Carcinogenesis 32:537
242. Lee WJ, Zhu BT (2006) Carcinogenesis 27:269
243. Morey Kinney SR, Zhang W, Pascual M, Greally JM, Gillard BM, Karasik E, Foster BA, Karpf AR (2009) Cancer Prev Res (Phila) 2:1065
244. Mittal A, Piyathilake C, Hara Y, Katiyar SK (2003) Neoplasia 5:555
245. Choi KC, Jung MG, Lee YH, Yoon JC, Kwon SH, Kang HB, Kim MJ, Cha JH, Kim YJ, Jun WJ, Lee JM, Yoon HG (2009) Cancer Res 69:583
246. Adlercreutz H (2002) Lancet Oncol 3:364
247. Wu AH, Ziegler RG, Nomura AM, West DW, Kolonel LN, Horn-Ross PL, Hoover RN, Pike MC (1998) Am J Clin Nutr 68:1437S
248. Majid S, Dar AA, Ahmad AE, Hirata H, Kawakami K, Shahryari V, Saini S, Tanaka Y, Dahiya AV, Khatri G, Dahiya R (2009) Carcinogenesis 30:662
249. King-Batoon A, Leszczynska JM, Klein CB (2008) Environ Mol Mutagen 49:36
250. Majid S, Kikuno N, Nelles J, Noonan E, Tanaka Y, Kawamoto K, Hirata H, Li LC, Zhao H, Okino ST, Place RF, Pookot D, Dahiya R (2008) Cancer Res 68:2736
251. Kikuno N, Shiina H, Urakami S, Kawamoto K, Hirata H, Tanaka Y, Majid S, Igawa M, Dahiya R (2008) Int J Cancer 123:552

Keap1–Nrf2 Signaling: A Target for Cancer Prevention by Sulforaphane

Thomas W. Kensler, Patricia A. Egner, Abena S. Agyeman, Kala Visvanathan, John D. Groopman, Jian-Guo Chen, Tao-Yang Chen, Jed W. Fahey, and Paul Talalay

Abstract Sulforaphane is a promising agent under preclinical evaluation in many models of disease prevention. This bioactive phytochemical affects many molecular targets in cellular and animal models; however, amongst the most sensitive is Keap1, a key sensor for the adaptive stress response system regulated through the transcription factor Nrf2. Keap1 is a sulfhydryl-rich protein that represses Nrf2

T.W. Kensler (✉)
Department of Environmental Health Sciences, Johns Hopkins Bloomberg School of Public Health, Baltimore, MD 21205, USA

Department of Biochemistry and Molecular Biology, Johns Hopkins Bloomberg School of Public Health, Baltimore, MD 21205, USA

Department of Pharmacology and Molecular Sciences, Johns Hopkins School of Medicine, Baltimore, MD 21205, USA

Department of Pharmacology & Chemical Biology, University of Pittsburgh, Pittsburgh, PA 15261, USA
e-mail: tkensler@jhsph.edu

P.A. Egner and J.D. Groopman
Department of Environmental Health Sciences, Johns Hopkins Bloomberg School of Public Health, Baltimore, MD 21205, USA

A.S. Agyeman
Department of Biochemistry and Molecular Biology, Johns Hopkins Bloomberg School of Public Health, Baltimore, MD 21205, USA

K. Visvanathan
Department of Epidemiology, Johns Hopkins Bloomberg School of Public Health, Baltimore, MD 21205, USA

J.-G. Chen and T.-Y. Chen
Qidong Liver Cancer Institute, Qidong, Jiangsu 226200, People's Republic of China

J.W. Fahey and P. Talalay
Department of Pharmacology and Molecular Sciences, Johns Hopkins School of Medicine, Baltimore, MD 21205, USA

signaling by facilitating the polyubiquitination of Nrf2, thereby enabling its subsequent proteasomal degradation. Interaction of sulforaphane with Keap1 disrupts this function and allows for nuclear accumulation of Nrf2 and activation of its transcriptional program. Enhanced transcription of Nrf2 target genes provokes a strong cytoprotective response that enhances resistance to carcinogenesis and other diseases mediated by exposures to electrophiles and oxidants. Clinical evaluation of sulforaphane has been largely conducted by utilizing preparations of broccoli or broccoli sprouts rich in either sulforaphane or its precursor form in plants, a stable β-thioglucose conjugate termed glucoraphanin. We have conducted a series of clinical trials in Qidong, China, a region where exposures to food- and air-borne carcinogens has been considerable, to evaluate the suitability of broccoli sprout beverages, rich in either glucoraphanin or sulforaphane or both, for their bioavailability, tolerability, and pharmacodynamic action in population-based interventions. Results from these clinical trials indicate that interventions with well characterized preparations of broccoli sprouts may enhance the detoxication of aflatoxins and air-borne toxins, which may in turn attenuate their associated health risks, including cancer, in exposed individuals.

Keywords Sulforaphane · Nrf2 · chemoprevention · DNA adducts · mercapturic acids · clinical trials

Contents

1 Introduction 164
2 Keap1–Nrf2 Signaling 167
3 Keap1 Is Targeted by Sulforaphane 168
4 Gene Expression Changes Evoked by Sulforaphane in Animal and Human Cells 169
5 Clinical Trials in Qidong with Broccoli Sprout Preparations 170
References 173

1 Introduction

Developing rational chemoprevention strategies requires well-characterized agents, suitable cohorts, and reliable intermediate biomarkers of cancer or cancer risk [1]. Sulforaphane is one promising agent under preclinical and clinical evaluation. Sulforaphane was isolated from broccoli guided by bioassays for the induction of the cytoprotective enzyme NQO1 [2]. The inducible expression of NQO1 is now recognized to be regulated principally through the Keap1–Nrf2–ARE signaling pathway [3]. This pathway in turn is an important modifier of susceptibility to electrophilic and oxidative stresses, factors central to the processes of chemical carcinogenesis and other chronic degenerative diseases [4]. Sulforaphane is a potent inducer of Nrf2 signaling and blocks the formation of dimethylbenz[a]anthracene-evoked mammary tumors in rats as well as other tumor types in various animal models [5, 6]. In some instances these protective effects are lost in Nrf2-disrupted mice [7, 8]. In addition to increasing cellular capacity for detoxifying electrophiles and oxidants, sulforaphane has been shown to induce apoptosis, inhibit cell cycle progression, and inhibit angiogenesis [9–11]. Collectively, these actions serve to impede tumor growth. However, not all of the molecular actions of sulforaphane are triggered at the same concentrations. For example, activation of Nrf2 signaling occurs at substantially lower concentrations than does induction of apoptosis [2, 12]. The overall potent and multimodal actions of sulforaphane make it appealing to use in both preventive and therapeutic settings.

Broccoli and other cruciferous vegetables (e.g., cabbage, kale, and Brussels sprouts), primary sources of sulforaphane, are widely consumed in many parts of the world. Epidemiological evidence from prospective cohort studies and retrospective case-control studies suggest that consumption of a diet rich in crucifers reduces the risk of several types of cancers as well as some chronic degenerative diseases [13, 14]. There is growing evidence that the protective effects of crucifers against disease may be attributable largely to their content of glucosinolates (β-thioglucose N-hydroxysulfates) [15]. Glucosinolates in plant cells are hydrolyzed to bioactive isothiocyanates by the β-thioglucosidase myrosinase [15]. Myrosinase is released from intracellular vesicles following crushing of the plant cells by chewing, food preparation, or damage by insects. This hydrolysis is also mediated in a less predictable manner by β-thioglucosidases in the microflora of the human gut [16]. Young broccoli plants are an especially good source of glucosinolates, with levels 20–50 times those found in mature market-stage broccoli [17]. The principal glucosinolate contained in broccoli is glucoraphanin, which is hydrolyzed by myrosinase to sulforaphane (see Fig. 1).

Human populations are continuously exposed to varying amounts of chemicals or manufacturing by-products that are carcinogenic in animal models; over 100 such compounds have been designated as human carcinogens by the International Agency for Research on Cancer [21] and the National Toxicology Program [22]. Exposures to these exogenous agents occur through the environmental vectors of food, water and air. In some cases the pathway to reducing cancer burden from these exposures is obvious – eliminate exposures. However, in some instances, exposures are largely unavoidable, such as exposures to aflatoxins and other mycotoxins in food, or require substantial behavioral changes (e.g., smoking cessation) or economic investments (e.g., clean air in developing megacities) that are exceedingly difficult to implement in individuals or populations. In these settings, effective, tolerable, low cost, and practical approaches to chemoprevention with foods rich in glucosinolates serving as precursors for anticarcinogenic isothiocyanates, such as glucoraphanin and its cognate isothiocyate sulforaphane in broccoli, may be especially desirable.

This chapter highlights recent studies on the mechanisms of action of sulforaphane as an inducer of Nrf2-regulated genes and their roles in attenuating or blocking carcinogenesis. These studies, in turn, have supported the development and conduct of a series of clinical trials in Qidong, China for the optimization of dose and formulation regimens seeking to reduce body burdens of environmental carcinogens in residents of this region. In Qidong, exposures to food-borne and air-borne toxins and carcinogens can be considerable. Heptatocellular carcinoma

Fig. 1 Glucoraphanin in broccoli is converted to sulforaphane either by plant myrosinases, or if the plant myrosinases have been denatured by cooking, by bacterial myrosinases in the human colon. Sulforaphane is passively absorbed and rapidly conjugated with glutathione by glutathione S-transferases (GSTs), then metabolized sequentially by γ-glutamyl-transpeptidase (GTP), cysteinyl-glycinease (GCase), and N-acetyltransferase (NAT). The conjugates are actively transported into the systemic circulation where the mercapturic acid and its precursors are urinary excretion products. Deconjugation may also occur to yield the parent isothiocyanate, sulforaphane. The mercapturic acid and cysteine conjugate forms are the major urinary metabolites of sulforaphane [18]. For the beverages used in the Qidong interventions enumerated in Table 1, sulforaphane was generated enterically from glucoraphanin through the action of thioglucosidases in the gut microflora (glucoraphanin-rich, GRR), or prereleased by treatment of aqueous broccoli sprout extract with myrosinase from the daikon plant *Raphanus sativus* (sulforaphane-rich, SFR)

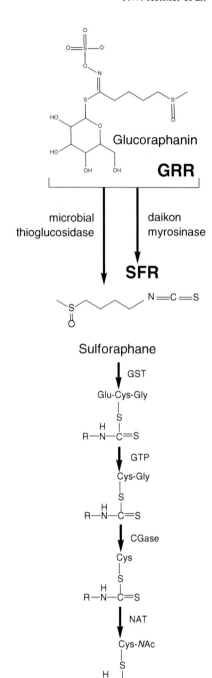

can account for up to 10% of the adult deaths in some rural townships there. Chronic infection with hepatitis B virus, coupled with exposure to aflatoxins, likely contributes to this high risk of liver cancer [23]. As vaccination programs and economic development take hold, risk factors for liver cancer are diminishing in Qidong; however, development is likely leading to increased exposures to air-borne chemicals with uncertain but potentially adverse health outcomes.

2 Keap1–Nrf2 Signaling

Environmental carcinogens typically undergo metabolic activation in target cells to form reactive electrophiles that damage DNA. Several completed clinical trials have attempted to reduce the burden of DNA damage imparted by environmental exposures to heterocyclic amines [24], tobacco smoke [25], and aflatoxins [19, 26, 27]. The end points for these trials were short-term biomarker modulations of carcinogen metabolism and/or DNA adducts and other forms of DNA damage. In these studies, modulation of these biomarkers is presumptive evidence for a cancer risk reduction, a concept that has been well validated in animal models [28]. Multiple strategies for modifying the bioactivation and/or detoxication of environmental carcinogens have been developed [4]. Disruption of Nrf2 signaling in mice leads to increased sensitivity to carcinogenesis by environmental agents [7, 29], increased burden of carcinogen-DNA adducts in target tissues [30–32], loss of chemopreventive efficacy of anticarcinogens such as sulforaphane, oltipraz, and CDDO-Im [7, 29, 32], and highlights a critical role for this adaptive stress response pathway as a critical determinant of susceptibility, and hence, a target for prevention.

The Keap1–Nrf2 signaling pathway provides a broad based cytoprotective response towards disruption of cellular homeostasis by extrinsic and intrinsic stresses. The current model of Keap1–Nrf2 interactions, as addressed in recent reviews [33, 34], involves the Kelch domains of a Keap1 homodimer functionally interacting with two different sites within the Neh2 domain of Nrf2, the ETGE, or high affinity "hinge" site and the DLG, the lower affinity "latch" site (see Fig. 2). Under normal cellular conditions, Tong et al. [35] propose that Nrf2 first interacts with the Keap1 dimer through the ETGE hinge interaction, tethering Nrf2 to the Keap1 homodimer, and subsequently the Cul3–Rbx1 complex which, following the stable interaction of Nrf2 to Keap1 through the DLG latch motif, leads to the appropriate orientation of proteins to facilitate the ubiquitination and subsequent proteasomal targeting as well as destruction of Nrf2. Upon cellular stress or pharmacologic induction, the ability of Keap1 to maintain both points of contact, the hinge and the latch, is thought to be disrupted by the alteration of the tertiary or quaternary structure of the Keap1 homodimer, accomplished via alterations of the many reactive cysteines within Keap1 through oxidation or covalent modification [36, 37]. The disruption of this efficient turnover of Nrf2 allows for the accumulation of the protein and permits Nrf2 to translocate into the nucleus. Once within the

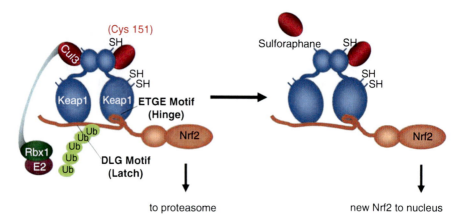

Fig. 2 Scheme of Keap1–Nrf2 interactions. Under homeostatic conditions, Nrf2 is bound by Keap1 through the "hinge" ETGE) and "latch" (DLG) domains of Nrf2. Upon association, Nrf2 is ubiquitinated by the Cul2/Rbx1/E2 ubiquitin ligase complex, marking it for proteasomal degradation. Induction of Nrf2 signaling by sulforaphane through thiocarbamylation at Cys 151may lead to disruption of the Cul3 association with Keap1 and abrogation of Nrf2 ubiquitination. Newly synthesized Nrf2 thereby escapes proteasomal degradation and translocates to the nucleus where it accumulates and activates the transcription of its target genes

nucleus, Nrf2 forms heterodimers with small Maf proteins, and drives the transcription of genes with a functional antioxidant response element (ARE) within their promoters [3, 38]. These genes include, but are not limited to, conjugation/detoxication proteins, antioxidative enzymes, anti-inflammation proteins, the proteasome, and cellular chaperones, creating a general cytoprotective response following pathway activation [39]. Recently, the response of Nrf2 has been broadened in scope, with studies documenting interactions between Nrf2 and Notch signaling [40], p53/p21 [41], p62 based autophagy [42, 43], aryl hydrocarbon receptor signaling [44], NF-κB [45, 46], and other processes [47]. These interactions provide the means to elicit the broad-based cell survival responses that now typify the pathway.

3 Keap1 Is Targeted by Sulforaphane

Sulforaphane is – or is amongst – the most potent naturally occurring inducers of Nrf2 signaling, exhibiting efficacy in the high nanomolar range in cell cultures. Its potency may reflect in part a capacity to accumulate in cells as an interchangeable conjugate with glutathione [48]. Keap1 is a cysteine-rich protein that serves as the sensor regulating activation of Nrf2 signaling by various chemical classes of anticarcinogens, all of which are thiol regents [49]. Hong et al. [50] observed that sulforaphane modified multiple Keap1 domains, whereas the model electrophiles but less potent pathway activators dexamethasone mesylate and biotinylated iodoacetic acid modified Keap1 preferentially in the central linker domain [49]. Some of the differences between sulforaphane modification patterns and those of other electrophiles probably reflect differences in electrophile chemistry. Dexamethasone mesylate and biotinylated iodoacetic acid are SN2 type electrophiles that alkylate by nucleophilic displacement of a leaving group. Thiols react with sulforaphane by addition to the isothiocyanate carbon to yield thionoacyl adducts. The acylation reaction occurs much more rapidly than does alkylation, although these adducts are subjected to dissociation and rearrangement. A follow-up analysis by Hu et al. [51] using a modified sample preparation protocol has determined C151 to be one of four cysteine residues preferentially modified by sulforaphane. These chemical mapping results are consistent with in vivo observations reported by multiple investigators in which C151 has also been determined to be the primary target for modification by sulforaphane [52, 53]. In cells in which cysteine 151 of Keap1 has been mutated to serine, nuclear accumulation and subsequent induction of Nrf2 target genes by sulforaphane are severely abrogated.

As depicted in Fig. 2, the Nrf2 signaling pathway is activated in response to the modification of Keap1 C151 by an increased amount of newly synthesized Nrf2 translocating to the nucleus, a result of decreased Keap1-mediated Nrf2 ubiquitination, and subsequent proteasomal degradation. This decrease in Nrf2 ubiquitination appears to arise from a diminished interaction between Keap1 and Cul3 upon the modification of C151, as shown by co-immunoprecipitation experiments in cells expressing mutant Keap1 (C151W) or treated with sulforaphane [36].

4 Gene Expression Changes Evoked by Sulforaphane in Animal and Human Cells

Extensive microarray-based studies have and continue to define the battery of Nrf2-regulated genes in the context of different species, tissues, cell types, and responses to small molecule activators of the pathway (reviewed in [33, 54]). These studies typically employ both genetic and pharmacologic perturbations of pathway activity to define the nature and range of induced or repressed genes. Several early studies focused on the comparative effects of sulforaphane or vehicle treatment in Nrf2-disrupted or wild-type mice in small intestine [55] and liver [56]. Patterns of elevated expression of Nrf2-regulated genes reflected those seen with other inducers such as 1,2-dithiole-3-thione [57] or with genetic upregulation via hepatic-specific disruption of Keap1 [58] in the liver. Families of genes elevated in response to sulforaphane include electrophile detoxication enzymes, enzymes involved in free radical metabolism, glutathione homeostasis, generation of reducing equivalents and lipid metabolism, solute transporters, subunits of the 26S proteasome, nucleotide excision repair proteins, and heat shock proteins. Bioavailability and Nrf2-dependent pharmacodynamic action of sulforaphane have been demonstrated in a number of extrahepatic tissues [59, 60]. More recent studies have

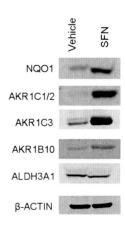

Fig. 3 Induction of Nrf2 target genes NQO1 and aldo-keto reductases (AKRs) following treatment of primary cultures of human mammosphere cultures. Western blots were conducted on cell isolates 48 h after treatment with 15 μM sulforaphane (SFN)

evaluated the Nrf2 transcriptional program in human cells [61, 62]. Recently, Agyeman et al. [63] analyzed the transcriptomic and proteomic changes in human breast epithelial MCF10A cells following sulforaphane treatment or Keap1 knockdown with siRNA using microarray and stable isotopic labeling with amino acids in culture, respectively. Strong concordance between the transcriptomic and proteomic profiles was observed. As seen in other studies with human cells, induction of aldo-keto reductase family members was most vigorous. Figure 3 demonstrates that aldo-keto reductases AKR1C1/2, AKR1C3, and AKR1B10, as well as the prototypic Nrf2-regulated enzyme NQO1, are substantively induced by sulforaphane following treatment of primary human mammary organoid cultures prepared from reduction mammaplasty specimens. Thus, an Nrf2 regulated response to sulforaphane in humans that recapitulates at least in part that observed in rodent models is evident.

5 Clinical Trials in Qidong with Broccoli Sprout Preparations

Extensive work by Talalay and colleagues has characterized the pharmacokinetics and safety in humans of ingestion of sulforaphane-rich (SFR) or glucoraphanin-rich (GRR) hot water extracts prepared from broccoli sprouts [16, 64, 65]. In many cases, freeze-dried standardized sprout extracts from specifically selected cultivars and seed sources grown in a prescribed manner were utilized to provide consistency of preparations across multiple studies. First and foremost, these studies have established the safety of these GRR and SFR preparations. Dose limiting factors center on taste, gastric irritation, and flatulence. Second, they have demonstrated a linear uptake and elimination of sulforaphane following administration of a wide range of doses as an SFR beverage. Third, bioavailability of sulforaphane was substantially better when administered as an SFR vs a GRR beverage. This latter result points to a limited capacity for the microbial thioglucosidases of the human gut to catalyze the conversion of glucoraphanin to sulforaphane. Subsequently, dozens

Table 1 Summary of clinical intervention trials with broccoli sprouts in Qidong

Agent	Dose and schedule	Size (duration)	Biomarker modulation	References
Broccoli sprout GRR	• 225 μmol GRR	12 (1 day)	Bioavailability study only: ~5% administered GR recovered in urine as SF metabolites	Unpublished
Broccoli sprout GRR	• Placebo, q.d. • 400 μmol GRR	200 (14 days)	9% decrease in urinary excretion of AFB-N^7-gua DNA adducts at 10 days; 10% decrease in pollutant PheT excretion	[19]
Broccoli sprout GRR ↔ SFR cross-over	• Run-in → (800 μmol) → wash-out → SFR (150 μmol) • Run-in → SFR → wash-out → GRR	50 (24 days)	Glucoraphanin and sulforaphane elimination pharmacokinetics; 20–50% increases in urinary excretion of mercapturic acid (NAC) conjugates of air pollutants: acrolein, ethylene oxide, crotonaldehyde, benzene	[18, 20]
Broccoli sprout GRR + SFR blend	• Placebo • GRR (600 μmol) + SFR (40 μmol)	291 (12 weeks)	Biomarker analyses in progress: primary endpoints are urinary biomarkers of food- and air-borne toxins and pollutants	Unpublished

of clinical trials are underway or completed utilizing broccoli or broccoli sprout preparations, as indicated by a review of the clinicaltrials.gov website. Summarized below and in Table 1 are the key findings in a series of four clinical trials we have conducted in Qidong, China with broccoli sprout derived beverages. All trials were approved by Institutional Review Boards in the United States and China.

Inasmuch as the initial hospital-based studies with broccoli sprout beverages were conducted in Baltimore amongst Caucasian and African-American participants, our first initiative in Qidong sought to address whether and to what extent the Chinese could convert, absorb, and excrete sulforaphane following administration of a GRR beverage. In 2002, 12 volunteers from the village of He Zuo in Qidong refrained from eating cruciferous and other green vegetables over a 4-day period. Extensive dietary logs were maintained and daily home visits to witness food preparation confirmed the absence of these vegetables from the diet. On the evening of the 3rd day, each volunteer consumed a GRR beverage containing 225 μmol glucoraphanin. Overnight, 12-h urine samples were collected during the run-in and post-intervention phases of the study. Using a cyclocondensation assay to measure sulforaphane and other isothiocyanate metabolites, average total excretion levels of 0.23, 0.32, 0.26, and 12.17 μmol of isothiocyanates were detected in the overnight voids. This greater than 40-fold increase reflects an excretion of sulforaphane metabolites as 5.4% of the administered dose of sulforaphane (in the form of its precursor glucoraphanin).

In 2003 a beverage formed from hot water infusions of 3-day old broccoli sprouts grown on site, containing defined concentrations of glucosinolates as the stable precursor of the sulforaphane, was evaluated for its ability to alter the disposition of aflatoxin. Exposures to aflatoxin, common in this community, likely arose from fungal contamination of their dietary staples. In this clinical study, also conducted in He Zuo, 200 healthy adults drank beverages containing either 400 or <3 μmole glucoraphanin nightly for 2 weeks. Urinary levels of aflatoxin-N^7-guanine, formed from depurination of the primary hepatic DNA adduct, were similar between the two intervention arms. A nonsignificant 9% decrease was seen in participants randomized to receive GRR compared to placebo beverage. However, measurement of urinary levels of sulforaphane metabolites indicated striking interindividual differences in bioavailability. This outcome may reflect individual differences in the rates of hydrolysis of glucoraphanin to sulforaphane by the intestinal microflora of the study participants. Accounting for this variability, a significant inverse association was observed for excretion of total sulforaphane metabolites and aflatoxin-N^7-guanine adducts in the 100 individuals receiving broccoli sprout glucosinolates [19]. This preliminary study illustrated the potential use of an inexpensive, easily implemented, food-based method for secondary prevention in a population at high risk for aflatoxin exposures.

One of several challenges in design of clinical chemoprevention trials is the selection of an adequate dose, type of formulation, and dose schedule of the intervention agent. A cross-over clinical trial was undertaken in He Zuo, Qidong in 2009 to compare the bioavailability and tolerability of sulforaphane from two broccoli sprout-derived beverages: one GRR and the other SFR (see Fig. 1). Sulforaphane was generated from glucoraphanin contained in the GRR beverage by gut microflora or formed by treatment of GRR with myrosinase from daikon sprouts to provide an SFR beverage [18]. Bulk amounts of freeze-dried powders of GRR and SFR were prepared in a commercial facility to provide a consistent composition throughout the study. Fifty healthy, eligible participants were requested to refrain from crucifer vegetable consumption and randomized into two treatment arms. The study design was as follows: 5-day run-in period, 7-day administration of beverages, 5-day washout period, and 7-day administration of the opposite intervention. Isotope dilution mass spectrometry was used to measure levels of glucoraphanin, sulforaphane, and sulforaphane thiol conjugates in urine samples collected daily throughout the study (see Fig. 1). Bioavailability, as measured by urinary excretion of sulforaphane and its metabolites, was substantially greater with the SFR (mean ~70%) than with GRR (mean ~5%) beverages. In addition, inter-individual variability in excretion was considerably lower with SFR than with GRR beverage. Elimination rates were considerably slower with GRR, allowing for achievement of steady-state dosing as opposed to bolus dosing with SFR [18].

An emerging problem in this region of China is outdoor air pollution. Analysis of urine samples for levels of phenanthrene tetraol, a metabolite of the polycyclic aromatic hydrocarbon and pollutant phenanthrene, from samples collected in the 2003 Qidong study indicated levels four to five times higher than measured in urine samples collected from urban residents of Minneapolis – St. Paul, Minnesota at the same time [19]. Urinary levels of phenanthrene tetraol remained high in the 2009 Qidong samples [20]. Therefore, urinary excretion of the mercapturic acids of the airborne toxins acrolein, crotonaldehyde, ethylene oxide, and benzene were also measured in urine samples from both pre- and post-interventions using liquid chromatography tandem mass spectrometry. Statistically significant increases of 20–50% in the levels of excretion of glutathione-derived conjugates of acrolein, crotonaldehyde, and benzene were seen in individuals receiving SFR, GRR, or both compared with their preintervention baseline values. No significant differences were seen between the effects of SFR vs GRR. Intervention with broccoli sprouts may enhance detoxication of airborne pollutants and attenuate their associated health risks [20].

Optimal dosing formulations in future studies might consider blends of sulforaphane and glucoraphanin as SFR and GRR mixtures to achieve peak concentrations for activation of some targets and prolonged inhibition of others implicated in the protective actions of sulforaphane. With that view in mind, a placebo-controlled intervention in 291 participants with a blend of 40 μmol SFR and 600 μmol GRR has been completed in early 2012 in He He, Qidong. This study will assess the impact of the broccoli sprout beverage on internal dose biomarkers of air pollution, and, in particular, evaluate the sustainability of the intervention over several months in terms of tolerability and efficacy. Although it is apparent that the Keap1–Nrf2 pathway can be activated in humans over the short term, it remains to be determined whether or not the pathway becomes refractory to repeated activation stimuli. Collectively, this series of clinical trials have defined paradigms for using biomarkers of exposures to environmental carcinogens as intermediate endpoints in the evaluation of agents for the prevention of chronic diseases. In particular, prevention trials of whole foods or simple extracts offer prospects for reducing an expanding global burden of cancer effectively with minimal cost, in contrast to promising isolated phytochemicals or pharmaceuticals [66].

Acknowledgments This work has been supported by USPHS grants P01 ES006052, R01 CA93780, R01 CA94076, Breast SPORE P50 CA088843, Center grant ES003819, Department of Defense W81XWH-08-1-0176, and the Prevent Cancer Foundation.

References

1. Kelloff GJ, Lieberman R, Steele VE et al (2001) Agents, biomarkers, and cohorts for chemopreventive agent development in prostate cancer. Urology 57:46–51
2. Zhang Y, Talalay P, Cho CG, Posner GH (1992) A major inducer of anticarcinogenic protective enzymes from broccoli: isolation and elucidation of structure. Proc Natl Acad Sci USA 89:2399–2403

3. Nioi P, McMahon M, Itoh K, Yamamoto M, Hayes JD (2003) Identification of a novel Nrf2-regulated antioxidant response element (ARE) in the mouse NAD(P)H:quinone oxidoreductase 1 gene: reassessment of the ARE consensus sequence. Biochem J 374:337–348
4. Kensler TW (1997) Chemoprevention by inducers of carcinogen detoxication enzymes. Environ Health Perspect 105(Suppl 4):965–970
5. Dinkova-Kostova AT (2007) Chemoprotection against cancer: an idea whose time has come. Altern Ther Health Med 13:S122–S127
6. Zhang Y, Kensler TW, Cho CG, Posner GH, Talalay P (1994) Anticarcinogenic activities of sulforaphane and structurally related synthetic norbornyl isothiocyanates. Proc Natl Acad Sci USA 91:3147–3150
7. Fahey JW, Haristoy X, Dolan PM et al (2002) Sulforaphane inhibits extracellular, intracellular, and antibiotic-resistant strains of Helicobacter pylori and prevents benzo[a]pyrene-induced stomach tumors. Proc Natl Acad Sci USA 99:7610–7615
8. Xu C, Huang MT, Shen G et al (2006) Inhibition of 7,12-dimethylbenz(a)anthracene-induced skin tumorigenesis in C57BL/6 mice by sulforaphane is mediated by nuclear factor E2-related factor 2. Cancer Res 66:8293–8296
9. Juge N, Mithen RF, Traka M (2007) Molecular basis for chemoprevention by sulforaphane: a comprehensive review. Cell Mol Life Sci 64:1105–1127
10. Higdon JV, Delage B, Williams DE, Dashwood RH (2007) Cruciferous vegetables and human cancer risk: epidemiologic evidence and mechanistic basis. Pharmacol Res 55:224–236
11. Zhang Y, Tang L (2007) Discovery and development of sulforaphane as a cancer chemopreventive phytochemical. Acta Pharmacol Sin 28:1343–1354
12. Gamet-Payrastre L, Li P, Lumeau S et al (2000) Sulforaphane, a naturally occurring isothiocyanate, induces cell cycle arrest and apoptosis in HT29 human colon cancer cells. Cancer Res 60:1426–1433
13. Riboli E, Norat T (2003) Epidemiologic evidence of the protective effects of fruit and vegetables on cancer risk. Am J Clin Nutr 78:559–695
14. World Cancer Research Fund/American Institute for Cancer Research (2007) Food, nutrition, physical activity and the prevention of cancer: a global perspective. American Institute for Cancer Research, Washington, DC
15. Fahey JW, Zalcmann AT, Talalay P (2001) The chemical diversity and distribution of glucosinolates and isothiocyanates among plants. Phytochemistry 56:5–51
16. Shapiro TA, Fahey JW, Wade KL et al (1998) Human metabolism and excretion of cancer chemoprotective glucosinolates and isothiocyanates of cruciferous vegetables. Cancer Epidemiol Biomarkers Prev 7:1091–1100
17. Fahey JW, Zhang Y, Talalay P (1997) Broccoli sprouts: an exceptionally rich source of inducers of enzymes that protect against chemical carcinogens. Proc Natl Acad Sci USA 94:10367–10372
18. Egner PA, Chen JG, Wang JB et al (2011) Bioavailability of sulforaphane from two broccoli sprout beverages: results of a short-term, cross-over clinical trial in Qidong, China. Cancer Prev Res 4:384–395
19. Kensler TW, Chen JG, Egner PA et al (2005) Effects of glucosinolate-rich broccoli sprouts on urinary levels of aflatoxin-DNA adducts and phenanthrene tetraols in a randomized clinical trial in He Zuo township, Qidong, People's Republic of China. Cancer Epidemiol Biomarkers Prev 14:2605–2613
20. Kensler TW, Ng D, Carmella SG et al (2012) Modulation of the metabolism of airborne pollutants by glucoraphanin-rich and sulforaphane-rich broccoli sprout beverages in Qidong, China. Carcinogenesis 33:101–107
21. International Agency for Research on Cancer (IARC) (2011) Agents classified by the IARC monographs, volumes 1–100. IARC Press, Lyon, France
22. US Department of Health and Human Services, Public Health Service, National Toxicology Program (2011) Report on Carcinogens, 12th edn

23. Kensler TW, Roebuck BD, Groopman JD, Wogan GN (2011) Aflatoxin: a 50-year odyssey of mechanistic and translational toxicology. Toxicol Sci 120(S1):S28–S48
24. Shaughnessy DT, Gangarosa LM, Schliebe B et al (2011) Inhibition of fried meat-induced colorectal DNA damage and altered systemic genotoxicity in humans by crucifera, chlorophyllin, and yogurt. PLoS One 6:e18707
25. Hecht SS, Carmella SG, Murphy SE (1999) Effects of watercress consumption on urinary metabolites of nicotine in smokers. Cancer Epidemiol Biomarkers Prev 8:907–913
26. Wang JS, Shen X, He X et al (1999) Protective alterations in phase 1 and 2 metabolism of aflatoxin B1 by oltipraz in residents of Qidong, People's Republic of China. J Natl Cancer Inst 91:347–354
27. Egner PA, Wang JB, Zhu YR et al (2001) Chlorophyllin intervention reduces aflatoxin-DNA adducts in individuals at high risk for liver cancer. Proc Natl Acad Sci USA 98:14601–14606
28. Kensler TW, Groopman JD, Wogan GN (1996) Use of carcinogen-DNA and carcinogen-protein adduct biomarkers for cohort selection and as modifiable end points in chemoprevention trials. IARC Sci Publ 139:237–248
29. Ramos-Gomez M, Kwak MK, Dolan PM et al (2001) Sensitivity to carcinogenesis is increased and chemoprotective efficacy of enzyme inducers is lost in nrf2 transcription factor-deficient mice. Proc Natl Acad Sci USA 98:3410–3415
30. Aoki Y, Sato H, Nishimura N et al (2001) Accelerated DNA adduct formation in the lung of the Nrf2 knockout mouse exposed to diesel exhaust. Toxicol Appl Pharmacol 173:154–160
31. Ramos-Gomez M, Dolan PM, Itoh K, Yamamoto M, Kensler TW (2003) Interactive effects of nrf2 genotype and oltipraz on benzo[a]pyrene-DNA adducts and tumor yield in mice. Carcinogenesis 24:461–467
32. Yates MS, Kwak MK, Egner PA et al (2006) Potent protection against aflatoxin-induced tumorigenesis through induction of Nrf2-regulated pathways by the triterpenoid 1-[2-cyano-3-,12-dioxooleana-1, 9(11)-dien-28-oyl]imidazole. Cancer Res 66:2488–2494
33. Hayes JD, McMahon M (2009) NRF2 and KEAP1 mutations: permanent activation of an adaptive response in cancer. Trends Biochem Sci 34:176–188
34. Taguchi K, Motohashi H, Yamamoto M (2011) Molecular mechanisms of the keap1-Nrf2 pathway in stress response and cancer evolution. Genes Cells 16:123–140
35. Tong KI, Katoh Y, Kusunoki H, Itoh K, Tanaka T, Yamamoto M (2006) Keap1 recruits Neh2 through binding to ETGE and DLG motifs: characterization of the two-site molecular recognition model. Mol Cell Biol 26:2887–2900
36. Zhang DD, Hannink M (2003) Distinct cysteine residues in Keap1 are required for Keap1-dependent ubiquitination of Nrf2 and for the stabilization of Nrf2 by chemopreventive agents and oxidative stress. Mol Cell Biol 23:8137–8151
37. Holland R, Hawkins AD, Eggler AL et al (2008) Prospective type 1 and type 2 disulfides of Keap1 protein. Chem Res Toxicol 21:2015–2060
38. Malhotra D, Portales-Casamar E, Singh A et al (2010) Global mapping of binding sites for Nrf2 identifies novel targets in cell survival response through ChIPSeq profiling and network analysis. Nucleic Acids Res 38:5718–5734
39. Kensler TW, Wakabayashi N, Biswal S (2007) Cell survival responses to environmental stresses via the Keap1-Nrf2-ARE pathway. Annu Rev Pharmacol Toxicol 47:89–116
40. Wakabayashi N, Shin S, Slocum SL, Agoston ES, Wakabayashi J, Kwak MK, Misra V, Biswal S, Yamamoto M, Kensler TW (2010) Regulation of notch1 signaling by nrf2: implications for tissue regeneration. Sci Signal 3(130):ra52
41. Chen W, Sun Z, Wang XJ et al (2009) Direct interaction between Nrf2 and p21(Cip1/WAF1) upregulates the Nrf2-mediated antioxidant response. Mol Cell 34:663–673
42. Komatsu M, Kurokawa H, Waguri S et al (2010) The selective autophagy substrate p62 activates the stress responsive transcription factor Nrf2 through inactivation of Keap1. Nat Cell Biol 12:213–223
43. Lau A, Wang XJ, Zhao F, Villeneuve NF, Wu T, Jiang T, Sun Z, White E, Zhang DD (2010) A noncanonical mechanism of activation by autophagy deficiency: direct interaction between Keap1 and p62. Mol Cell Biol 30:3275–3285

44. Shin S, Wakabayashi N, Misra V et al (2007) NRF2 modulates aryl hydrocarbon receptor signaling: influence on adipogenesis. Mol Cell Biol 27:7188–7197
45. Li W, Khor TO, Xu C et al (2008) Activation of Nrf2-antioxidant signaling attenuates NFKappaB-inflammatory response and elicits apoptosis. Biochem Pharmacol 76:1485–1489
46. Song MY, Kim EK, Moon WS et al (2009) Sulforaphane protects against cytokine- and streptozotocin-induced beta-cell damage by suppressing the NF-kappaB pathway. Toxicol Appl Pharmacol 235:57–67
47. Wakabayashi N, Slocum SL, Skoko JJ, Shin S, Kensler TW (2010) When NRF2 talks, who's listening? Antioxid Redox Signal 13:1649–1663
48. Zhang Y (2000) Role of glutathione in the accumulation of anticarcinogenic isothiocyanates and their glutathione conjugates by murine hepatoma cells. Carcinogenesis 21:1175–1182
49. Dinkova-Kostova AT, Holtzclaw WD, Cole RN et al (2002) Direct evidence that sulfhydryl groups of Keap1 are the sensors regulating induction of phase 2 enzymes that protect against carcinogens and oxidants. Proc Natl Acad Sci USA 99:11908–11913
50. Hong F, Sekhar KR, Freeman ML, Liebler DC (2005) Identification of sensor cysteines in human Keap1 modified by the cancer chemopreventive agent sulforaphane. Chem Res Toxicol 18:1917–1926
51. Hu C, Eggler AL, Mesecar AD, van Breemen RB (2011) Modification of keap1 cysteine residues by sulforaphane. Chem Res Toxicol 24:515–521
52. Kobayashi M, Li L, Iwamoto N et al (2009) The antioxidant defense system Keap1-Nrf2 comprises a multiple sensing mechanism for responding to a wide range of chemical compounds. Mol Cell Biol 29:493–502
53. McMahon M, Lamont DJ, Beattie KA, Hayes JD (2010) Keap1 perceives stress via three sensors for the endogenous signaling molecules nitric oxide, zinc and alkenals. Proc Natl Acad Sci USA 107:18838–18843
54. Kwak MK, Kensler TW (2010) Targeting Nrf2 signaling for cancer chemoprevention. Toxicol Appl Pharmacol 244:66–76
55. Thimmulappa RK, Mai KH, Srisuma S et al (2002) Identification of Nrf2-regulated genes induced by the chemopreventive agent sulforaphane by oligonucleotide array. Cancer Res 62:5196–5203
56. Hu R, Xu C, Shen G et al (2006) Gene expression profiles induced by cancer chemopreventive isothiocyanate sulforaphane in the liver of C57BL/6J mice and C57BL6J/Nrf2(−/−) mice. Cancer Lett 243:170–192
57. Kwak MK, Wakabayashi N, Itoh K et al (2003) Modulation of gene expression by cancer chemopreventive dithiolethiones through the Keap1-Nrf2 pathway. Identification of novel gene clusters for cell survival. J Biol Chem 278:8135–8145
58. Yates MS, Tran QT, Dolan PM et al (2009) Genetic versus chemoprotective activation of Nrf2 signaling: overlapping yet distinct gene expression profiles between Keap1 knockout and triterpenoid-treated mice. Carcinogenesis 30:1024–1031
59. Cornblatt BS, Ye L, Dinkova-Kostova AT et al (2007) Preclinical and clinical evaluation of sulforaphane for chemoprevention in the breast. Carcinogenesis 28:1485–1490
60. Clarke JD, Hsu A, Williams DE et al (2011) Metabolism and distribution of sulforaphane in Nrf2 knockout and wild-type mice. Pharm Res 28:3171–3179
61. Devling TW, Lindsay CD, McLellan LI et al (2005) Utility of siRNA against Keap1 as a strategy to stimulate a cancer chemopreventive phenotype. Proc Natl Acad Sci USA 102:7280–7285
62. Jeong WS, Keum YS, Chen C et al (2005) Differential expression and stability of endogenous nuclear factor E2-related factor 2 (Nrf2) by natural chemopreventive compounds. J Biochem Mol Biol 38:167–176
63. Agyeman AS, Chaerkaedy R, Shaw PG et al (2012) Transcriptomic and proteomic profiling of KEAP1 disrupted and sulforaphane-treated human breast epithelial cells reveals common expression profiles. Breast Cancer Res Treat 132:175–187

64. Shapiro TA, Fahey JW, Dinkova-Kostova AT et al (2006) Safety, tolerance and metabolism of broccoli sprout glucosinolates and isothiocyanates: a clinical phase I study. Nutr Cancer 55:53–62
65. Ye L, Dinkova-Kostova AT, Wade KL et al (2002) Quantitative determination of dithiolcarbamates in human plasma, serum, erythrocytes and urine: pharmacokinetics of broccoli sprout isothiocyanates in humans. Clin Chim Acta 316:43–53
66. Fahey JW, Talalay P, Kensler TW (2012) Notes from the field: "green" chemoprevention as frugal medicine. Cancer Prev Res 5:179–188

Chemoprotection Against Cancer by Isothiocyanates: A Focus on the Animal Models and the Protective Mechanisms

Albena T. Dinkova-Kostova

Abstract The isothiocyanates are among the most extensively studied chemoprotective agents. They are derived from glucosinolate precursors by the action of β-thioglucosidase enzymes (myrosinases). The Cruciferae family represents a rich source of glucosinolates. Notably, nearly all of the biological activities of glucosinolates, in both plants and animals, are attributable to their cognate hydrolytic products, and the isothiocyanates are prominent examples. In contrast to their relatively inert glucosinolate precursors, the isothiocyanates are endowed with high chemical reactivity, especially with sulfur-centered nucleophiles, such as protein cysteine residues. There are numerous examples of the chemoprotective effects of isothiocyanates in a number of animal models of experimental carcinogenesis at various organ sites and against carcinogens of several different types. It is becoming increasingly clear that this efficient protection is due to multiple mechanisms, including induction of cytoprotective proteins through the Keap1/Nrf2/ARE pathway, inhibition of proinflammatory responses through the NFκB pathway, induction of cell cycle arrest and apoptosis, effects on heat shock proteins, and inhibition of angiogenesis and metastasis. Because the isothiocyanates affect the function of transcription factors and ultimately the expression of networks of genes, such protection is comprehensive and long-lasting.

Keywords Glucosinolate · Keap1 · NFκB · NQO1 · Nrf2 · Sulforaphane · Phenethyl isothiocyanate

A.T. Dinkova-Kostova (✉)
Division of Cancer Research, Medical Research Institute, University of Dundee, Dundee, Scotland DD1 9SY, United Kingdom

Departments of Medicine and Pharmacology and Molecular Sciences, Johns Hopkins University School of Medicine, Baltimore, MD 21205, USA
e-mail: a.dinkovakostova@dundee.ac.uk

Contents

1 Introduction 180
2 Protective Effects in Animal Models of Carcinogenesis 182
3 Inhibition of Tumor Growth in Xenograft Models 186
4 Protective Mechanisms 186
5 Concluding Remarks and Future Perspectives 192
References 192

1 Introduction

The isothiocyanates are a diverse family of biologically active phytochemicals which are derived from glucosinolate precursors. Glucosinolates are S-β-thioglucoside N-hydroxysulfates (Fig. 1) that are particularly abundant in cruciferous (Brassicacea) plants. Depending on the origin of their side chain, there are three different types of glucosinolates: (1) aromatic, originating from Phe or Tyr; (2) aliphatic, originating from Ala, Leu, Ile, Met, or Val; and (3) indole, derived from Trp [1–3]. In the intact plant, the glucosinolates are always accompanied by β-thioglucosidase enzymes, known as myrosinases (EC 3.2.1.147). However, under physiological conditions, the myrosinases are physically separated from their substrates. Curiously, certain parts of the plant may contain extraordinary high concentrations of glucosinolates. Thus, in the root of field-grown canola (*Brassica napus*), two cell layers located under the outermost periderm layer contain 100 times higher concentrations of glucosinolates than in whole roots [4]. Similarly, in *Arabidopsis thaliana*, the flower stalk has specialized S-(sulfur-rich) cells with concentrations of glucosinolates exceeding 130 mmol/L. The S-cells are located between the phloem and the endodermis, whereas myrosinase is present in the adjacent phloem parenchyma cells [5, 6]. Enzyme and substrate come in contact upon damage of the plant tissue such as during injury or chewing, resulting in rapid hydrolysis of the glucosinolates to give rise to a variety of highly reactive compounds (Fig. 1) that are essential for plant defense against herbivores and pathogens [7], and also have beneficial effects in mammals [8]. The importance of this reaction for plant defense is emphasized by its name, "the mustard oil bomb" [9], and by the extraordinary changes in cell composition and the extreme degree of metabolic specialization that take place during differentiation of the S-cells which are accompanied by degradation of a number of organelles [6]. The isothiocyanates represent one of the major types of products of the myrosinase reaction and are largely responsible for most of the biological activities associated with the glucosinolates.

In contrast to their relatively inert precursors, the isothiocyanates are characterized by high chemical reactivity. The central carbon atom of the isothiocyanate (–N=C=S) group is highly electrophilic and reacts avidly with sulfur-, nitrogen-, and oxygen-centered nucleophiles (Fig. 2). As such nucleophiles are integral components of amino acids, it is perhaps not surprising that one of the

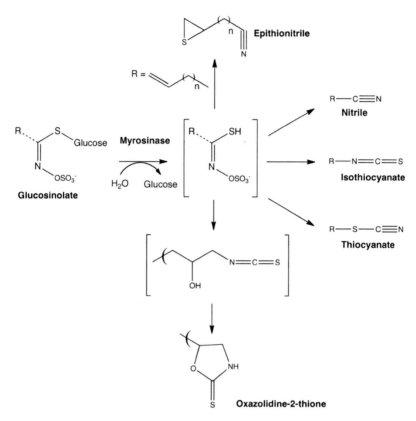

Fig. 1 The glucosinolates are hydrolyzed by the catalytic action of myrosinases to give unstable aglucones and liberate glucose. Depending on the structure of the glucosinolate side chain (R) and the reaction conditions, a variety of final products can be formed, including epithionitriles, nitriles, isothiocyanates, thiocyanites, and oxazolidine-2-thiones. Modified from [1]

major cellular targets of isothiocyanates are proteins and peptides (reviewed in [10, 11]). Probably the most common in cells is the reaction of isothiocyanates with cysteine residues in proteins and glutathione, leading to the formation of thiocarbamate products. Because of the high (millimolar) concentration of glutathione in tissues, the conjugation reaction of isothiocyanates to glutathione is a common occurrence, and it also represents the first step in the metabolism of isothiocyanates in biological systems. This reaction is further facilitated by the enzymatic activity of glutathione transferases (GSTs) which also lower the pK_a value of the cysteine residue of glutathione such that, when bound to the enzyme, it exists as the thiolate anion even at physiological pH, and is thus primed for nucleophilic attack on the electrophilic substrate [12].

Another possibility is an alkylation reaction of isothiocyanates with the α-amino groups in N-terminal residues of proteins. The products of this reaction, as well as of the reaction of the isothiocyanates with the ε-amino groups of lysines, are known

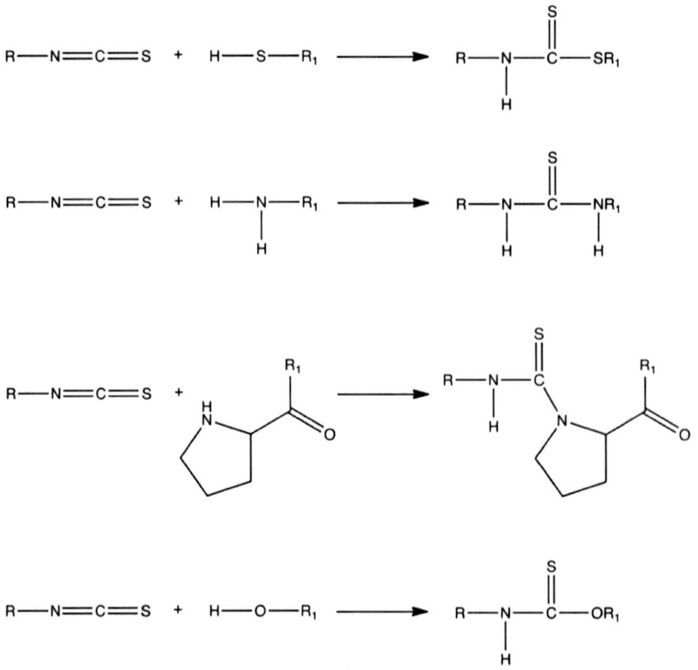

Fig. 2 The central carbon of the isothiocyanate (–N=C=S) group is electrophilic and reacts readily with sulfur-, nitrogen-, and oxygen-centered nucleophiles. The most common reactions of isothiocyanates with cellular proteins are: conjugation with sulfhydryl groups, such as the sulfhydryl group of cysteine, alkylation with α-amino groups in N-terminal residues and the ε-amino group of lysine, reactions with the secondary amine in proline, and, although not occurring at physiological conditions, reactions with hydroxyl group-containing residues, such as tyrosine. Modified from [10]

as thioureas. The isothiocyanates can also react with secondary amines, especially those which, due to their surrounding amino acid environment, have low pK_a values and are therefore highly reactive. Under certain conditions, reactions of isothiocyanates with hydroxyl group-containing residues (e.g., tyrosine), are also possible.

2 Protective Effects in Animal Models of Carcinogenesis

Nearly 50 years ago, it was reported that feeding of α-naphthyl isothiocyanate (Fig. 3) to Wistar rats dose-dependently reduced the development of liver tumors caused by the chemical carcinogens 3′-methyl 4-dimethylaminoazobenzene, ethionine, and N-2-fluorenylacetamide [13, 14]. In the late 1970s, Lee Wattenberg demonstrated that benzyl isothiocyanate, phenyl isothiocyanate, and phenethyl isothiocyanate (Fig. 3) inhibited the carcinogenic effects of polycyclic aromatic hydrocarbons using the 9,10-dimethyl-1,2-benzanthracene (DMBA)-induced

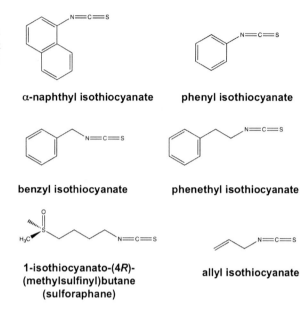

Fig. 3 Chemical structures of isothiocyanates which have been shown to protect against tumor development in animal models of carcinogenesis

mammary carcinogenesis model in Sprague–Dawley rats [15–19]. Benzyl isothiocyanate also effectively inhibited the formation of benzo[a]pyrene-induced forestomach and pulmonary adenomas in ICR/Ha mice [15]. In a series of extensive studies, Fung-Lung Chung and his colleagues showed that chemically-induced lung carcinogenesis is inhibited by orally administered isothiocyanates [20–28]. In male F344 rats treated with the tobacco-derived nitrosamine carcinogen 4-(methylnitrosamino)-1-(3-pyridyl)-1-butanone (NNK), phenethyl isothiocyanate inhibited the methylation and pyridyloxobutylation of DNA in lung and the formation of pulmonary adenomas and carcinomas [20]. In a structure–activity study, female A/J mice received four daily doses by gavage of phenyl-$(CH_2)_n$-isothiocyanate ($n = 0$–6) before carcinogen administration (a single dose of NNK, i.p.). Four months later, when the experiment was terminated, it was found that isothiocyanates with $n \geq 2$ reduced formation of pulmonary tumors, whereas phenyl isothiocyanate and benzyl isothiocyanate had no effect [21–23, 28]. However, if phenethyl isothiocyanate was given 1 week after the carcinogen and then continued till the end of the experiment, the protective effect was lost [23], indicating that the timing when the protective agents are given is critical. Importantly, this finding also suggested that the isothiocyanate was perhaps able to alter the metabolism of the carcinogen. Pretreatment with phenethyl isothiocyanate at a dose of 5 μmol or with 6-phenylhexyl isothiocyanate at a dose of 0.2 μmol p.o., either once or for 4 consecutive days with the final (or single) administration occurring 2 h prior to a single i.p. injection of the carcinogen, resulted in significant reductions of tumor multiplicity regardless of whether the isothiocyanate was administered one or four times [26]. In contrast, post-treatment was without effect [23], again suggesting that protection against carcinogenesis is due to inhibition of the metabolic activation of the carcinogen. In strong support of this conclusion, it was found that phenethyl isothiocyanate effectively inhibited the NNK-induced O^6-methylguanine formation in the lungs of F344 rats [20] and A/J mice [21].

More recently, reduction in carcinogen-DNA adduct formation by phenethyl isothiocyanate was observed following administration to rats of low doses of the radiolabeled heterocyclic amines 2-amino-1-methyl-6-phenyl-imidazo[4,5-b]-pyridine (PhIP) and 2-amino-3-methylimidazo[4,5-f]quinoline (IQ) [29]. PhIP adducts were quantified by accelerator mass spectrometry in the liver, colon, prostate, and blood plasma and IQ adducts in the liver and blood plasma. It was shown that phenethyl isothiocyanate decreased the formation of DNA adducts in tissues and albumin adduct in blood, and elevated the activity of phase 2 enzymes in liver. Phenethyl isothiocyanate was also shown to be highly effective in protection against the development of tumors of the esophagus caused by N-nitrosobenzylmethylamine (NMBA) in male F344 rats [30–33]. Importantly, phenethyl isothiocyanate significantly inhibited tumor incidence and multiplicity when given before and during, but not following, NMBA treatment [33]. In a structure–activity study, 3-phenylpropyl isothiocyanate was identified as an especially effective inhibitor, reducing the incidence and multiplicity of NMBA-induced esophageal tumors by >95% [34].

Yuesheng Zhang and colleagues have evaluated the ability of sulforaphane [1-isothiocyanato-(4R)-(methylsulfinyl)butane] (Fig. 3) and three synthetic norbornyl analogs (exo-2-acetyl-exo-6-isothiocyanatonorbornane, endo-2-acetyl-exo-6-isothiocyanatonorbornane, and exo-2-acetyl-exo-5-isothiocyanatonorbornane) to block the formation of mammary tumors in Sprague–Dawley rats treated with single doses of 9,10-dimethyl-1,2-benzanthracene (DMBA) [35]. They found that when sulforaphane and exo-2-acetyl-exo-6-isothiocyanatonorbornane were administered p.o. at doses of 75 or 150 μmol per day for 5 days around the time of exposure to the carcinogen, the development of mammary tumors was delayed and their incidence, multiplicity, and weight were reduced. The analogs endo-2-acetyl-exo-6-isothiocyanatonorbornane and exo-2-acetyl-exo-5-isothiocyanatonorbornane were less effective.

Feeding sulforaphane at 7.5 μmol/day from 7 days before until 2 days after the last dose of the carcinogen benzo[a]pyrene inhibited the development of stomach carcinogenesis in mice [36]. Sulforaphane and phenethyl isothiocyanate also reduced the formation of azoxymethane-induced colonic aberrant crypt foci in rats at 20 or 50 μmol/day, respectively p.o. for 3 days before the carcinogen, or 5 or 20 μmol, respectively three times/week for 8 weeks after the carcinogen [37]. The malignant progression of lung adenomas induced by tobacco carcinogens in mice was also inhibited in animals given 1.5 or 3 mmol/kg diet of either sulforaphane or phenethyl isothiocyanate during weeks 21–42 after administration of the carcinogen [38]. Dietary sulforaphane or benzyl isothiocyanate inhibited the development of pancreatic tumors when administered before or during the initiation stage in Syrian hamsters treated with N-nitroso-bis(2-oxopropyl)amine, but had no effect when administered post-initiation [39].

The development of intestinal adenomas in mice in which the *apc* tumor suppressor gene is truncated (a condition that makes them genetically predisposed to multiple intestinal neoplasia) was inhibited by feeding sulforaphane in the diet at doses of 6 μmol/mouse daily for 10 weeks [40], or 300 ppm (~4.25 μmol/mouse) or 600 ppm (~8.5 μmol/mouse) for 3 weeks [41]. Dietary supplementation of phenethyl isothiocyanate at 0.05% of the diet for 3 weeks was also protective in this model [42]. In a transgenic mouse model of prostate cancer (TRAMP, transgenic adenocarcinoma of mouse prostate, which has similar disease progression to human prostate carcinogenesis from histologic prostatic intraepithelial neoplasia to well-differentiated and poorly differentiated carcinoma, and distant site metastasis), orally-administered sulforaphane (6 μmol/mouse, three times per week, for 17–19 weeks, beginning at 6 weeks of age), or sulforaphane-rich broccoli sprouts had a significant inhibitory effects on prostate tumorigenesis and pulmonary metastasis [43, 44].

Pretreatment with sulforaphane (at daily doses of 10 or 40 μmol/kg, p.o., for 5 days) was shown to inhibit DNA damage in the mouse bladder following exposure to 4-aminobiphenyl (ABP), a major human bladder carcinogen from tobacco smoke [45]. Furthermore, dietary administration to rats of a freeze-dried aqueous extract of isothiocyanate-containing broccoli sprouts (70, 25, and 5% of sulforaphane, iberin, and erucin, respectively) significantly and dose-dependently inhibited bladder cancer development induced by *N*-butyl-*N*-(4-hydroxybutyl) nitrosamine [46]. Remarkably, the concentrations of isothiocyanates in the urine were two to three orders of magnitude higher than in plasma, indicating that the bladder epithelium is the most exposed tissue to orally-administered isothiocyanates. In an orthotopic bladder cancer model, *N*-acetyl-*S*-(*N*-allylthiocarbamoyl)cysteine, the major urinary metabolite of allyl isothiocyanate, was administered orally at 10 μmol/kg body weight, daily, for 3 weeks, to female F344 rats [47]. This treatment inhibited tumor growth by 40% and reduced muscle invasion by 49%.

In the two-stage chemical skin carcinogenesis model (a single dose of DMBA as initiator followed by multiple doses of the phorbol ester 12-*O*-tetradecanoylphorbol-13-acetate [TPA] as promoter) sulforaphane protected hairless mice against the development of skin tumors when administered topically twice a week at levels of 1, 5, or 10 μmol/mouse during the promotion stage [48]. In C57BL/6 mice, pretreatment with 100 nmol of sulforaphane topically for 14 days before DMBA application decreased tumor incidence and multiplicity [49]. Topical application of broccoli sprout extracts (containing the equivalent of 1 μmol of sulforaphane) 5 days a week, for 11 weeks, reduced by 50% tumor incidence, multiplicity, and burden in SKH-1 hairless mice that had been rendered "high-risk" for skin cancer development by prior chronic exposure (20 weeks) to low doses (30 mJ/cm^2) of UVB radiation [50]. Incorporation of glucoraphanin-rich broccoli sprout powder in the mouse diet had a similar effect in this model [51].

3 Inhibition of Tumor Growth in Xenograft Models

In addition to their protective effects against tumor development in animal models of carcinogenesis, the isothiocyanates have been shown to inhibit the growth of human tumor cells in xenograft models. Thus, reduction of the growth of PC-3 human prostate cancer xenografts was demonstrated following i.p. injections of 10 µmol allyl isothiocyanate, three times per week beginning on the day of implantation of the tumor cells, and on day 26 post-implantation the tumor size in the treated mice was 1.7-fold smaller than the tumor size in the control animals [52]. Similarly, i.p. administration of phenethyl isothiocyanate (5 µmol, three times per week for 28 days), beginning 1 day before tumor implantation [53] or dietary intervention with 8 µmol per gram of diet of the N-acetylcysteine conjugate of phenethyl isothiocyanate (a metabolite of this isothiocyanate) [54] reduced the tumor volume of PC-3 xenografts. Sulforaphane in the diet at a daily dose of 7.5 µmol per animal for 21 days also suppressed (by 40%) the growth of PC-3 xenografts [55]. Dietary administration of phenethyl isothiocyanate (100–150 mg/kg body weight/day) inhibited the growth of the androgen-dependent LNCaP human prostate cancer xenografts [56]. Treatment with sulforaphane (50 mg/kg, i.p., 5 times per week for 26 days, or a total of 20 injections) reduced tumor growth by 50% in orthotopically (right thoracic mammary fat pad)-transplanted human breast cancer KPL-1 cells in female athymic BALB/c mice [57]. In a pancreatic cancer xenograft model, co-treatment with sulforaphane enhanced the antitumor effect of the 17-allylamino 17-demethoxygeldanamycin (17-AAG), an Hsp90 inhibitor, resulting in inhibition of tumor growth by more than 70% [58]. In mice with primary human colorectal cancer cell xenografts, sulforaphane treatment (400 µmol/kg, s.c., daily, for 3 weeks) decreased the mean tumor weight by 70% compared with vehicle-treatment [59]. Significant reduction in tumor volume was also observed by sulforaphane treatment (0.75 mg, s.c., daily, for 2 weeks) in a subcutaneous tumor xenograft model of human Barrett esophageal adenocarcinoma in mice [60].

4 Protective Mechanisms

Most cancer-causing xenobiotics are procarcinogenic and are converted to the ultimate carcinogens by metabolism. Therefore one possible mechanism by which the isothiocyanates exert their protective effects is by modulating the biotransformation of the procarcinogens. Indeed, Paul Talalay and his colleagues showed that isothiocyanates, as well as some dietary antioxidants, were causing alterations in the metabolism of carcinogens by (1) reducing the activation of carcinogens by inhibiting phase 1 drug metabolizing enzymes (mostly cytochrome P450s) that are largely responsible for converting procarcinogens (e.g., polycyclic aromatic hydrocarbons) to highly reactive electrophilic species and (2) inducing the

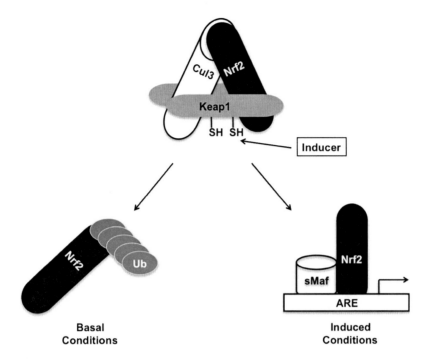

Fig. 4 The Keap1/Nrf2/ARE pathway. Under basal conditions, dimeric Keap1 (*gray*) binds and targets transcription factor Nrf2 (*black*) for ubiquitination and proteasomal degradation via association with Cullin 3 (Cul3, *white*)-based E3 ubiquitin ligase. Under induced conditions, inducers, such as isothiocyanates, bind and chemically modify reactive cysteine residues of Keap1 leading to loss of its ability to target Nrf2 for ubiquitination and degradation. As a result, Nrf2 accumulates, forms a heterodimer with a small Maf transcription factor, and the complex binds to the antioxidant response elements (*ARE*) in the promoter region of cytoprotective genes, enhancing their transcription

gene expression of cytoprotective (phase 2) enzymes [61–65]. These seminal studies provided an explanation for the mechanism of action of the protective agents [66].

The genes encoding phase 2 and other cytoprotective proteins, such as NAD(P) H:quinone oxidoreductase 1 (NQO1), glutathione *S*-transferases (GSTs), heme oxygenase 1, thioredoxin reductase, and aldo-keto reductases, share common transcriptional regulation via the Keap1/Nrf2/ARE pathway (Fig. 4) (reviewed in [67–85]). Isothiocyanates enhance the expression of cytoprotective genes by reacting with specific cysteine residues of the protein sensor Kelch-like ECH-associated protein 1 (Keap1) leading to loss of its ability to target transcription factor NF-E2-related factor 2 (Nrf2) for ubiquitination and proteasomal degradation, and subsequently allowing Nrf2 to undergo nuclear translocation, bind to the antioxidant response elements (AREs, specific sequences that are present in the promoter regions of cytoprotective genes), and activate their transcription. Direct cysteine modifications of Keap1 by sulforaphane have been demonstrated using purified recombinant protein and ectopically-expressed Keap1 isolated from cells exposed to this isothiocyanate [86–88].

The fact that it was possible to achieve protection against a variety of carcinogens at several different organ sites by administering compounds from edible plants that have been present in the human diet for centuries provided a strong impetus for the search of new and more potent protectors. A highly quantitative bioassay system was developed based on the ability of a potential protective agent to elevate the activity of NQO1 in murine Hepa1c1c7 cells [89–91]. The most potent inducer activity that was identified among a large series of extracts from edible plants that belong to ten different plant families covering almost the entire spectrum of vegetables commonly consumed in Europe and the USA was that of broccoli (*Brassica oleracea italica*) [90]. Activity-guided fractionation led to the isolation of the isothiocyanate sulforaphane as the principal inducer [92, 93]. In a structure activity study sulforaphane and various analogs differing in the oxidation state of sulfur and the number of methylene groups, $CH_3-SO_m-(CH_2)_n-NCS$, where $m = 0$, 1, or 2 and $n = 3$, 4, or 5, were evaluated for their ability to induce NQO1 in the Hepa1c1c7 bioassay system. It was found that sulforaphane was the most potent inducer, and the presence of oxygen on sulfur enhanced potency.

In CD-1 mice, daily doses of 15 µmol of sulforaphane (and its sulfide and sulfone analogs), p.o., for 5 days resulted in induction of NQO1 and GST activities in liver, forestomach, glandular stomach, small intestine, and lung [92]. In the rat, induction of NQO1 and GST activities was reported in liver, colon, and pancreas when the animals were given daily doses of 200, 500, or 1,000 µmol/kg/day [94] or 40 µmol/kg/day of sulforaphane, p.o., for 5 days [95]. Especially striking was the magnitude of induction in bladder [95, 96]. Feeding sulforaphane at a dose of 3 µmol/g diet for 14 days induced the activities of NQO1 and GST in the small intestine in wild-type mice, whereas an identical treatment was without effect in mice that lack transcription factor Nrf2 [97].

The ability of phenethyl isothiocyanate to induce cytoprotective enzymes in vivo has also been reported. Thus, in rats, hepatic mRNA and enzyme activity of GST were elevated in a dose-dependent manner following treatment with phenethyl isothiocyanate (0, 3.16, 10, 31.6, 100, and 200 mg/kg/day, p.o., for 3 days) [98]. The highest dose also doubled the levels of glutathione. Importantly, pretreatment with phenethyl isothiocyanate at a dose of 100 mg/kg enhanced the biliary excretion of glutathione conjugate of acetaminophen twofold; however, treatment with a dose of 200 mg/kg was without effect. A detailed structure–activity study examined several alkyl–aryl isothiocyanates for the ability to induce GST and NQO1 in various organs in female Sprague–Dawley rats [99]. The compounds, i.e., 1-benzyl-, 1-phenylethyl-, 2-phenylethyl-, 3-phenylpropyl-, 4-phenylbutyl-, 1-methyl-3-phenylpropyl-, 4-methylbenzyl-, 4-chlorobenzyl-, 2-methoxybenzyl-, 3-methoxybenzyl-, 4-methoxybenzyl-, 3,4,5-trimethoxybenzyl-, and cyclohexylmethyl isothiocyanate, were administered p.o. at daily doses of 250 µmol/kg/day for 5 days. It was found that the most inducible organ was the bladder with ratios of treated over control values ranging from ~two- to fourfold for GST and ~four- to eightfold for NQO1. The most effective inducers were 1-phenylethyl-, 2-methoxybenzyl-, and cyclohexylmethyl isothiocyanate. 1-Phenylethyl isothiocyanate was more effective that 1-benzyl isothiocyanate, but further increase in the length of the alkyl chain decreased the efficacy. The presence of a methoxyl group at position 3 or 4 on the aromatic ring, or a chloro group at position 4, all led to a decrease in activity, whereas a methyl group at position 4 had little effect. Allyl isothiocyanate at doses as low as 10 μmol/kg/day induced the enzyme activity of NQO1 and GST in bladder [95, 100]. Induction in many other organs, i.e., liver, kidney, lung, spleen, urinary bladder, glandular and nonglandular stomach, duodenum, jejunum, ileum, cecum, and colon was observed at high doses. Global gene expression profiling has confirmed that sulforaphane and phenethyl isothiocyanate modulate numerous cytoprotective genes and signaling pathways in mice, rats and humans [101–105].

We have determined the enzyme activity of NQO1 in homogenates prepared from 3-mm skin punch biopsies of healthy human volunteers who received a single topical application to their skin of 100 nmol of sulforaphane [106]. Despite large interindividual variations in basal activity levels, NQO1 was increased by around twofold 24 h after the application of the extract. Importantly, the NQO1 activity remained higher than that of the placebo-treated sites even when the biopsies were performed 72 h after the application of the extract, emphasizing the long-lasting effect of the treatment. Three repeated applications (at 24-h intervals) of an extract containing 50 nmol of sulforaphane were also effective in inducing NQO1, and, when this dose was tripled, induction reached ~4.5-fold [106].

The ability of isothiocyanates to inhibit proinflammatory responses represents another mechanism by which these compounds exhibit their protective effects. Sulforaphane, phenethyl-, and hexyl isothiocyanate inhibit proinflammatory responses (i.e., lipopolysaccharide- and interferon-γ-mediated elevation of inducible nitric oxide synthase [iNOS] and cyclooxygenase 2 [COX-2]) [50, 107–113]. Notably, this anti-inflammatory activity is only partially dependent on transcription factor Nrf2 [111]. Inhibition of the NFkB pathway (Fig. 5) by isothiocyanates has been demonstrated in both cells and animals [107, 108, 114–123]. Phenethyl isothiocyanate was reported to reduce significantly the carageenin-induced edema in the rat paw [124]. The acute and chronic symptoms of ulcerative colitis in mice were improved by oral administration of phenethyl isothiocyanate, and there was less intestinal bleeding and inflammatory infiltrate, a lower degree of mucosal inflammation, and better preservation of goblet cells [125]. Topical application of sulforaphane reduced inflammation resulting from exposure to ultraviolet (UVB) radiation in the skin of SKH-1 hairless [126] and C57BL/6 mice [127]. Similarly, UVB-induced skin thickening, COX-2 protein levels, and hyperplasia were suppressed by feeding sulforaphane for 14 days to HR-1 hairless mice [128]. Gene expression of COX-2 was also reduced in polyps of Apc^{Min} mice that had been fed sulforaphane [129]. Interestingly, it was recently discovered that, by direct covalent binding to the N-terminal proline residue, several isothiocyanates potently and irreversibly inhibit the tautomerase activity of the proinflammatory cytokine macrophage migration inhibitory factor (MIF) [130–133].

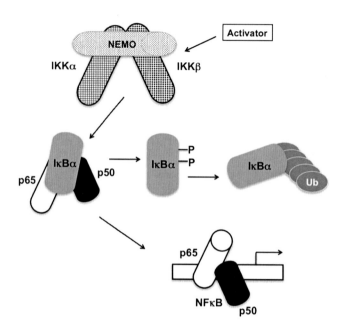

Fig. 5 The NFκB pathway. Under basal conditions, IκBα (*gray*) negatively regulates transcription factor NFκB by forming a complex with both subunits (p65, *white*, and p50, *black*) of NFκB, thus retaining the transcription factor in the cytoplasm. Activators of the pathway stimulate a kinase complex (IKKα, IKKβ, NEMO) leading to phosphorylation of IκBα, followed by ubiquitination and proteasomal degradation of the inhibitor. As a result, NFκB translocates to the nucleus where the p50-p65 heterodimer binds specific DNA sequences of the promoter regions of its target genes. Isothiocyanates can inhibit this pathway by binding to critical cysteine residues of NFκB or components of the kinase complex, such as IKKβ

In addition to inducing cytoprotective proteins and inhibiting proinflammatory responses, the isothiocyanates have also been shown to cause cell cycle arrest and apoptosis in a number of experimental systems (for reviews see [134–139]). This mechanism undoubtedly contributes to the antitumor activity of these compounds. Thus, BALB/c mice that were injected subcutaneously with F3II cells and subsequently injected daily intravenously with small amounts of sulforaphane (15 nmol/day for 13 days) developed significantly smaller tumors (approximately 60% less in weight) than vehicle-treated controls [140]. Western blot analysis revealed significantly reduced PCNA and elevated PARP fragmentation in samples from mice that received sulforaphane. Inhibition of carcinogenesis involved perturbation of mitotic microtubules and early M-phase block associated with Cdc2 kinase activation, indicating that cells arrest prior to metaphase exit.

The isothiocyanates can inhibit angiogenesis and metastasis, two processes that are critical for the growth and dissemination of solid tumors. Thus, sulforaphane administered intravenously (100 nmol/day, for 7 days) to female BALB/c mice inhibited endothelial cell response to vascular endothelial growth factor (VEGF) in a subcutaneous VEGF-impregnated Matrigel plug model [141]. In C57BL/6 mice, sulforaphane administered intraperitoneally (0.5 mg/kg) inhibited formation of tumor nodules in lungs caused by intravenous injection of Bl6-F 10 melanoma cells [142]. The protective effect of allyl isothiocyanate and phenethyl isothiocyanate (25 mg per animal per day for 5 days, i.p.) on the serum cytokine profiles of C57BL/6 mice following an intradermal injection of Bl6-F 10 melanoma cells have also been reported [143]. Both allyl isothiocyanate and phenethyl isothiocyanate are highly and equally potent in downregulating VEGF and proinflammatory cytokines such as interleukin IL-1β, IL-6, granulocyte macrophage colony-stimulating factor (GM-CSF), and tumor necrosis factor alpha (TNFα). Serum nitric oxide levels were also reduced [144]. In contrast, the levels of the antiangiogenic IL-2 and tissue inhibitor of metalloproteinases (TIMP)-1 were elevated. Importantly, tumor-directed capillary formation was inhibited in the skin of the animals that received the isothiocyanates. In an LNCaP human prostate cancer cell xenograft model, dietary administration of phenethyl isothiocyanate reduced the expression of tumor platelet/endothelial cell adhesion molecule (PECAM-1/CD31), a marker of angiogenesis, and reduced tumor cell growth [56].

Another property of the isothiocyanates that could contribute to their chemoprotective effects is their immunomodulatory activity. Thus, in BALB/c mice, five daily doses of sulforaphane at 0.5 mg per animal per day, administered intraperitoneally, elevated the total white blood cell count, the bone marrow cellularity, and the phagocytic activity of macrophages [145, 146]. Oral administration of sulforaphane (9 μmol per mouse per day for 11 days) reversed the age-associated decrease of contact hypersensitivity and TH_1 immunity through induction of cytoprotective enzymes and glutathione biosynthesis [147]. Similar to the effects of sulforaphane, treatment with five doses of allyl isothiocyanate or with phenyl isothiocyanate (25 μg/dose/animal, i.p.) was found to enhance the total white blood cell count, the bone marrow cellularity, as well as the alpha-esterase positive cell number, and when combined with treatment with the antigen sheep red blood cells produced an enhancement in the circulating antibody titer and the number of plaque forming cells in the spleen [148]. Curiously, at low concentration of fetal bovine serum (at or below 1%) in the cell culture medium, sulforaphane was recently shown to raise the phagocytosis activity of RAW 264.7 cells [149].

Overexpression of histone deacetylase (HDAC) enzymes has been implicated in protecting cancer cells, and HDAC inhibitors have been found to cause growth arrest, differentiation, and apoptosis [150]. Sulforaphane, erucin, benzyl-phenylbutyl-, phenylhexyl-, and phenethyl isothiocyanate inhibit the activity of HDAC in human cell lines established from colon, prostate, pancreatic, and breast cancer, and in leukemia cells [151–157]. Sulforaphane incorporation in the diet inhibited HDAC activity and elevated global histone acetylation, with specific increases at the bax and the p21 promoter regions, in polyp tissue from Apc^{Min} mice and in PC-3 xenografts [40, 55]. Inhibition of HDAC activity was also observed in circulating peripheral blood mononuclear cells obtained from human subjects after consumption of broccoli sprouts [55].

Sulforaphane was recently shown to upregulate the heat shock response through activation of heat shock transcription factor 1 (HSF1) [158]. The levels of heat shock proteins 70 (Hsp70) and 27 (Hsp27) were increased and the proteasomal activity was elevated in an Hsp27-dependent manner. An independent study reported that sulforaphane disrupted the Hsp90-p50(Cdc37) interaction [58]. Furthermore, a synergistic activity between sulforaphane and the Hsp90 inhibitor 17-allylamino 17-demethoxygeldanamycin (17-AAG) was observed in downregulating several Hsp90 client proteins such as mutant p53, Raf-1, and Cdk4 [58]. Global gene expression profiling on liver samples obtained from C57BL/6 mice that had received a single dose of sulforaphane (90 mg/kg, p.o.) showed an increase in the gene expression of several heat shock proteins [102]. Taken together, these findings reveal another mechanism by which the isothiocyanates could contribute to inhibition of carcinogenesis.

5 Concluding Remarks and Future Perspectives

There is now a wealth of convincing evidence for the chemoprotective effects of isothiocyanates in experimental models of carcinogenesis, including models of high-risk genetic predisposition. Notably, the anticarcinogenic activity of isothiocyanates spans all of the major stages of the multistage process of carcinogenesis. In addition, there have been major advances in understanding the multiple mechanisms by which the isothiocyanates exert their protective effects. Induction of cytoprotective enzymes, inhibition of proinflammatory responses, immunomodulatory activity, and alterations of signaling pathways are all contributing factors. This mechanistic diversity makes this class of phytochemicals particularly effective chemoprotective agents. We have witnessed many successes using animal models of carcinogenesis and employing defined dosing regimens and routes of administration of these protective agents. The challenges ahead are to be able to translate the laboratory findings to human populations. Data are already available on the safety, pharmacokinetics, and efficacy of sulforaphane- and glucoraphanin-rich broccoli preparations in human subjects [55, 105, 106, 126, 133, 159–174], and a number of studies in high risk populations are currently in progress.

Acknowledgments The author is very grateful to Research Councils UK and Cancer Research UK (C20953/A10270) for financial support.

References

1. Halkier BA, Gershenzon J (2006) Biology and biochemistry of glucosinolates. Annu Rev Plant Biol 57:303–333
2. Fahey JW, Zalcmann AT, Talalay P (2001) The chemical diversity and distribution of glucosinolates and isothiocyanates among plants. Phytochemistry 56:5–51
3. Sønderby IE, Geu-Flores F, Halkier BA (2010) Biosynthesis of glucosinolates – gene discovery and beyond. Trends Plant Sci 15:283–290
4. McCully ME, Miller C, Sprague SJ et al (2008) Distribution of glucosinolates and sulphur-rich cells in roots of field-grown canola (*Brassica napus*). New Phytol 180:193–205

5. Koroleva OA, Davies A, Deeken R et al (2000) Identification of a new glucosinolate-rich cell type in Arabidopsis flower stalk. Plant Physiol 124:599–608
6. Koroleva OA, Gibson TM, Cramer R et al (2010) Glucosinolate-accumulating S-cells in Arabidopsis leaves and flower stalks undergo programmed cell death at early stages of differentiation. Plant J 64:456–469
7. Wittstock U, Gershenzon J (2002) Constitutive plant toxins and their role in defense against herbivores and pathogens. Curr Opin Plant Biol 5:300–307
8. Talalay P, Fahey JW (2001) Phytochemicals from cruciferous plants protect against cancer by modulating carcinogen metabolism. J Nutr 131:3027S–3033S
9. Ratzka A, Vogel H, Kliebenstein DJ et al (2002) Disarming the mustard oil bomb. Proc Natl Acad Sci USA 99:11223–11228
10. Mi L, Di Pasqua AJ, Chung FL (2011) Proteins as binding targets of isothiocyanates in cancer prevention. Carcinogenesis 32:1405–1413
11. Zhang Y (2012) The molecular basis that unifies the metabolism, cellular uptake and chemopreventive activities of dietary isothiocyanates. Carcinogenesis 33:2–9
12. Winayanuwattikun P, Ketterman AJ (2005) An electron-sharing network involved in the catalytic mechanism is functionally conserved in different glutathione transferase classes. J Biol Chem 280:31776–31782
13. Sidransky H, Ito N, Verney E (1966) Influence of alpha-naphthyl-isothiocyanate on liver tumorigenesis in rats ingesting ethionine and N-2-fluorenylacetamide. J Natl Cancer Inst 37:677–686
14. Lacasagne A, Hurst L, Xuong MD (1970) Inhibition by 2-naphtylisothiocyanates of hepatocarcinogenesis induced by p-dimethylaminoazobenzene (DAB) in rats. C R Seances Soc Biol Fil 164:230–233
15. Wattenberg LW (1977) Inhibition of carcinogenic effects of polycyclic hydrocarbons by benzyl isothiocyanate and related compounds. J Natl Cancer Inst 58:395–398
16. Wattenberg LW (1981) Inhibition of carcinogen-induced neoplasia by sodium cyanate, tert-butyl isocyanate, and benzyl isothiocyanate administered subsequent to carcinogen exposure. Cancer Res 41:2991–2994
17. Wattenberg LW (1987) Inhibitory effects of benzyl isothiocyanate administered shortly before diethylnitrosamine or benzo[a]pyrene on pulmonary and forestomach neoplasia in A/J mice. Carcinogenesis 8:1971–1973
18. Wattenberg LW (1990) Inhibition of carcinogenesis by naturally-occurring and synthetic compounds. Basic Life Sci 52:155–166
19. Wattenberg LW (1990) Inhibition of carcinogenesis by minor anutrient constituents of the diet. Proc Nutr Soc 49:173–183
20. Morse MA, Wang CX, Stoner GD et al (1989) Inhibition of 4-(methylnitrosamino)-1-(3-pyridyl)-1-butanone-induced DNA adduct formation and tumorigenicity in the lung of F344 rats by dietary phenethyl isothiocyanate. Cancer Res 49:549–553
21. Morse MA, Amin SG, Hecht SS et al (1989) Effects of aromatic isothiocyanates on tumorigenicity, O^6-methylguanine formation, and metabolism of the tobacco-specific nitrosamine 4-(methylnitrosamino)-1-(3-pyridyl)-1-butanone in A/J mouse lung. Cancer Res 49:2894–2897
22. Morse MA, Eklind KI, Amin SG et al (1989) Effects of alkyl chain length on the inhibition of NNK-induced lung neoplasia in A/J mice by arylalkyl isothiocyanates. Carcinogenesis 10:1757–1759
23. Morse MA, Reinhardt JC, Amin SG et al (1990) Effect of dietary aromatic isothiocyanates fed subsequent to the administration of 4-(methylnitrosamino)-1-(3-pyridyl)-1-butanone on lung tumorigenicity in mice. Cancer Lett 49:225–230
24. Morse MA, Eklind KI, Hecht SS et al (1991) Structure-activity relationships for inhibition of 4-(methylnitrosamino)-1-(3-pyridyl)-1-butanone lung tumorigenesis by arylalkyl isothiocyanates in A/J mice. Cancer Res 51:1846–1850

25. Morse MA, Eklind KI, Hecht SS et al (1991) Inhibition of tobacco-specific nitrosamine 4-(N-nitrosomethylamino)-1-(3-pyridyl)-1-butanone (NNK) tumorigenesis with aromatic isothiocyanates. IARC Sci Publ 105:529–534
26. Morse MA, Eklind KI, Amin SG et al (1992) Effect of frequency of isothiocyanate administration on inhibition of 4-(methylnitrosamino)-1-(3-pyridyl)-1-butanone-induced pulmonary adenoma formation in A/J mice. Cancer Lett 62:77–81
27. Hecht SS, Morse MA, Eklind KI et al (1991) A/J mouse lung tumorigenesis by the tobacco-specific nitrosamine 4-(methylnitrosamino)-1-(3-pyridyl)-1-butanone and its inhibition by arylalkyl isothiocyanates. Exp Lung Res 17:501–511
28. Chung FL, Morse MA, Eklind KI (1992) New potential chemopreventive agents for lung carcinogenesis of tobacco-specific nitrosamine. Cancer Res 52:2719s–2722s
29. Dingley KH, Ubick EA, Chiarappa-Zucca ML et al (2003) Effect of dietary constituents with chemopreventive potential on adduct formation of a low dose of the heterocyclic amines PhIP and IQ and phase II hepatic enzymes. Nutr Cancer 46:212–221
30. Stoner GD, Morrissey DT, Heur YH et al (1991) Inhibitory effects of phenethyl isothiocyanate on N-nitrosobenzylmethylamine carcinogenesis in the rat esophagus. Cancer Res 51:2063–2068
31. Morse MA, Zu H, Galati AJ et al (1993) Dose-related inhibition by dietary phenethyl isothiocyanate of esophageal tumorigenesis and DNA methylation induced by N-nitrosomethylbenzylamine in rats. Cancer Lett 72:103–110
32. Wilkinson JT, Morse MA, Kresty LA et al (1995) Effect of alkyl chain length on inhibition of N-nitrosomethylbenzylamine-induced esophageal tumorigenesis and DNA methylation by isothiocyanates. Carcinogenesis 16:1011–1015
33. Siglin JC, Barch DH, Stoner GD (1995) Effects of dietary phenethyl isothiocyanate, ellagic acid, sulindac and calcium on the induction and progression of N-nitrosomethylbenzylamine-induced esophageal carcinogenesis in rats. Carcinogenesis 16:1101–1106
34. Stoner GD, Adams C, Kresty LA et al (1998) Inhibition of N'-nitrosonornicotine-induced esophageal tumorigenesis by 3-phenylpropyl isothiocyanate. Carcinogenesis 19:2139–2143
35. Zhang Y, Kensler TW, Cho CG et al (1994) Anticarcinogenic activities of sulforaphane and structurally related synthetic norbornyl isothiocyanates. Proc Natl Acad Sci USA 91:3147–3150
36. Fahey JW, Haristoy X, Dolan PM et al (2002) Sulforaphane inhibits extracellular, intracellular, and antibiotic-resistant strains of Helicobacter pylori and prevents benzo[a]pyrene-induced stomach tumors. Proc Natl Acad Sci USA 99:7610–7615
37. Chung FL, Conaway CC, Rao CV et al (2000) Chemoprevention of colonic aberrant crypt foci in Fischer rats by sulforaphane and phenethyl isothiocyanate. Carcinogenesis 21:2287–2291
38. Conaway CC, Wang CX, Pittman B et al (2005) Phenethyl isothiocyanate and sulforaphane and their N-acetylcysteine conjugates inhibit malignant progression of lung adenomas induced by tobacco carcinogens in A/J mice. Cancer Res 65:8548–8557
39. Kuroiwa Y, Nishikawa A, Kitamura Y et al (2006) Protective effects of benzyl isothiocyanate and sulforaphane but not resveratrol against initiation of pancreatic carcinogenesis in hamsters. Cancer Lett 241:275–280
40. Myzak MC, Dashwood WM, Orner GA et al (2006) Sulforaphane inhibits histone deacetylase in vivo and suppresses tumorigenesis in Apc-minus mice. FASEB J 20:506–508
41. Hu R, Khor TO, Shen G et al (2006) Cancer chemoprevention of intestinal polyposis in ApcMin/+ mice by sulforaphane, a natural product derived from cruciferous vegetable. Carcinogenesis 27:2038–2046
42. Khor TO, Cheung WK, Prawan A et al (2008) Chemoprevention of familial adenomatous polyposis in Apc(Min/+) mice by phenethyl isothiocyanate (PEITC). Mol Carcinog 47:321–325
43. Singh SV, Warin R, Xiao D et al (2009) Sulforaphane inhibits prostate carcinogenesis and pulmonary metastasis in TRAMP mice in association with increased cytotoxicity of natural killer cells. Cancer Res 69:2117–2125

44. Keum YS, Khor TO, Lin W et al (2009) Pharmacokinetics and pharmacodynamics of broccoli sprouts on the suppression of prostate cancer in transgenic adenocarcinoma of mouse prostate (TRAMP) mice: implication of induction of Nrf2, HO-1 and apoptosis and the suppression of Akt-dependent kinase pathway. Pharm Res 26:2324–2331
45. Ding Y, Paonessa JD, Randall KL et al (2010) Sulforaphane inhibits 4-aminobiphenyl-induced DNA damage in bladder cells and tissues. Carcinogenesis 31:1999–2003
46. Munday R, Mhawech-Fauceglia P, Munday CM et al (2008) Inhibition of urinary bladder carcinogenesis by broccoli sprouts. Cancer Res 68:1593–1600
47. Bhattacharya A, Li Y, Geng F et al (2012) The principal urinary metabolite of allyl isothiocyanate, N-acetyl-S-(N-allylthiocarbamoyl)cysteine, inhibits the growth and muscle invasion of bladder cancer. Carcinogenesis 33:394–398
48. Gills JJ, Jeffery EH, Matusheski NV et al (2006) Sulforaphane prevents mouse skin tumorigenesis during the stage of promotion. Cancer Lett 236:72–79
49. Xu C, Huang MT, Shen G et al (2006) Inhibition of 7,12-dimethylbenz(a)anthracene-induced skin tumorigenesis in C57BL/6 mice by sulforaphane is mediated by nuclear factor E2-related factor 2. Cancer Res 66:8293–8296
50. Dinkova-Kostova AT, Jenkins SN, Fahey JW et al (2006) Protection against UV-light-induced skin carcinogenesis in SKH-1 high-risk mice by sulforaphane-containing broccoli sprout extracts. Cancer Lett 240:243–252
51. Dinkova-Kostova AT, Fahey JW, Benedict AL et al (2010) Dietary glucoraphanin-rich broccoli sprout extracts protect against UV radiation-induced skin carcinogenesis in SKH-1 hairless mice. Photochem Photobiol Sci 9:597–600
52. Srivastava SK, Xiao D, Lew KL et al (2003) Allyl isothiocyanate, a constituent of cruciferous vegetables, inhibits growth of PC-3 human prostate cancer xenografts in vivo. Carcinogenesis 24:1665–1670
53. Khor TO, Keum YS, Lin W et al (2006) Combined inhibitory effects of curcumin and phenethyl isothiocyanate on the growth of human PC-3 prostate xenografts in immunodeficient mice. Cancer Res 66:613–621
54. Chiao JW, Wu H, Ramaswamy G et al (2004) Ingestion of an isothiocyanate metabolite from cruciferous vegetables inhibits growth of human prostate cancer cell xenografts by apoptosis and cell cycle arrest. Carcinogenesis 25:1403–1408
55. Myzak MC, Tong P, Dashwood WM et al (2007) Sulforaphane retards the growth of human PC-3 xenografts and inhibits HDAC activity in human subjects. Exp Biol Med (Maywood) 232:227–234
56. Hudson TS, Perkins SN, Hursting SD et al (2012) Inhibition of androgen-responsive LNCaP prostate cancer cell tumor xenograft growth by dietary phenethyl isothiocyanate correlates with decreased angiogenesis and inhibition of cell attachment. Int J Oncol 40:1113–1121
57. Kanematsu S, Yoshizawa K, Uehara N et al (2011) Sulforaphane inhibits the growth of KPL-1 human breast cancer cells in vitro and suppresses the growth and metastasis of orthotopically transplanted KPL-1 cells in female athymic mice. Oncol Rep 26:603–608
58. Li Y, Zhang T, Schwartz SJ et al (2011) Sulforaphane potentiates the efficacy of 17-allylamino 17-demethoxygeldanamycin against pancreatic cancer through enhanced abrogation of Hsp90 chaperone function. Nutr Cancer 63:1151–1159
59. Chen MJ, Tang WY, Hsu CW et al (2012) Apoptosis induction in primary human colorectal cancer cell lines and retarded tumor growth in SCID mice by sulforaphane. Evid Based Complement Alternat Med 2012:415231
60. Qazi A, Pal J, Maitah M et al (2010) Anticancer activity of a broccoli derivative, sulforaphane, in Barrett adenocarcinoma: potential use in chemoprevention and as adjuvant in chemotherapy. Transl Oncol 3:389–399
61. Benson AM, Batzinger RP, Ou SY et al (1978) Elevation of hepatic glutathione S-transferase activities and protection against mutagenic metabolites of benzo(a)pyrene by dietary antioxidants. Cancer Res 38:4486–4495

62. Benson AM, Cha YN, Bueding E et al (1979) Elevation of extrahepatic glutathione S-transferase and epoxide hydratase activities by 2(3)-tert-butyl-4-hydroxyanisole. Cancer Res 39:2971–2977
63. Benson AM, Hunkeler MJ, Talalay P (1980) Increase of NAD(P)H:quinone reductase by dietary antioxidants: possible role in protection against carcinogenesis and toxicity. Proc Natl Acad Sci USA 77:5216–5220
64. Benson AM, Barretto PB (1985) Effects of disulfiram, diethyldithiocarbamate, bisethylxanthogen, and benzyl isothiocyanate on glutathione transferase activities in mouse organs. Cancer Res 45:4219–4223
65. Benson AM, Barretto PB, Stanley JS (1986) Induction of DT-diaphorase by anticarcinogenic sulfur compounds in mice. J Natl Cancer Inst 76:467–473
66. Talalay P, Batzinger RP, Benson AM et al (1978) Biochemical studies on the mechanisms by which dietary antioxidants suppress mutagenic activity. Adv Enzyme Regul 17:23–36
67. Hayes JD, McMahon M (2001) Molecular basis for the contribution of the antioxidant responsive element to cancer chemoprevention. Cancer Lett 174:103–113
68. Kwak MK, Wakabayashi N, Kensler TW (2004) Chemoprevention through the Keap1-Nrf2 signaling pathway by phase 2 enzyme inducers. Mutat Res 555:133–148
69. Motohashi H, Yamamoto M (2004) Nrf2-Keap1 defines a physiologically important stress response mechanism. Trends Mol Med 10:549–557
70. Nguyen T, Yang CS, Pickett CB (2004) The pathways and molecular mechanisms regulating Nrf2 activation in response to chemical stress. Free Radic Biol Med 37:433–441
71. Kobayashi M, Yamamoto M (2005) Molecular mechanisms activating the Nrf2-Keap1 pathway of antioxidant gene regulation. Antioxid Redox Signal 7:385–394
72. Kobayashi M, Yamamoto M (2006) Nrf2-Keap1 regulation of cellular defense mechanisms against electrophiles and reactive oxygen species. Adv Enzyme Regul 46:113–140
73. Zhang DD (2006) Mechanistic studies of the Nrf2-Keap1 signaling pathway. Drug Metab Rev 38:769–789
74. Kensler TW, Wakabayashi N, Biswal S (2007) Cell survival responses to environmental stresses via the Keap1-Nrf2-ARE pathway. Annu Rev Pharmacol Toxicol 47:89–116
75. Eggler AL, Gay KA, Mesecar AD (2008) Molecular mechanisms of natural products in chemoprevention: induction of cytoprotective enzymes by Nrf2. Mol Nutr Food Res 52:S84–S94
76. Dinkova-Kostova AT, Talalay P (2008) Direct and indirect antioxidant properties of inducers of cytoprotective proteins. Mol Nutr Food Res 52:S128–S138
77. Osburn WO, Kensler TW (2008) Nrf2 signaling: an adaptive response pathway for protection against environmental toxic insults. Mutat Res 659:31–39
78. Surh YJ, Kundu JK, Na HK (2008) Nrf2 as a master redox switch in turning on the cellular signaling involved in the induction of cytoprotective genes by some chemopreventive phytochemicals. Planta Med 74:1526–1539
79. Nguyen T, Nioi P, Pickett CB (2009) The Nrf2-antioxidant response element signaling pathway and its activation by oxidative stress. J Biol Chem 284:13291–13295
80. Li W, Kong AN (2009) Molecular mechanisms of Nrf2-mediated antioxidant response. Mol Carcinog 48:91–104
81. Hayes JD, McMahon M (2009) NRF2 and KEAP1 mutations: permanent activation of an adaptive response in cancer. Trends Biochem Sci 34:176–188
82. Hayes JD, McMahon M, Chowdhry S et al (2010) Cancer chemoprevention mechanisms mediated through the Keap1-Nrf2 pathway. Antioxid Redox Signal 13:1713–1748
83. Kwak MK, Kensler TW (2010) Targeting NRF2 signaling for cancer chemoprevention. Toxicol Appl Pharmacol 244:66–76
84. Villeneuve NF, Lau A, Zhang DD (2010) Regulation of the Nrf2-Keap1 antioxidant response by the ubiquitin proteasome system: an insight into cullin-ring ubiquitin ligases. Antioxid Redox Signal 13:1699–1712

85. Slocum SL, Kensler TW (2011) Nrf2: control of sensitivity to carcinogens. Arch Toxicol 85:273–284
86. Dinkova-Kostova AT, Holtzclaw WD, Cole RN et al (2002) Direct evidence that sulfhydryl groups of Keap1 are the sensors regulating induction of phase 2 enzymes that protect against carcinogens and oxidants. Proc Natl Acad Sci USA 99:11908–11913
87. McMahon M, Lamont DJ, Beattie KA et al (2010) Keap1 perceives stress via three sensors for the endogenous signaling molecules nitric oxide, zinc, and alkenals. Proc Natl Acad Sci USA 107:18838–18843
88. Hu C, Eggler AL, Mesecar AD et al (2011) Modification of Keap1 cysteine residues by sulforaphane. Chem Res Toxicol 24:515–521
89. Prochaska HJ, Santamaria AB (1988) Direct measurement of NAD(P)H:quinone reductase from cells cultured in microtiter wells: a screening assay for anticarcinogenic enzyme inducers. Anal Biochem 169:328–336
90. Prochaska HJ, Santamaria AB, Talalay P (1992) Rapid detection of inducers of enzymes that protect against carcinogens. Proc Natl Acad Sci USA 89:2394–2398
91. Fahey JW, Dinkova-Kostova AT, Stephenson KK et al (2004) The "Prochaska" microtiter plate bioassay for inducers of NQO1. Methods Enzymol 382:243–258
92. Zhang Y, Talalay P, Cho CG et al (1992) A major inducer of anticarcinogenic protective enzymes from broccoli: isolation and elucidation of structure. Proc Natl Acad Sci USA 89:2399–2403
93. Zhang Y, Tang L (2007) Discovery and development of sulforaphane as a cancer chemopreventive phytochemical. Acta Pharmacol Sin 28:1343–1354
94. Matusheski NV, Jeffery EH (2001) Comparison of the bioactivity of two glucoraphanin hydrolysis products found in broccoli, sulforaphane and sulforaphane nitrile. J Agric Food Chem 49:5743–5749
95. Munday R, Munday CM (2004) Induction of phase II detoxification enzymes in rats by plant-derived isothiocyanates: comparison of allyl isothiocyanate with sulforaphane and related compounds. J Agric Food Chem 52:1867–1871
96. Zhang Y, Munday R, Jobson HE et al (2006) Induction of GST and NQO1 in cultured bladder cells and in the urinary bladders of rats by an extract of broccoli (Brassica oleracea italica) sprouts. J Agric Food Chem 54:9370–9376
97. McMahon M, Itoh K, Yamamoto M et al (2001) The Cap'n'Collar basic leucine zipper transcription factor Nrf2 (NF-E2 p45-related factor 2) controls both constitutive and inducible expression of intestinal detoxification and glutathione biosynthetic enzymes. Cancer Res 61:3299–3307
98. Seo KW, Kim JG, Park M et al (2000) Effects of phenethylisothiocyanate on the expression of glutathione S-transferases and hepatotoxicity induced by acetaminophen. Xenobiotica 30:535–545
99. Munday R, Zhang Y, Munday CM et al (2008) Structure–activity relationships and organ specificity in the induction of GST and NQO1 by alkyl–aryl isothiocyanates. Pharm Res 25:2164–2170
100. Munday R, Munday CM (2002) Selective induction of phase II enzymes in the urinary bladder of rats by allyl isothiocyanate, a compound derived from Brassica vegetables. Nutr Cancer 44:52–59
101. Thimmulappa RK, Mai KH, Srisuma S et al (2002) Identification of Nrf2-regulated genes induced by the chemopreventive agent sulforaphane by oligonucleotide microarray. Cancer Res 62:5196–5203
102. Hu R, Xu C, Shen G et al (2006) Gene expression profiles induced by cancer chemopreventive isothiocyanate sulforaphane in the liver of C57BL/6J mice and C57BL/6J/Nrf2 (−/−) mice. Cancer Lett 243:170–192
103. Reen RK, Dombkowski AA, Kresty LA et al (2007) Effects of phenylethyl isothiocyanate on early molecular events in N-nitrosomethylbenzylamine-induced cytotoxicity in rat esophagus. Cancer Res 67:6484–6492

104. Stoner GD, Dombkowski AA, Reen RK et al (2008) Carcinogen-altered genes in rat esophagus positively modulated to normal levels of expression by both black raspberries and phenylethyl isothiocyanate. Cancer Res 68:6460–6467
105. Traka M, Gasper AV, Melchini A et al (2008) Broccoli consumption interacts with GSTM1 to perturb oncogenic signalling pathways in the prostate. PLoS One 3:e2568
106. Dinkova-Kostova AT, Fahey JW, Wade KL et al (2007) Induction of the phase 2 response in mouse and human skin by sulforaphane-containing broccoli sprout extracts. Cancer Epidemiol Biomarkers Prev 16:847–851
107. Heiss E, Herhaus C, Klimo K et al (2001) NFκB is a molecular target for sulforaphane-mediated anti-inflammatory mechanisms. J Biol Chem 276:32008–32015
108. Heiss E, Gerhäuser C (2005) Time-dependent modulation of thioredoxin reductase activity might contribute to sulforaphane-mediated inhibition of NFκB binding to DNA. Antioxid Redox Signal 7:1601–1611
109. Woo KJ, Kwon TK (2007) Sulforaphane suppresses lipopolysaccharide-induced cyclooxygenase-2 (COX-2) expression through the modulation of multiple targets in COX-2 gene promoter. Int Immunopharmacol 7:1776–1783
110. Lin W, Wu RT, Wu T (2008) Sulforaphane suppressed LPS-induced inflammation in mouse peritoneal macrophages through Nrf2 dependent pathway. Biochem Pharmacol 76:967–973
111. Liu H, Dinkova-Kostova AT, Talalay P (2008) Coordinate regulation of enzyme markers for inflammation and for protection against oxidants and electrophiles. Proc Natl Acad Sci USA 105:15926–15931
112. Prawan A, Saw CL, Khor TO et al (2009) Anti-NF-κB and anti-inflammatory activities of synthetic isothiocyanates: effect of chemical structures and cellular signaling. Chem Biol Interact 179:202–211
113. Cheung KL, Khor TO, Kong AN (2009) Synergistic effect of combination of phenethyl isothiocyanate and sulforaphane or curcumin and sulforaphane in the inhibition of inflammation. Pharm Res 26:224–231
114. Wu L, Noyan Ashraf MH, Facci M et al (2004) Dietary approach to attenuate oxidative stress, hypertension, and inflammation in the cardiovascular system. Proc Natl Acad Sci USA 101:7094–7099
115. Xu C, Shen G, Chen C et al (2005) Suppression of NF-κB and NF-κB-regulated gene expression by sulforaphane and PEITC through IκBα, IKK pathway in human prostate cancer PC-3 cells. Oncogene 24:4486–4495
116. Kallifatidis G, Rausch V, Baumann B et al (2009) Sulforaphane targets pancreatic tumour-initiating cells by NF-B-induced antiapoptotic signalling. Gut 58:949–963
117. Moon DO, Kim MO, Kang SH et al (2009) Sulforaphane suppresses TNF-α-mediated activation of NF-κB and induces apoptosis through activation of reactive oxygen species-dependent caspase-3. Cancer Lett 274:132–142
118. Song MY, Kim EK, Moon WS et al (2009) Sulforaphane protects against cytokine- and streptozotocin-induced beta-cell damage by suppressing the NF-κB pathway. Toxicol Appl Pharmacol 235:57–67
119. Shan Y, Wu K, Wang W et al (2009) Sulforaphane down-regulates COX-2 expression by activating p38 and inhibiting NF-κB-DNA-binding activity in human bladder T24 cells. Int J Oncol 34:1129–1134
120. Brandenburg LO, Kipp M, Lucius R et al (2010) Sulforaphane suppresses LPS-induced inflammation in primary rat microglia. Inflamm Res 59:443–450
121. Kivelä AM, Mäkinen PI, Jyrkkänen HK et al (2010) Sulforaphane inhibits endothelial lipase expression through NF-κB in endothelial cells. Atherosclerosis 213:122–128
122. Jeong SI, Choi BM, Jang SI (2010) Sulforaphane suppresses TARC/CCL17 and MDC/CCL22 expression through heme oxygenase-1 and NF-κB in human keratinocytes. Arch Pharm Res 33:1867–1876

123. Negi G, Kumar A, Sharma SS (2011) Nrf2 and NF-κB Modulation by sulforaphane counteracts multiple manifestations of diabetic neuropathy in rats and high glucose-induced changes. Curr Neurovasc Res 8:294–304
124. Dey M, Ribnicky D, Kurmukov AG et al (2006) In vitro and in vivo anti-inflammatory activity of a seed preparation containing phenethylisothiocyanate. J Pharmacol Exp Ther 317:326–333
125. Dey M, Kuhn P, Ribnicky D et al (2010) Dietary phenethylisothiocyanate attenuates bowel inflammation in mice. BMC Chem Biol 10:4
126. Talalay P, Fahey JW, Healy ZR et al (2007) Sulforaphane mobilizes cellular defenses that protect skin against damage by UV radiation. Proc Natl Acad Sci USA 104:17500–17505
127. Saw CL, Huang MT, Liu Y et al (2011) Impact of Nrf2 on UVB-induced skin inflammation/photoprotection and photoprotective effect of sulforaphane. Mol Carcinog 50:479–486
128. Shibata A, Nakagawa K, Yamanoi H et al (2010) Sulforaphane suppresses ultraviolet B-induced inflammation in HaCaT keratinocytes and HR-1 hairless mice. J Nutr Biochem 21:702–709
129. Khor TO, Hu R, Shen G et al (2006) Pharmacogenomics of cancer chemopreventive isothiocyanate compound sulforaphane in the intestinal polyps of ApcMin/+ mice. Biopharm Drug Dispos 27:407–420
130. Brown KK, Blaikie FH, Smith RA et al (2009) Direct modification of the proinflammatory cytokine macrophage migration inhibitory factor by dietary isothiocyanates. J Biol Chem 284:32425–32433
131. Cross JV, Rady JM, Foss FW et al (2009) Nutrient isothiocyanates covalently modify and inhibit the inflammatory cytokine macrophage migration inhibitory factor (MIF). Biochem J 423:315–321
132. Ouertatani-Sakouhi H, El-Turk F, Fauvet B et al (2009) A new class of isothiocyanate-based irreversible inhibitors of macrophage migration inhibitory factor. Biochemistry 48:9858–9870
133. Healy ZR, Liu H, Holtzclaw WD et al (2011) Inactivation of tautomerase activity of macrophage migration inhibitory factor by sulforaphane: a potential biomarker for anti-inflammatory intervention. Cancer Epidemiol Biomarkers Prev 20:1516–1523
134. Fimognari C, Nüsse M, Berti F et al (2004) Isothiocyanates as novel cytotoxic and cytostatic agents: molecular pathway on human transformed and non-transformed cells. Biochem Pharmacol 68:1133–1138
135. Gamet-Payrastre L (2006) Signaling pathways and intracellular targets of sulforaphane mediating cell cycle arrest and apoptosis. Curr Cancer Drug Targets 6:135–145
136. Juge N, Mithen RF, Traka M (2007) Molecular basis for chemoprevention by sulforaphane: a comprehensive review. Cell Mol Life Sci 64:1105–1127
137. Antosiewicz J, Ziolkowski W, Kar S et al (2008) Role of reactive oxygen intermediates in cellular responses to dietary cancer chemopreventive agents. Planta Med 74:1570–1579
138. Clarke JD, Dashwood RH, Ho E (2008) Multi-targeted prevention of cancer by sulforaphane. Cancer Lett 269:291–304
139. Cheung KL, Kong AN (2010) Molecular targets of dietary phenethyl isothiocyanate and sulforaphane for cancer chemoprevention. AAPS J 12:87–97
140. Jackson SJ, Singletary KW (2004) Sulforaphane: a naturally occurring mammary carcinoma mitotic inhibitor, which disrupts tubulin polymerization. Carcinogenesis 25:219–227
141. Jackson SJ, Singletary KW, Venema RC (2007) Sulforaphane suppresses angiogenesis and disrupts endothelial mitotic progression and microtubule polymerization. Vascul Pharmacol 46:77–84
142. Thejass P, Kuttan G (2006) Antimetastatic activity of sulforaphane. Life Sci 78:3043–3050
143. Thejass P, Kuttan G (2007) Allyl isothiocyanate (AITC) and phenyl isothiocyanate (PITC) inhibit tumour-specific angiogenesis by downregulating nitric oxide (NO) and tumour necrosis factor-alpha (TNF-α) production. Nitric Oxide 16:247–257

144. Thejass P, Kuttan G (2007) Immunomodulatory activity of Sulforaphane, a naturally occurring isothiocyanate from broccoli (*Brassica oleracea*). Phytomedicine 14:538–545
145. Thejass P, Kuttan G (2007) Inhibition of endothelial cell differentiation and proinflammatory cytokine production during angiogenesis by allyl isothiocyanate and phenyl isothiocyanate. Integr Cancer Ther 6:389–399
146. Thejass P, Kuttan G (2007) Modulation of cell-mediated immune response in B16F-10 melanoma-induced metastatic tumor-bearing C57BL/6 mice by sulforaphane. Immunopharmacol Immunotoxicol 29:173–186
147. Kim HJ, Barajas B, Wang M et al (2008) Nrf2 activation by sulforaphane restores the age-related decrease of T(H)1 immunity: role of dendritic cells. J Allergy Clin Immunol 121:1255–1261.e7
148. Manesh C, Kuttan G (2003) Effect of naturally occurring isothiocyanates on the immune system. Immunopharmacol Immunotoxicol 25:451–459
149. Suganuma H, Fahey JW, Bryan KE et al (2011) Stimulation of phagocytosis by sulforaphane. Biochem Biophys Res Commun 405:146–151
150. Marks P, Rifkind RA, Richon VM et al (2001) Histone deacetylases and cancer: causes and therapies. Nat Rev Cancer 1:194–202
151. Myzak MC, Karplus PA, Chung FL et al (2004) A novel mechanism of chemoprotection by sulforaphane: inhibition of histone deacetylase. Cancer Res 64:5767–5774
152. Myzak MC, Hardin K, Wang R et al (2006) Sulforaphane inhibits histone deacetylase activity in BPH-1, LnCaP and PC-3 prostate epithelial cells. Carcinogenesis 27:811–819
153. Dashwood RH, Ho E (2007) Dietary histone deacetylase inhibitors: from cells to mice to man. Semin Cancer Biol 17:363–369
154. Ma X, Fang Y, Beklemisheva A et al (2006) Phenylhexyl isothiocyanate inhibits histone deacetylases and remodels chromatins to induce growth arrest in human leukemia cells. Int J Oncol 28:1287–1293
155. Beklemisheva AA, Fang Y, Feng J et al (2006) Epigenetic mechanism of growth inhibition induced by phenylhexyl isothiocyanate in prostate cancer cells. Anticancer Res 26:1225–1230
156. Pledgie-Tracy A, Sobolewski MD, Davidson NE (2007) Sulforaphane induces cell type-specific apoptosis in human breast cancer cell lines. Mol Cancer Ther 6:1013–1021
157. Wang LG, Liu XM, Fang Y et al (2008) De-repression of the p21 promoter in prostate cancer cells by an isothiocyanate via inhibition of HDACs and c-Myc. Int J Oncol 33:375–380
158. Gan N, Wu YC, Brunet M et al (2010) Sulforaphane activates heat shock response and enhances proteasome activity through up-regulation of Hsp27. J Biol Chem 285:35528–35536
159. Shapiro TA, Fahey JW, Wade KL et al (1998) Human metabolism and excretion of cancer chemoprotective glucosinolates and isothiocyanates of cruciferous vegetables. Cancer Epidemiol Biomarkers Prev 7:1091–1100
160. Getahun SM, Chung FL (1999) Conversion of glucosinolates to isothiocyanates in humans after ingestion of cooked watercress. Cancer Epidemiol Biomarkers Prev 8:447–451
161. Conaway CC, Getahun SM, Liebes LL et al (2000) Disposition of glucosinolates and sulforaphane in humans after ingestion of steamed and fresh broccoli. Nutr Cancer 38:168–178
162. Shapiro TA, Fahey JW, Wade KL et al (2001) Chemoprotective glucosinolates and isothiocyanates of broccoli sprouts: metabolism and excretion in humans. Cancer Epidemiol Biomarkers Prev 10:501–508
163. Ye L, Dinkova-Kostova AT, Wade KL et al (2002) Quantitative determination of dithiocarbamates in human plasma, serum, erythrocytes and urine: pharmacokinetics of broccoli sprout isothiocyanates in humans. Clin Chim Acta 316:43–53
164. Kensler TW, Chen JG, Egner PA et al (2005) Effects of glucosinolate-rich broccoli sprouts on urinary levels of aflatoxin-DNA adducts and phenanthrene tetraols in a randomized clinical trial in He Zuo township, Qidong, People's Republic of China. Cancer Epidemiol Biomarkers Prev 14:2605–2613

165. Gasper AV, Al-Janobi A, Smith JA et al (2005) Glutathione S-transferase M1 polymorphism and metabolism of sulforaphane from standard and high-glucosinolate broccoli. Am J Clin Nutr 82:1283–1291
166. Shapiro TA, Fahey JW, Dinkova-Kostova AT et al (2006) Safety, tolerance, and metabolism of broccoli sprout glucosinolates and isothiocyanates: a clinical phase I study. Nutr Cancer 55:53–62
167. Cornblatt BS, Ye L, Dinkova-Kostova AT et al (2007) Preclinical and clinical evaluation of sulforaphane for chemoprevention in the breast. Carcinogenesis 28:1485–1490
168. Riedl MA, Saxon A, Diaz-Sanchez D (2009) Oral sulforaphane increases phase II antioxidant enzymes in the human upper airway. Clin Immunol 130:244–251
169. Egner PA, Chen JG, Wang JB et al (2011) Bioavailability of sulforaphane from two broccoli sprout beverages: results of a short-term, cross-over clinical trial in Qidong, China. Cancer Prev Res (Phila) 4:384–395
170. Cramer JM, Jeffery EH (2011) Sulforaphane absorption and excretion following ingestion of a semi-purified broccoli powder rich in glucoraphanin and broccoli sprouts in healthy men. Nutr Cancer 63:196–201
171. Li F, Hullar MA, Beresford SA et al (2011) Variation of glucoraphanin metabolism in vivo and ex vivo by human gut bacteria. Br J Nutr 106:408–416
172. Harvey CJ, Thimmulappa RK, Sethi S et al (2011) Targeting Nrf2 signaling improves bacterial clearance by alveolar macrophages in patients with COPD and in a mouse model. Sci Transl Med 3:78ra32
173. Clarke JD, Hsu A, Riedl K et al (2011) Bioavailability and inter-conversion of sulforaphane and erucin in human subjects consuming broccoli sprouts or broccoli supplement in a cross-over study design. Pharmacol Res 64:456–463
174. Kensler TW, Ng D, Carmella SG et al (2012) Modulation of the metabolism of airborne pollutants by glucoraphanin-rich and sulforaphane-rich broccoli sprout beverages in Qidong, China. Carcinogenesis 33:101–107

Human Cancer Chemoprevention: Hurdles and Challenges

Vaqar Mustafa Adhami and Hasan Mukhtar

Abstract Cancer is considered a disease of aging since the risk for developing the disease considerably increases with age. It is estimated that 77% of all cancers are diagnosed in people who fall within the age group of 55 or older. Also, it takes several years from initiation to the development of detectable cancer. One advantage of the long latency is that it provides numerous opportunities for intervention. While intervention approaches cannot be geared towards a whole population, they can nevertheless be directed towards a defined group of people who have a greater relative risk for developing the disease. The idea of cancer prevention through the use of nontoxic agents, preferably from dietary sources, has therefore emerged as an appropriate strategy for controlling the disease. An important aspect of chemoprevention is that agents can be designed for intervention at any stage during the multistage process of carcinogenesis. This process of slowing the progression of cancer is applicable to many cancers with long latency, including prostate cancer. Over the past two decades we have put considerable effort into identifying dietary substances in the form of extracts and pure compounds that can be used for the prevention of prostate and other cancers. Although cancer chemoprevention has proven to be a successful strategy in animals and, to some extent, we can say that the mission has been accomplished, its application to humans has met with limited success. This chapter will discuss various challenges associated with chemoprevention of cancer with the focus on studies with green tea and prostate cancer.

Keywords Cancer · Chemoprevention · Clinical · Green tea · Prostate · Trials

---

V.M. Adhami and H. Mukhtar (✉)
Department of Dermatology, University of Wisconsin-Madison, Madison, WI 53706, USA
e-mail: hmukhtar@dermatology.wisc.edu

Contents

1 Introduction 204
2 Animal and Human Cancer Chemoprevention Studies 205
3 Green Tea and Prostate Cancer Chemoprevention: An Appropriate Model to Illustrate Differences Between Human and Animal Studies 206
   3.1 In Vitro Studies with Green Tea 207
   3.2 In Vivo Studies with Green Tea 208
   3.3 Epidemiologic and Clinical Studies with Green Tea 210
4 Clinical Successes with Cancer Chemoprevention 212
5 Complexities Associated with Cancer Chemoprevention 213
6 Conclusions 216
References 217

Abbreviations

CAM	Complementary and alternative medicine
CI	Confidence interval
COMT	Catechol-O-methyltransferase
DNA	Deoxyribose nucleic acid
EGC	Epigallocatechin
EGCG	(−)-Epigallocatechin gallate
FRET	Fluorescence resonance energy transfer
GSTM1	Glutathione S-transferase M1 gene
GTCs	Green tea catechins
GTP	Green tea polyphenols
HRPC	Hormone refractory prostate cancer
IGF	Insulin-like growth factor
JPHC	Japanese Public Health Center
ODC	Ornithine decarboxylase
PIN	Prostatic intraepithelial neoplasia
PKC	Protein kinase C
PSA	Prostate specific antigen
SELECT	Selenium and vitamin E cancer prevention trial
TRAMP	Transgenic adenocarcinoma of the mouse prostate

1 Introduction

According to estimates of the American Cancer Society, one-third of cancer deaths expected to occur in the year 2012 will be related to life style factors such as nutrition, obesity and lack of physical activity and therefore can be prevented [1]. Besides modifying life style as a primary cancer prevention method, another approach to decrease the incidence of cancer is through chemoprevention, a means of cancer management in which the progression of the disease can be manipulated through administration of natural and/or synthetic compounds [2–8].

The history of cancer prevention dates back a few hundred years, but its recognition and application is recent [9]. Lee W. Wattenberg demonstrated prevention of chemical carcinogenesis and coined the term "chemoprophylaxis" [10]. Michael B. Sporn coined the widely used and familiar term "cancer chemoprevention" to define the use of agents to reverse, suppress, or prevent the carcinogenic process to invasive cancer [11, 12]. We believe that chemopreventive intervention is only possible during the process of cancer development and unlikely when cancer is already established. Cancer chemoprevention if directed at the right population only has the potential to delay the process of cancer development and therefore we define cancer chemoprevention as "slowing the process of carcinogenesis." This concept of slowing the progression of the disease could apply to most other solid malignancies including cancers of the breast, colon, lung, bladder, prostate, and others.

Since chemopreventive protocols are expected to run over long periods of time, it is practical that, for human use, only those agents that are nontoxic and widely acceptable should be advocated. Naturally occurring compounds that are part of our diet fit very well into this category and have been extensively studied in crude and chemically defined forms. Extracts have been analyzed to identify active ingredients such as epigallocatechin gallate (EGCG) from green tea and curcumin from turmeric. Some of the compounds have also been subjected to modification to enhance their bioactivity [13]. However, the practice of chemical modification in principle eliminates the natural form of the compound and makes it synthetic. While these modified compounds and other synthetic agents could be more potent, they nevertheless qualify as drugs and have to be tested extensively for lack of toxicity before they are approved by the regulatory agencies for wide public use. It is also for these reasons that we promote the use of agents from dietary sources as they are nature's gift molecules with potential cancer preventive properties. Many of the compounds identified from natural sources are antioxidant, exhibit anti-inflammatory activity, and possess antiproliferative properties. Because many of the natural agents form part of our diet they are considered nontoxic and humans have acquired the ability to consume them without any known side effects. Cancer chemopreventive agents from natural and especially dietary sources are often cost effective and have wide human acceptance compared to synthetic compounds that can only be marketed as drugs and come with a wide variety of off target and debilitating side effects.

2 Animal and Human Cancer Chemoprevention Studies

Cancer chemoprevention research relies greatly on the availability of animal models. Chemically induced animal models of carcinogenesis accelerated research on cancer chemoprevention and later, with the availability of spontaneous, transgenic and xenograft mouse models, further led to a surge in chemoprevention studies and identification of many agents both synthetic and natural [11–20]. Preclinical studies using animals to model chemoprevention of human cancer have been largely successful and promising. Human cancer chemoprevention studies began earnestly after the discovery that retinoids enhance susceptibility to chemical carcinogenesis [21]. However, many human clinical trials have not been successful and have yielded disappointing results. This conundrum of success in animals and disappointment in humans is not without reason and many arguments could be made. We have to admit that both human trials and preclinical animal studies are not optimally planned and executed. Another reason is the obvious differences between human and animal studies. Animal studies are generally well optimized and conducted in genetically identical populations excluding issues related to genetic variability. Human chemoprevention trials on the other hand cannot be properly optimized. Humans enrolled in clinical trials usually belong to diverse genetic backgrounds and have wide-ranging food habits and other life style factors such as smoking, alcoholism, and cooking methods which could affect chemopreventive outcomes.

Humans consume a variety of food items that contain both carcinogens and chemopreventive ingredients. We believe that humans are already protected to varying degrees because of their diversified food habits, and by our lifestyle we have attained some level of chemoprevention that preempts the possibility of observing large effects from human chemoprevention trials. Even people considered to have poor food habits consume a sizeable portion of chemopreventive ingredients such as fruits, vegetables, tea, and red wine. Thus, to appreciate the outcome of human chemoprevention trials we may have to lower our expectations and settle for modest to moderate effects.

3 Green Tea and Prostate Cancer Chemoprevention: An Appropriate Model to Illustrate Differences Between Human and Animal Studies

Prostate cancer is the second most frequently diagnosed cancer in males, being next only to skin cancer. An estimated 28,170 prostate cancer related deaths are projected for the year 2012, making it the second-leading cause of cancer deaths in men in the United States [1]. Because prostate cancer develops slowly over a period of decades and is commonly diagnosed in men over the age of 50, it is considered an ideal disease for chemopreventive intervention [4–8]. Guided by data derived from epidemiological, clinical, and laboratory studies, many agents and their molecular targets for prostate cancer chemoprevention have been identified. The fact that prostate cancer – like many other cancers – exhibits aberrations in different molecular events, blocking or inhibiting only one event may not be sufficient to prevent the onset and progression of the disease. Efforts are ongoing for a better understanding of the disease so that novel approaches for its prevention and treatment could be developed.

Asian populations are usually considered to be at a lower risk for the development of prostate cancer [22]. However, interestingly, after migrating to the west these populations acquire trends and risks for development of prostate cancer similar to their western counterparts [22, 23]. These facts point strongly towards a role for environmental and dietary factors in the development of prostate cancer. There is also evidence from geographic and epidemiological data that suggests an increase in the incidence of prostate cancer possibly due to adoption of a western lifestyle [24, 25]. Also, there is epidemiological evidence that suggests populations with higher consumption of selenium, vitamin E, green tea, fruits, and tomatoes have lower risk for the development of prostate cancer [26]. These observations have led to an increase in research on the use of natural agents for the prevention of prostate cancer and currently several natural agents are being studied for their cancer chemopreventive potential.

The beverage tea has been studied extensively and it has emerged as an agent having antimutagenic and anticancer effects [4–7]. Green tea, a popular beverage in many countries, is made from the leaves of an evergreen shrub *Camellia sinensis*. Although native to China, its consumption rapidly spread across the globe and currently the plant is grown in many countries around the world and demand for tea has been growing each year. The method of processing tea leaves determines the type of tea produced. Green tea undergoes the least amount of oxidation and retains most of the chemical composition of the original tea leaves. Black tea, obtained through complete oxidation of tea leaves, constitutes about 78% of the tea produced in the world and is predominantly consumed in Western and some Asian countries. Green tea contains flavanols, flavandiols, flavanoids, and phenolic acids. Most of the green tea polyphenols are flavanols, known as catechins, such as catechin, gallocatechin, epicatechin, epicatechin-3-gallate, epigallocatechin, and epigallocatechin-3-gallate. The polyphenolic components in tea have been observed to be antioxidant in nature and possess the ability to prevent oxidant-induced cellular damage [27]. Studies conducted in many organ specific animal bioassay systems have shown that tea and its polyphenolic constituents are capable of affording protection against a variety of cancer types. Although the majority of the studies conducted have used green tea, a limited number of studies have also shown the anticancer efficacy of black tea.

3.1 In Vitro Studies with Green Tea

Androgens play an important role in the development of the prostate gland, are considered as major stimuli for inducing neoplastic transformation, and therefore constitute a potential target for prostate cancer prevention [28]. Both EGCG and EGC have been found to inhibit 5-α-reductase, the enzyme that converts testosterone to its active metabolite 5-α-dihydroxytestosterone [29]. EGCG inhibits growth of androgen-responsive LNCaP cells and the expression of androgen regulated prostate specific antigen (PSA) and hK2 genes. An Sp1 binding site in the androgen receptor gene promoter was identified as a target for tea polyphenols as treatments with EGCG decreased the expression, DNA binding activity, and transactivation of Sp1 protein [30]. PSA secretion was significantly decreased in a dose-dependent as well as time-dependent manner when LNCaP cells were treated with EGCG [31]. Activity of ornithine decarboxylase (ODC), an androgen regulated molecule that is up regulated in prostate cancer, was significantly inhibited when LNCaP cells were treated with GTP [30]. GTP also inhibited testosterone induced colony formation in LNCaP cells in a dose-dependent manner [32]. We recently provided evidence that EGCG is a direct antagonist of androgen action [33]. In silico modeling and FRET-based competition assay showed that EGCG physically interacted with the ligand-binding domain of androgen receptor by replacing a high-affinity labeled ligand [33].

Induction of apoptosis of cancerous cells has emerged as a therapeutic modality against cancer by naturally derived bioactive agents from diet [34]. EGCG treatment resulted in an induction of apoptosis in several human cancer cells including human prostate cancer cells DU145, LNCaP, and PC-3 regardless of their androgen or p53 status [35, 36]. In subsequent experiments we observed that EGCG-mediated cell cycle dysregulation and apoptosis is mediated through modulation of cyclin kinase inhibitor (cki)-cyclin-cyclin-dependent kinase (cdk) pathway and via a concurrent effect on two important transcription factors p53 and NF-kappa B, causing a change in the ratio of Bax/Bcl-2 that favors apoptosis [37–39]. Further studies showed that these effects were p53-dependent and involved the function of both p21 and Bax such that down-regulation of either conferred a growth advantage to the cells [40]. The ubiquitin–proteasome system plays a vital role in degradation of cellular proteins and hence allows tumor cell survival while proteasome inhibitors induce tumor growth arrest [41]. EGCG also inhibited the chymotrypsin-like activity of the proteasome in vitro in several tumor and transformed cell lines resulting in the accumulation of p27/Kip1 and IkB-α, an inhibitor of transcription factor NF-κB, leading to cell cycle arrest [42, 43]. Using cDNA microarray, we identified a set of 25 genes that showed a significant response after treatment with EGCG (12 μM, for 12 h). Expression of 16 genes was significantly increased and 9 genes were found to be significantly repressed by EGCG treatment [44]. Repression of PKC-α was found to be most prominent, suggesting that inhibition of PKC-α gene expression could inhibit cancer cell proliferation [45, 46]. The cDNA microarray also identified induction of receptor-type protein tyrosine phosphatase-λ gene expression, a tumor suppressor gene candidate frequently deleted in some human cancers [44].

3.2 In Vivo Studies with Green Tea

Green tea consumption has consistently been shown to prevent or delay prostate cancer in rodents (Table 1) [13–19]. EGCG (daily 1 mg/mouse, i.p.) treatment to athymic nude mice implanted with androgen-insensitive PC-3 and androgen-sensitive LNCaP 104-R cells resulted in reduction in the initial tumor growth of both cell types by 20–30% [58]. Similar findings were observed when nude mice

Table 1 Preclinical and clinical trials with green against prostate cancer

Trial (ref #) & type	Study design	Outcome
Gupta et al. [47] Preclinical	TRAMP mice; 0.1% GTP in drinking water starting at 8 weeks of age	Excellent
Jatoi et al. [48] Clinical	Patients with androgen-independent metastatic prostate cancer; (1 g/glass six times/day)	Poor
Adhami et al. [49] Preclinical	TRAMP mice; 0.1% GTP in drinking water starting at 8 weeks of age	Excellent
Caporali et al. [50] Preclinical	TRAMP mice: 0.3% GTC in drinking water starting at 8 weeks of age	Excellent
Choan et al. [51] Clinical	Patients with hormone refractory prostate cancer; (250-mg capsules twice daily)	Poor
Bettuzzi et al. [52] Clinical	Volunteers with high grade PIN lesions; (200-mg capsules thrice daily)	Excellent
Harper et al. [53] Preclinical	TRAMP mice; 0.06% EGCG in drinking water starting at 5 weeks of age	Excellent; efficacy depends on stage at the time of initiation
Brausi et al. [54] Clinical	Volunteers with High Grade PIN lesions; (200-mg capsules thrice daily)	Excellent
Adhami et al. [55] Preclinical	TRAMP mice; 0.1% GTP in drinking water starting at 8 weeks of age	Excellent; efficacy depends on stage at the time of initiation
McLarty et al. [56] Clinical	Patients scheduled for prostatectomy; (1.3 g daily green tea in capsules)	Good
Nguyen et al. [57] Clinical	Patients scheduled for prostatectomy; (800 mg polyphenon E daily)	Promising but not statistically significant

with xenografts were fed with a nutrient mixture containing green tea extract [59]. Employing androgen-responsive CWR22Rv1 prostate cancer tumor xenografts implanted in athymic nude mice, we demonstrated that treatment with GTP and EGCG not only resulted in significant inhibition of growth of implanted tumors but also a reduction in the serum PSA levels [60]. Furthermore, GTP (0.01% or 0.05% w/v) given after establishment of CWR22Rv1 tumors caused a significant regression of tumors, suggesting therapeutic effects of GTP at human achievable concentrations. Using the transgenic adenocarcinoma of the mouse prostate (TRAMP), a model in which progressive forms of human disease occur spontaneously [61], we showed that oral infusion of GTP at a human achievable dose significantly inhibits prostate cancer development and increases overall survival in these mice [47]. In a follow-up study we demonstrated that continuous GTP infusion for 24 weeks to these mice resulted in substantial reduction in the levels of IGF-I and significant increase in the levels of IGFBP-3, suggesting that IGF-I/IGFBP-3 signaling pathway is a prime pathway for GTP-mediated inhibition of prostate cancer [49]. These effects of green tea on the development of prostate cancer in TRAMP were subsequently corroborated by Caporali et al. [50] who demonstrated progressive accumulation of clusterin mRNA and protein in the prostate gland, suggesting a possible role for clusterin as a novel tumor-suppressor gene in the prostate. To identify the efficacy of green tea at different stages of prostate cancer development, TRAMP mice received oral infusion of GTP (0.1% in drinking water) at ages representing different stages of the disease. Tumor free survival and median life expectancy was highest in animals in which intervention was initiated early compared with animals where treatment was started later [55]. Our studies suggested that chemopreventive potential of GTP decreases with advancing stage of the disease. These observations were corroborated by Harper et al. [53] who observed that EGCG suppresses early stage – but not late stage – prostate cancer in TRAMP mice.

3.3 Epidemiologic and Clinical Studies with Green Tea

Epidemiologic evidence suggests that regular use of green tea in the Asian population in general is inversely associated with the risk of several types of human cancers including prostate cancer, compared to those in Western societies [24, 25]. The Japan Public Health Center (JPHC)-based Prospective Study established in 1990 and 1993 enrolled 49,920 men aged 40–69 years [62]. The subjects completed a questionnaire that included their green tea consumption habit and were followed until the end of year 2004. Green tea consumption was not associated with localized prostate cancer; however, its consumption was associated with a dose-dependent decrease in the risk of advanced prostate cancer. The multivariate relative risk was 0.52 (95% confidence interval: 0.28, 0.96) for men drinking five or more cups/day compared with less than one cup/day (p(trend) = 0.01). The study concluded that green tea may be associated with a decreased risk of advanced prostate cancer. In another case–control study conducted in Hangzhou, China the possible joint effects of lycopene and green tea on prostate cancer risk were examined [63]. A total of 130 prostate cancer patients and 274 hospital controls were enrolled and information on tea and dietary intakes, and possible confounders, was collected using a structured questionnaire. Prostate cancer risk was reduced with increased consumption of green tea. The protective effect of green tea was significant (odds ratio 0.14, 95% CI: 0.06–0.35) for the highest quartile relative to the lowest after adjusting for total vegetables and fruits intakes and other potential confounding factors. This study suggested that habitual drinking of tea could lead to a reduced risk of prostate cancer in Chinese men. The study also suggested that tea, together with other dietary ingredients, could have a stronger preventive effect than either component taken separately.

To explore whether green tea consumption had an etiological association with prostate cancer, a case–control study was conducted in Hangzhou, southeast China during 2001–2002 [64]. Among the cases, 55.4% were tea drinkers compared to 79.9% for the controls. Almost all the tea consumed was green tea. The prostate cancer risk declined with increasing frequency, duration, and quantity of green tea consumption. The adjusted odds ratios (OR), relative to non-tea drinkers, were 0.28

(95% CI = 0.17–0.47) for tea drinking, 0.12 (95% CI = 0.06–0.26) for drinking tea over 40 years, 0.09 (95% CI = 0.04–0.21) for those consuming more than 1.5 kg of tea leaves yearly, and 0.27 (95% CI = 0.15–0.48) for those drinking more than three cups (1 L) daily. The dose response relationships were also significant, suggesting that green tea is protective against prostate cancer. These observations are further supported by facts that suggest Asian men migrating to the United States and their subsequent US born generations acquire a higher clinical incidence of prostate cancer [23].

Several clinical studies have explored the effects of green tea consumption in prostate cancer patients (Table 1). Jatoi et al. [48] reported a phase II trial that explored green tea's antineoplastic effects in patients with androgen independent prostate carcinoma. The study conducted by the North Central Cancer Treatment Group evaluated 42 patients who were asymptomatic and had manifested progressive PSA elevation with hormone therapy. Patients were instructed to take 6 g of green tea per day orally in six divided doses. Tumor response, defined as 50% decline in baseline PSA value, occurred in a single patient or 2% of the cohort and was not sustained beyond 2 months. The study concluded that green tea carries limited antineoplastic activity, as defined by a decline in PSA levels, among patients with androgen independent prostate carcinoma. This poorly planned and executed trial of green tea in human prostate cancer patients initiated uncalled-for discussion about lack of preventive effects of green tea for human prostate cancer.

Another study evaluated the efficacy and toxicity of green tea, prescribed as a complementary and alternative medicine (CAM) formulation on hormone refractory prostate cancer (HRPC). Nineteen patients with HRPC were enrolled into the study and prescribed green tea extract capsules at a dose level of 250 mg twice daily [51]. Nine patients had progressive disease within 2 months of starting therapy and another six patients developed progressive disease after additional 1–4 months of therapy. Based on the observations it was concluded that green tea, as CAM therapy, was found to have minimal clinical activity against HRPC.

In another study Bettuzzi et al. [52] conducted a proof-of-principle clinical trial to assess the efficacy of green tea catechins (GTCs) for the chemoprevention of prostate cancer in volunteers with high grade prostatic intraepithelial neoplasia (HG-PIN) based on observations that 30% of men with HG-PIN develop prostate cancer within 1 year after repeated biopsy. Sixty volunteers with HG-PIN enrolled in this double-blind, placebo-controlled study received daily treatment that consisted of three GTC capsules, 200 mg each. After 1 year, only 1 tumor was diagnosed among the 30 GTCs-treated men whereas 9 cancers were found among the 30 placebo-treated men. GTCs-treated men showed PSA values that were constantly lower with respect to placebo-treated ones. This study concluded that GTCs are safe and very effective for treating premalignant lesions before prostate cancer develops. In a follow-up of the same study, it was observed that the inhibition of prostate cancer progression achieved in these subjects after 1 year of GTCs administration was long-lasting [54]. The mean follow-up from the end of GTCs dosing was 23.3 months for placebo-arm and 19.1 months for GTCs-arm. Overall, treatment with GTCs led to an almost 80% reduction in prostate cancer diagnosis, from 53% to 11% [54].

McLarty et al. [56] determined the effects of short-term supplementation with green tea on serum biomarkers in patients with prostate cancer. Twenty-six men with positive prostate biopsies and scheduled for radical prostatectomy were given daily doses of polyphenon E, which contained 800 mg of EGCG until time of radical prostatectomy. Treatment with green tea resulted in a significant reduction in serum levels of PSA in men with prostate cancer, with no elevation of liver enzymes. These findings supported a potential role for Polyphenon E in the treatment or prevention of prostate cancer.

A randomized, double-blind, placebo-controlled trial of polyphenon E was conducted in men with prostate cancer scheduled to undergo radical prostatectomy to determine its effects on systemic and tissue biomarkers of prostate carcinogenesis [57]. Patients received polyphenon E (containing 800 mg EGCG) or placebo daily for 3–6 weeks before surgery. Polyphenon E intervention resulted in favorable but not statistically significant changes in serum PSA, serum IGF, and oxidative DNA damage in blood leukocytes. The proportion of subjects who had a decrease in Gleason score between biopsy and surgical specimens was greater in those on polyphenon E but was not statistically significant. Although changes observed with green tea administration were not statistically significant, the study suggested that prostate cancer preventive activity, if occurring, may be through indirect means and/or that the activity may need to be evaluated with longer intervention durations, repeated dosing, or in patients at earlier stages of the disease.

The outcome of laboratory and clinical studies with green tea has been undisputed and unanimous. Green tea supplementation was associated with significant prevention of disease progression when intervention was started early and relative efficacy found to be dependent on the stage of the disease at the time of intervention [53, 55]. Human clinical trials with green tea conducted in advanced prostate cancer patients without any preclinical evidence yielded disappointing results [48, 51]. An obvious conclusion from these studies is that before planning to conduct clinical trials in human patients more efforts should be devoted to experiments in the rodent model. The failure of the SELECT trial for prostate cancer is an example that explains this point [65]. This ambitious undertaking should have been verified first in relevant mouse model(s) of human prostate cancer.

4 Clinical Successes with Cancer Chemoprevention

Because of many recent successful trials, cancer chemoprevention is being recognized as a practical strategy for the management of cancer. There is evidence from preclinical studies and encouraging clinical trials, and many proof-of-principle studies support cancer chemopreventive approaches. In a randomized clinical trial of rofecoxib, a cyclooxygenase (Cox)-2 inhibitor, adenoma recurrence was less frequent for rofecoxib subjects than for those randomized to placebo [66]. Rofecoxib also conferred a reduction in risk of advanced adenomas. The chemopreventive effect was more pronounced in the first year than in the subsequent 2 years. In this randomized trial, rofecoxib significantly reduced the risk of colorectal adenomas, but also had serious toxicity [66].

Encouraged by the findings that tamoxifen as an adjuvant therapy decreased contralateral breast cancer incidence, it was suggested that the drug might play a role in breast cancer prevention. To test this hypothesis, the National Surgical Adjuvant Breast and Bowel Project initiated the Breast Cancer Prevention Trial in 1992 [67]. Women at increased risk for breast cancer were randomly assigned to receive placebo or 20 mg/day tamoxifen for 5 years. Tamoxifen reduced the risk of invasive breast cancer by 49%, of noninvasive breast cancer by 50%, and the occurrence of estrogen receptor-positive tumors by 69%. The study concluded that tamoxifen decreases the incidence of invasive and noninvasive breast cancer and recommended its use as a breast cancer preventive agent in women at increased risk for the disease.

The Prostate Cancer Prevention Trial enrolled 18,882 men to determine the effect of finasteride relative to placebo on prostate cancer risk [68]. Men (55 years and older) with a PSA level of <3.0 ng/mL and normal digital rectal examination findings were randomized to receive finasteride 5 mg daily or placebo. Finasteride significantly reduced the risk of prostate cancer risk relative to placebo across multiple Gleason scores including the most frequently detected intermediate-grade and high-grade cancers [68].

A recent clinical trial suggested the usefulness of Cox-2 inhibitors for prevention of non-melanoma skin cancers [69]. Patients with actinic keratoses were given 200 mg of celecoxib or placebo orally twice daily for 9 months. Eleven months post randomization, incidence of non-melanoma skin cancer was lower in the celecoxib arm than in the placebo arm. The study concluded that celecoxib may be effective for prevention of squamous cell and basal cell carcinomas in individuals who have extensive actinic damage and are at high risk for development of non-melanoma skin cancers.

5 Complexities Associated with Cancer Chemoprevention

Some of the encouraging data from clinical trials need to be verified more stringently in larger human trials. In spite of the promising data obtained from many of these cancer chemoprevention trials, there are several issues and complexities especially related to studies involving dietary ingredients that put hurdles on the future of chemoprevention. Differences in human genomes that influence nutrient metabolism impact the way a particular cancer-fighting molecule is metabolized in the body. Wu et al. [70] examined the interrelationships between tea intake, catechol-O-methyltransferase (COMT) genotype, and breast cancer risk in 589 incident cases and 563 population-based controls from a population-based case–control study of breast cancer in Chinese-American, Japanese-American, and Filipino-American women in Los Angeles County. Risk of breast cancer was influenced significantly by intake of tea, particularly green tea. However, the inverse association between tea intake and breast cancer risk was observed only among individuals who possessed at least one low-activity COMT allele. Among women who carried at least one low activity COMT allele, tea drinkers showed a significantly reduced risk of breast cancer compared with non-tea drinkers after adjustment for relevant demographic, menstrual, reproductive, and dietary factors [70]. This risk reduction was observed in relation to both green tea and black tea intake. In contrast, risk of breast cancer did not differ between tea drinkers and non-tea drinkers among those who were homozygous for the high activity COMT allele. Tea catechins reduced breast cancer risk in this study of Asian-American women and was strongest among persons who had the low activity COMT alleles, suggesting these individuals were less efficient in eliminating tea catechins and may derive the most benefit from these compounds [70].

In another study, suppression of COMT activity in human breast cancer cells increased the proteasome-inhibitory potency of EGCG and therefore enhanced its tumor cell growth-inhibitory activity [71]. When breast cancer cells containing high COMT activity were tested, the diminished COMT activity apparently increased the effectiveness of EGCG via augmented proteasome inhibition and apoptosis induction. This study supports the notion that COMT inhibition may increase the anticancer properties of tea polyphenols and the combination may serve as a novel approach or supplemental treatment for breast cancer chemotherapy.

Glutathione S-transferase M1 gene (GSTM1) affects sulforaphane metabolism and the faster it happens the less benefit we get from eating broccoli. Gasper et al. [72] compared sulforaphane metabolism in GSTM1-null and GSTM1-positive subjects after they consumed standard broccoli and high-glucosinolate broccoli (super broccoli). Sixteen subjects were recruited into a randomized, three-phase crossover dietary trial of standard broccoli, super broccoli, and water. GSTM1-null subjects had slightly higher, but statistically significant sulforaphane metabolite concentrations in plasma, a greater rate of urinary excretion of sulforaphane metabolites during the first 6 h after broccoli consumption, and a higher percentage of sulforaphane excretion 24 h after ingestion than did GSTM1-positive subjects. These observations suggest that GSTM1 genotypes have a significant impact on the metabolism of sulforaphane and that the differences in metabolism may explain the greater protection that GSTM1-positive persons gain from consuming broccoli.

Apart from human genetic makeup, the quality and content of human intestinal microbial flora varies considerably which influences phytonutrient metabolism. Depending on the type of intestinal flora, phytonutrients metabolism is known to vary considerably between individuals having similar dietary habits [73]. Some phytonutrients are difficult to access being present in seasonal and expensive food items, some may be present in small quantities in bulky foods, and others pose significant problems with oral bioavailability [74]. Hopefully, with the help of additional research and help from the food industry, such obstacles can be overcome.

When it comes to cancer prevention, the age at which chemopreventive intervention is initiated is critical. There is epidemiological evidence to suggest that higher consumption of soy foods is associated with lower breast cancer risk. However, this is true only for Asian women who start eating soy products starting early in life compared to their western counterparts who start eating later in life [75]. We and others have observed that green tea efficacy depends on the stage of the disease at which it is initiated [53, 55]. Our studies have indicated that the chemopreventive potential of green tea polyphenols decreases with advancing stage of the disease and have highlighted the need to design appropriate chemoprevention clinical trials taking these observations into consideration.

The unsatisfactory outcome of cancer chemoprevention in many human trials is certainly of concern but should not dissuade us from looking into ways of making chemoprevention a success story. Unlike cancer chemoprevention, even minor advances in cancer chemotherapy become big news and modest effects are considered significant. Success from chemoprevention studies is hardly noticed and, in general, expectations from chemoprevention are too high. But there are several other factors that need to be seriously deliberated if chemoprevention is to have a future in controlling human cancer. The lack of interest shown by the pharmaceutical industry towards cancer chemoprevention is indeed of concern. Recent data on breast cancer chemoprevention with exemestane, an aromatase inhibitor, showed a 65% tumor risk reduction compared with placebo [76]. Despite this promising outcome, the pharmaceutical industry has shown a lack of interest in following up on these findings. Many support organizations, such as the National Breast Cancer Coalition, do not consider cancer chemoprevention a viable approach and describe programs and efforts to market chemopreventive agents as irresponsible [77].

Other issues to consider are the lack of awareness of care givers and the public's attitude towards cancer chemoprevention. Primary physicians may not be aware of the opportunities cancer chemoprevention provides for a high risk group resulting in few patients enrolling for chemoprevention trials. The patient population at high risk for developing cancer is uncertain about the outcome of cancer chemopreventive interventions, worrying about side effects and long term use. Recent data suggest that the number of women who use tamoxifen for chemoprevention fell from 120,000 in 2000 to 60,000 in 2005, even though more than 2,000,000 women are eligible for treatment. Lastly, cancer chemoprevention, like other fields of research, is dependent on federal money for support. However, the share devoted to cancer prevention research over the past decade has seen a steady decline, even though overall the National Cancer Institute's budget has increased.

To realize the potential of cancer chemoprevention, several strategies aimed at identifying the right patient population and the right agent need to be investigated simultaneously. Based on cell culture and animal data, many signaling molecules and biomarkers have been identified that could serve as excellent targets for chemoprevention. This knowledge also advocates the use of synthetic agents for cancer chemoprevention and further advocates the use of natural agents, as most of them have been documented to exert multitargeted effects. We need to establish risk factors and gene signatures for risk factors to identify a high risk population that will benefit from chemoprevention. This high risk population would have to be profiled individually to classify responders from nonresponders. Because individuals differ based on their genetic profile, a one size fits all approach seems inappropriate for cancer chemoprevention and therefore smarter prevention trials need to be undertaken. Based on this information, we may be able to custom tailor specific chemopreventive cocktails for effective prevention of many cancers.

A tumor cell uses multiple pathways to survive and therefore agents that interfere with a single pathway are unlikely to succeed. Either multiple agents with distinct targets in combination or single agents with multiple targets need to be developed. Nanotechnology could help overcome issues related to bioavailability and at the same time help to deliver sustained levels of bioactive agents and reduce toxicity. Cancer chemopreventive regimes usually run over a long course and have been observed to produce unexpected and serious side effects. A recently proposed short-term intermittent approach could help offset toxicity issues with agents that have to be taken long-term [78]. Designing and genetically modifying commonly used food items to contain cancer fighting ingredients such as anthocyanin-rich tomatoes is a smart way to move forward in cancer prevention but, before that, additional data in animal models would be needed. In addition the use of probiotics to manipulate and tweak intestinal microbial flora could help escape the unnecessary molecular degradation of bioactive phytonutrients.

6 Conclusions

Chemoprevention could be an important tool for controlling cancer based on evidence generated during the last few decades [79]. While considerable improvements in diagnosis and treatment have improved overall survival, cancer continues to remain a public health concern. Many nontoxic dietary ingredients are showing promise for the management of cancer and it is increasingly appreciated that many such molecules are nature's gifts endowed with the power to prevent cancer in the human population. Based on many studies, and as outlined in this review, there is an urgent need for more in-depth clinical studies to identify categorically and develop agents for the prevention of cancer. Because of a complex nature of the disease, it is also worthwhile to conduct studies with a combination of agents.

Cancer chemoprevention is a viable approach only during the process of carcinogenesis but when cancer is established its prevention is next to impossible. Thus "slowing the process of carcinogenesis" concept appears to be a viable approach for cancer control and appears to be valid for most solid malignancies. Rapid progress is being made in the field in terms of identifying agents that are target specific and at the same time less toxic. Specific recommendations for cancer prevention need to be based on individual genotypes. This will require many more years of rigorous case controlled studies. According to many advocates a promotional campaign about cancer chemoprevention awareness needs to be started on the same scale as statins for cardiovascular health. Primary care physicians and the public need to be educated on the benefits of cancer chemoprevention. Funding agencies and the pharmaceutical industry must realize the benefits of this approach in terms of cost and effort. We need to lower our expectations and settle for moderate effects from cancer chemopreventive trials. If modeled carefully, cancer chemoprevention could be a viable approach for high risk populations.

References

1. Siegel R, Naishadham D, Jemal A (2012) Cancer statistics, 2012. CA Cancer J Clin 62:10–29
2. Sporn MB, Suh N (2000) Chemoprevention of cancer. Carcinogenesis 21:525–530
3. Boone CW, Bacus JW, Bacus JV et al (1997) Properties of intraepithelial neoplasia relevant to the development of cancer chemopreventive agents. J Cell Biochem 28–29:1–20
4. Adhami VM, Ahmad N, Mukhtar H (2003) Molecular targets for green tea in prostate cancer prevention. J Nutr 133:2417S–2424S
5. Siddiqui IA, Adhami VM, Saleem M et al (2006) Beneficial effects of tea and its polyphenols against prostate cancer. Mol Nutr Food Res 50:130–143
6. Syed DN, Khan N, Afaq F, Mukhtar H (2007) Chemoprevention of prostate cancer through dietary agents: progress and promise. Cancer Epidemiol Biomarkers Prev 16:2193–2203
7. Khan N, Afaq F, Mukhtar H (2008) Cancer chemoprevention through dietary antioxidants: progress and promise. Antioxid Redox Signal 10:475–510
8. Klein EA (2005) Chemoprevention of prostate cancer. Crit Rev Oncol Hematol 54:1–10, Comfortable
9. Lippman SM, Hawk ET (2009) Cancer prevention: from 1727 to milestones of the past 100 years. Cancer Res 69:5269–5284
10. Wattenberg LW (1966) Chemoprophylaxis of carcinogenesis: a review. Cancer Res 26:1520–1526
11. Sporn MB, Dunlop NM, Newton DL et al (1976) Prevention of chemical carcinogenesis by vitamin A and its synthetic analogs (retinoids). Fed Proc 35:1332–1338
12. Sporn MB (1976) Approaches to prevention of epithelial cancer during the preneoplastic period. Cancer Res 36:2699–2702
13. Gescher A, Pastorino U, Plummer SM et al (1998) Suppression of tumour development by substances derived from the diet – mechanisms and clinical implications. Br J Clin Pharmacol 45:1–12
14. Wattenberg LW (1985) Chemoprevention of cancer. Cancer Res 45:1–8
15. Ip C, Ip MM (1981) Chemoprevention of mammary tumorigenesis by a combined regimen of selenium and vitamin A. Carcinogenesis 2:915–918
16. Ip C (1981) Prophylaxis of mammary neoplasia by selenium supplementation in the initiation and promotion phases of chemical carcinogenesis. Cancer Res 41:4386–4390
17. Croft WA, Croft MA, Paulus KP et al (1981) 13-cis-Retinoic acid: effect on urinary bladder carcinogenesis by N-[4-(5-nitro-2-furyl)-2-thiazolyl]-formamide in Fischer rats. Cancer Lett 12:355–360
18. Thompson HJ, Chasteen ND, Meeker LD (1984) Dietary vanadyl(IV) sulfate inhibits chemically-induced mammary carcinogenesis. Carcinogenesis 5:849–851
19. McCormick DL, Wilson AM (1986) Combination chemoprevention of rat mammary carcinogenesis by indomethacin and butylated hydroxytoluene. Cancer Res 46:3907–3911
20. Wattenberg LW, Bueding E (1986) Inhibitory effects of 5-(2-pyrazinyl)-4-methyl-1,2-dithiol-3-thione (Oltipraz) on carcinogenesis induced by benzo[a]pyrene, diethylnitrosamine and uracil mustard. Carcinogenesis 7:1379–1381
21. Gunby P (1978) Retinoid chemoprevention trial begins against bladder cancer. JAMA 240:609–610
22. Sim HG, Cheng CW (2005) Changing demography of prostate cancer in Asia. Eur J Cancer 41:834–845
23. Angwafo FF (1998) Migration and prostate cancer: an international perspective. J Natl Med Assoc 90:S720–S723
24. Boyle P, Severi G (1999) Epidemiology of prostate cancer chemoprevention. Eur Urol 35:370–376
25. Adlercreutz H (1990) Western diet and Western diseases: some hormonal and biochemical mechanisms and associations. Scand J Clin Lab Invest 201:3–23

26. Schuurman AG, Goldbohm RA, Brants HA et al (2002) A prospective cohort study on intake of retinol, vitamins C and E, and carotenoids and prostate cancer risk (Netherlands). Cancer Causes Control 13:573–582
27. Katiyar S, Mukhtar H (1996) Tea in chemoprevention of cancer. Int J Oncol 8:221–238
28. Wilding G (1995) Endocrine control of prostate cancer. Cancer Surv 23:43–62
29. Liao S, Hiipakka RA (1995) Selective inhibition of steroid 5 alpha-reductase isozymes by tea epicatechin-3-gallate and epigallocatechin-3-gallate. Biochem Biophys Res Commun 214:833–838
30. Ren F, Zhang S, Mitchell SH et al (2000) Tea polyphenols down-regulate the expression of the androgen receptor in LNCaP prostate cancer cells. Oncogene 19:1924–1932
31. Gupta S, Ahmad N, Mukhtar H (1999) Prostate cancer chemoprevention by green tea. Semin Urol Oncol 17:70–76
32. Gupta S, Ahmad N, Mohan RR et al (1999) Prostate cancer chemoprevention by green tea: in vitro and in vivo inhibition of testosterone-mediated induction of ornithine decarboxylase. Cancer Res 59:2115–2120
33. Siddiqui IA, Asim M, Hafeez BB et al (2011) Green tea polyphenol EGCG blunts androgen receptor function in prostate cancer. FASEB J 25:1198–1207
34. Khan N, Afaq F, Mukhtar H (2006) Apoptosis by dietary factors: the suicide solution for delaying cancer growth. Carcinogenesis 28:233–239
35. Ahmad N, Feyes DK, Nieminen AL et al (1997) Green tea constituent epigallocatechin-3-gallate and induction of apoptosis and cell cycle arrest in human carcinoma cells. J Natl Cancer Inst 89:1881–1886
36. Gupta S, Ahmad N, Nieminen AL et al (2000) Growth inhibition, cell-cycle dysregulation, and induction of apoptosis by green tea constituent (−)-epigallocatechin-3-gallate in androgen-sensitive and androgen-insensitive human prostate carcinoma cells. Toxicol Appl Pharmacol 164:82–90
37. Gupta S, Hussain T, Mukhtar H (2003) Molecular pathway for (−)-epigallocatechin-3-gallate-induced cell cycle arrest and apoptosis of human prostate carcinoma cells. Arch Biochem Biophys 410:177–185
38. Hastak K, Gupta S, Ahmad N et al (2003) Role of p53 and NF-kappaB in epigallocatechin-3-gallate-induced apoptosis of LNCaP cells. Oncogene 22:4851–4859
39. Gupta S, Hastak K, Afaq F et al (2004) Essential role of caspases in epigallocatechin-3-gallate-mediated inhibition of nuclear factor kappa B and induction of apoptosis. Oncogene 23:2507–2522
40. Hastak K, Agarwal MK, Mukhtar H et al (2005) Ablation of either p21 or Bax prevents p53-dependent apoptosis induced by green tea polyphenol epigallocatechin-3-gallate. FASEB J 19:789–791
41. Dou QP, Li B (1999) Proteasome inhibitors as potential novel anticancer agents. Drug Resist Updat 2:215–223
42. Nam S, Smith DM, Dou QP (2001) Ester bond-containing tea polyphenols potently inhibit proteasome activity in vitro and in vivo. J Biol Chem 276:13322–13330
43. Smith DM, Wang Z, Kazi A et al (2002) Synthetic analogs of green tea polyphenols as proteasome inhibitors. Mol Med 8:382–392
44. Wang SI, Mukhtar H (2002) Gene expression profile in human prostate LNCaP cancer cells by epigallocatechin-3-gallate. Cancer Lett 182:43–51
45. Livne E, Fishman DD (1997) Linking protein kinase C to cell-cycle control. Eur J Biochem 248:1–9
46. Fishman DD, Segal S, Livneh E (1998) The role of protein kinase C in G1 and G2/M phases of the cell cycle. Int J Oncol 12:181–186
47. Gupta S, Hastak K, Ahmad N et al (2001) Inhibition of prostate carcinogenesis in TRAMP mice by oral infusion of green tea polyphenols. Proc Natl Acad Sci USA 98:10350–10355
48. Jatoi A, Ellison N, Burch PA et al (2003) A phase II trial of green tea in the treatment of patients with androgen independent metastatic prostate carcinoma. Cancer 97:1442–1446

49. Adhami VM, Siddiqui IA, Ahmad N et al (2004) Oral consumption of green tea polyphenols inhibits insulin-like growth factor-I-induced signaling in an autochthonous mouse model of prostate cancer. Cancer Res 64:8715–8722
50. Caporali A, Davalli P, Astancolle S et al (2004) The chemopreventive action of catechins in the TRAMP mouse model of prostate carcinogenesis is accompanied by clusterin overexpression. Carcinogenesis 25:2217–2224
51. Choan E, Segal R, Jonker D (2005) A prospective clinical trial of green tea for hormone refractory prostate cancer: an evaluation of the complementary/alternative therapy approach. Urol Oncol 23:108–113
52. Bettuzzi S, Brausi M, Rizzi F et al (2006) Chemoprevention of human prostate cancer by oral administration of green tea catechins in volunteers with high-grade prostate intraepithelial neoplasia: a preliminary report from a one-year proof-of-principle study. Cancer Res 66:1234–1240
53. Harper CE, Patel BB, Wang J et al (2007) Epigallocatechin-3-gallate suppresses early stage, but not late stage prostate cancer in TRAMP mice: mechanisms of action. Prostate 67:1576–1589
54. Brausi M, Rizzi F, Bettuzzi S (2008) Chemoprevention of human prostate cancer by green tea catechins: two years later. A follow-up update. Eur Urol 54:472–473
55. Adhami VM, Siddiqui IA, Sarfaraz S et al (2009) Effective prostate cancer chemopreventive intervention with green tea polyphenols in the TRAMP model depends on the stage of the disease. Clin Cancer Res 15:1947–1953
56. McLarty J, Bigelow RL, Smith M et al (2009) Tea polyphenols decrease serum levels of prostate-specific antigen, hepatocyte growth factor, and vascular endothelial growth factor in prostate cancer patients and inhibit production of hepatocyte growth factor and vascular endothelial growth factor in vitro. Cancer Prev Res (Phila) 2:673–682
57. Nguyen MM, Ahmann FR, Nagle RB et al (2012) Randomized, double-blind, placebo-controlled trial of polyphenon E in prostate cancer patients before prostatectomy: evaluation of potential chemopreventive activities. Cancer Prev Res (Phila) 5:290–298
58. Liao S, Umekita Y, Guo J et al (1995) Growth inhibition and regression of human prostate and breast tumors in athymic mice by tea epigallocatechin gallate. Cancer Lett 96:239–243
59. Roomi MW, Ivanov V, Kalinovsky T et al (2005) In vivo antitumor effect of ascorbic acid, lysine, proline and green tea extract on human prostate cancer PC-3 xenografts in nude mice: evaluation of tumor growth and immunohistochemistry. In Vivo 19:179–183
60. Siddiqui IA, Zaman N, Aziz MH et al (2006) Inhibition of CWR22Rv1 tumor growth and PSA secretion in athymic nude mice by green and black teas. Carcinogenesis 27:833–839
61. Gingrich JR, Barrios RJ, Kattan MW et al (1997) Androgen-independent prostate cancer progression in the TRAMP model. Cancer Res 57:4687–4691
62. Kurahashi N, Sasazuki S, Iwasaki M et al (2008) Green tea consumption and prostate cancer risk in Japanese men: a prospective study. Am J Epidemiol 167:71–77
63. Jian L, Lee AH, Binns CW (2007) Tea and lycopene protect against prostate cancer. Asia Pac J Clin Nutr 16:453–457
64. Jian L, Du CJ, Lee AH et al (2005) Do dietary lycopene and other carotenoids protect against prostate cancer? Int J Cancer 113:1010–1014
65. Lippman SM, Klein EA, Goodman PJ et al (2009) Effect of selenium and vitamin E on risk of prostate cancer and other cancers: the selenium and vitamin E cancer prevention trial (SELECT). JAMA 301:39–51
66. Baron JA, Sandler RS, Bresalier RS et al (2006) A randomized trial of rofecoxib for the chemoprevention of colorectal adenomas. Gastroenterology 131:1674–1682
67. Fisher B, Costantino JP, Wickerham DL et al (1998) Tamoxifen for prevention of breast cancer: report of the National Surgical Adjuvant Breast and Bowel Project P-1 Study. J Natl Cancer Inst 90:1371–1388

68. Kaplan SA, Roehrborn CG, Meehan AG et al (2009) Evidence that finasteride reduces risk of most frequently detected intermediate- and high-grade (Gleason score 6 and 7) cancer. Urology 73:935–939
69. Elmets CA, Viner JL, Pentland AP et al (2010) Chemoprevention of nonmelanoma skin cancer with celecoxib: a randomized, double-blind, placebo-controlled trial. J Natl Cancer Inst 102:1835–1844
70. Wu AH, Tseng CC, Van Den Berg D et al (2003) Tea intake, COMT genotype, and breast cancer in Asian-American women. Cancer Res 63:7526–7529
71. Landis-Piwowar K, Chen D, Chan TH et al (2010) Inhibition of catechol-Omicron-methyltransferase activity in human breast cancer cells enhances the biological effect of the green tea polyphenol (−)-EGCG. Oncol Rep 24(2):563–569
72. Gasper AV, Al-Janobi A, Smith JA et al (2005) Glutathione S-transferase M1 polymorphism and metabolism of sulforaphane from standard and high-glucosinolate broccoli. Am J Clin Nutr 82:1283–1291
73. Lampe JW (2009) Is equol the key to the efficacy of soy foods? Am J Clin Nutr 89:1664S–16647S
74. DeWeerdt S (2011) Food: the omnivore's labyrinth. Nature 471:S22–S24
75. Maskarinec G, Noh JJ (2004) The effect of migration on cancer incidence among Japanese in Hawaii. Ethn Dis 14:431–439
76. Goss PE, Ingle JN, Alés-Martínez JE et al (2011) Exemestane for breast-cancer prevention in postmenopausal women. N Engl J Med 364:2381–2391
77. Schmidt C (2011) The breast cancer chemoprevention debate. J Natl Cancer Inst 103:1646–1647
78. Wu X, Lippman SM (2011) An intermittent approach for cancer chemoprevention. Nat Rev Cancer 11:879–885
79. Wu X, Patterson S, Hawk E (2011) Chemoprevention – history and general principles. Best Pract Res Clin Gastroenterol 25:445–459

Personalizing Lung Cancer Prevention Through a Reverse Migration Strategy

Kathryn A. Gold, Edward S. Kim, Ignacio I. Wistuba, and Waun K. Hong

Abstract Lung cancer is the deadliest cancer in the United States and worldwide. Tobacco use is the one of the primary causes of lung cancer and smoking cessation is an important step towards prevention, but patients who have quit smoking remain at risk for lung cancer. Finding pharmacologic agents to prevent lung cancer could potentially save many lives. Unfortunately, despite extensive research, there are no known effective chemoprevention agents for lung cancer. Clinical trials in the past, using agents without a clear target in an unselected population, have shown pharmacologic interventions to be ineffective or even harmful. We propose a new approach to drug development in the chemoprevention setting: reverse migration, that is, drawing on our experience in the treatment of advanced cancer to bring agents, biomarkers, and study designs into the prevention setting. By identifying molecular drivers of lung neoplasia and using matched targeted agents, we hope to personalize therapy to each individual to develop more effective, tolerable chemoprevention. Also, advances in risk modeling, using not only clinical characteristics but also biomarkers, may help us to select patients better for chemoprevention

K.A. Gold (✉) and E.S. Kim
Department of Thoracic/Head and Neck Medical Oncology, The University of Texas M. D. Anderson Cancer Center, 1515 Holcombe Blvd, Houston, TX 77030, USA
e-mail: KAGold@mdanderson.org

I.I. Wistuba
Department of Thoracic/Head and Neck Medical Oncology, The University of Texas M. D. Anderson Cancer Center, 1515 Holcombe Blvd, Houston, TX 77030, USA

Department of Pathology, The University of Texas M. D. Anderson Cancer Center, 1515 Holcombe Blvd, Houston, TX 77030, USA

W.K. Hong
Department of Thoracic/Head and Neck Medical Oncology, The University of Texas M. D. Anderson Cancer Center, 1515 Holcombe Blvd, Houston, TX 77030, USA

Division of Cancer Medicine, The University of Texas M. D. Anderson Cancer Center, 1515 Holcombe Blvd, Houston, TX 77030, USA

efforts, thus sparing patients at low risk for cancer the potential toxicities of treatment. Our institution has experience with biomarker-driven clinical trials, as in the recently reported *B*iomarker-integrated *A*pproaches of *T*argeted *T*herapy for *L*ung Cancer *E*limination (BATTLE) trial, and we now propose to bring this trial design into the prevention setting.

Keywords Chemoprevention · Lung cancer · Targeted therapies

Contents

1 Introduction	222
2 Basics of Chemoprevention	223
2.1 History of Lung Cancer Chemoprevention	225
3 Molecular Biology of Lung Cancer	226
3.1 KRAS	227
3.2 EGFR	227
3.3 EML4-ALK	227
3.4 Other Molecular Changes	228
4 Personalizing Treatment of Lung Cancer	229
5 Personalizing Prevention of Lung Cancer	229
5.1 Personalizing Tertiary Chemoprevention	231
5.2 Personalizing Primary and Secondary Chemoprevention	232
6 Conclusions	234
References	235

1 Introduction

Lung cancer is the most common cause of cancer death, both in the United States and worldwide [1]. Once lung cancer is diagnosed, outcomes are poor, with only 15.6% of patients surviving 5 years after diagnosis (http://seer.cancer.gov/statfacts/html/lungb.html). Lung cancer prevention is an attractive goal. Smoking cessation is an important step towards this goal, but the risk of lung cancer remains elevated even after a patient has quit smoking. About half of all lung cancers are diagnosed in patients who have already quit smoking; therefore, tobacco cessation alone is not sufficient. Lung cancer chemoprevention is a promising field, though one that has met with only limited success.

Chemoprevention refers to the use of any agent, either synthetic, biologic, or natural, to suppress, reverse, or prevent carcinogenesis [2, 3]. Unfortunately, clinical trials of chemopreventive agents for lung cancer have been largely negative or even harmful [4–8]. To improve outcomes, we must change our approach to drug development. Historically, there have been two main approaches to development of therapeutics for chemoprevention. First is the development of agents, usually natural agents, that were identified in epidemiologic studies as potentially important in the development of cancer. Examples include beta-carotene [4, 5] and selenium [9]. The second approach has been to study agents developed for different indications in the setting of chemoprevention. Examples of this approach include the cyclo-oxygenase-2 (COX-2) inhibitors, initially developed for arthritis and later studied for the prevention of colon cancer [10–12]. We propose a new approach: reverse migration, that is, importing ideas and therapies developed in advanced cancer into the setting of chemoprevention [13].

2 Basics of Chemoprevention

Several concepts that are important in chemoprevention are the ideas of field cancerization and multistep carcinogenesis. "Field cancerization" was first described in 1953 with the study of histologically abnormal tissue surrounding oral cancer [14]. Slaughter et al. hypothesized that an injury from a toxin occurs at multiple locations within a field such as the aerodigestive tract, and carcinogenesis may occur at multiple sites. This field effect is responsible for the high rates of recurrence of squamous cell carcinoma of the oral cavity following local treatment. Though the initial observations of field cancerization focused on histologic changes, we now know that molecular changes can be found in histologically normal epithelium adjacent to tumors [15–17]. Thus, a premalignant lesion in one area of the lung implies increased risk of cancer throughout the lungs.

Multistep carcinogenesis was first described in 1938 [18]. Serial changes in the lungs of smokers were described on a histologic level by Auerbach et al., with a progression from hyperplasia to metaplasia to dysplasia to carcinoma in situ to invasive cancer [19, 20]. The earliest events in carcinogenesis are at the genomic level – additional events are necessary to induce phenotypical changes in the tissue.

Since carcinogenesis occurs in multiple steps and in multiple locations, we have opportunities to "detour carcinogenesis," that is, to take steps to prevent the progression to invasive cancer in patients at risk for malignancy [21]. By understanding the molecular progression leading to cancer, we hope to identify mutations that drive cancer. Data suggests that in a single patient, primary and metastatic tumors are very similar genetically [22, 23]; it is not unreasonable to think that premalignancy will also share characteristics with more advanced tumors. Agents that are effective against these drivers in the metastatic setting may also be useful in prevention, as part of a reverse migration approach.

Prevention efforts can target different groups of patients. With primary prevention, the focus is on healthy individuals who are at high risk; for example, current and former smokers. The goal of secondary prevention is to prevent progression to cancer in patients with premalignant lesions, such as intraepithelial neoplasia. Tertiary prevention aims to prevent the development of recurrent or second primary tumors in patients who have a history of cancer.

There have been notable successes in chemoprevention, as described in Table 1 [3, 10, 24–33]. Successful trials often involve known molecular targets that can be effectively inhibited by drugs; for example, hormone receptors in breast and

Table 1 Successes in chemoprevention

	Intervention	Population	Target	Endpoint	Outcome
BCPT [25]	Tamoxifen	Women >60	Estrogen receptor	Invasive breast cancer	49% decrease in invasive breast cancer
PCPT [27]	Finasteride	Men ≥55	Testosterone production	Prostate cancer	25% reduction in prostate cancer
FUTURE II [28]	Vaccine	Women >15, <26	HPV	Premalignant cervical lesions	98% decrease in HPV 16/18 associated lesions
Baron et al. [32]	Aspirin	Patients with adenomas	Inflammation	Adenomas	19% decrease in adenomas

BCPT Breast cancer prevention trial, HPV Human papilloma virus, PCPT Prostate cancer prevention trial

prostate cancer [25, 27], and inflammation in colon cancer [30]. Alternatively, other successful trials have used vaccines to target viruses known to be involved in carcinogenesis, such as the human papilloma virus in cervical cancer [28] and hepatitis B in hepatocellular carcinoma [29]. Many negative trials have used agents identified from epidemiologic studies, without clear targets, such as beta-carotene [4, 5] or selenium [9].

2.1 History of Lung Cancer Chemoprevention

There have been extensive efforts in chemoprevention of nonsmall cell lung cancer (NSCLC), with many clinical trials using agents selected based on epidemiologic data. An influential 1981 review discussed data correlating intake of beta-carotene, which is partially converted to vitamin A in the body, with lower risk of cancer [34]. Based on this data, a number of trials tested vitamin supplementation as a cancer prevention strategy.

The alpha-tocopherol and beta-carotene (ATBC) trial randomized male smokers from Finland to either vitamin E, beta-carotene, a combination of both, or placebo [4]. Unexpectedly, beta-carotene supplementation was associated with an 18% increase in the risk of lung cancer, as well as a significant increase in overall mortality. Vitamin E had no significant effect on incidence of cancer or mortality. Patients in the study who consumed higher dietary amounts of beta-carotene and vitamin E, however, had lower risks of cancer, suggesting the supplementation may have different effects than dietary intake. Another large trial, the beta-carotene and retinol efficacy trial (CARET), was stopped after an interim analysis following the publication of the ATBC trial. This trial randomized patients to receive a combination of retinol and beta-carotene vs placebo. The patients in the active treatment group had a higher risk of lung cancer and all-cause mortality [5]. Interestingly, the increase in lung cancer incidence was seen in current smokers; in former smokers, a trend towards decreased lung cancer incidence with supplementation was noted.

Patients with surgically treated lung cancer are at a high risk for second primary tumors, with rates reported as high as 3% per year [35, 36]. This group has been extensively studied to determine if any chemopreventive agent can reduce risk. For example, in the Intergroup Lung Trial, patients with resected early stage NSCLC were randomized to receive either isotretinoin, a synthetic vitamin A derivative, or placebo [7]. There were no significant differences between the arms with respect to second primary tumors, recurrence, or survival, though subgroup analysis revealed increased mortality in the treatment arm for active smokers, and a trend towards benefit in never smokers. Other studies in this setting, using selenium [6] or a combination of retinol and N-acetylcysteine [8], have also been negative.

Table 2 Major phase III studies in lung cancer chemoprevention

	Intervention	Population	N	Endpoint	Outcome
ATBC [4]	β-Carotene, α-tocopherol	Male smokers	29,133	Lung cancer	Harmful
CARET [5]	β-Carotene, retinol	Current and former smokers	18,314	Lung cancer	Harmful
Lung Intergroup Trial [7]	Isotretinoin	Resected NSCLC	1,166	Second primary tumor (SPT)	Negative[a]
EUROSCAN [8]	Retinol, NAC	Resected NSCLC or HNSCC	2,592	SPT	Negative
Intergroup Selenium Study [6]	Selenium	Resected NSCLC	1,772	SPT	Negative

ATBC Alpha-tocopherol, beta-carotene cancer prevention study, *CARET* Carotene and retinol efficacy trial, *NAC* N-acetylcysteine, *NSCLC* non-small cell lung cancer, *HNSCC* head and neck squamous cell carcinoma
[a]Harm in current smokers

Currently, there are no proven chemoprevention agents for lung cancer (Table 2). To improve our chances of developing effective chemoprevention, we need a better understanding of the biology of lung cancer.

3 Molecular Biology of Lung Cancer

There are many different types of lung cancer, and different molecular pathways leading to each type. About 20% of lung cancer is small cell lung cancer, and the rest is NSCLC. In the United States, the most common subtype of NSCLC is adenocarcinoma, and squamous cell carcinoma is the second most common histology [37].

Adenocarcinomas classically originate in the peripheral airways. Though most patients diagnosed with adenocarcinoma have a history of cigarette smoking, this is the most common type of lung cancer in nonsmokers. The proportion of adenocarcinomas has been increasing over the past few decades – the reasons for this increase are unknown, but might include changing smoking habits or the increased use of filtered cigarettes. Atypical adenomatous hyperplasia is a precursor lesion for a subset of adenocarcinomas [38]; however, the precursor lesions of adenocarcinoma have been less extensively studied than those of squamous cell carcinoma.

Squamous cell carcinomas usually arise in the proximal airways. There is a strong association between squamous cell carcinoma and smoking. The precursor lesions to squamous cell carcinoma have been well described, and include squamous metaplasia and dysplasia [20, 39].

We now understand the mutations that drive certain subsets of these tumors.

3.1 KRAS

KRAS is a GTPase which is an early component of multiple cell signaling pathways. Mutations in *KRAS* are found in 20–30% of adenocarcinomas, and are very rarely found in squamous cell carcinoma [40, 41]. Mutations are often seen in atypical adenomatous hyperplasia, a precursor to adenocarcinoma [38, 41]. These mutations are more common in current and former smokers than nonsmokers [42] and they are associated with resistance to epidermal growth factor receptor (EGFR) inhibition [43]. There are no effective targeted therapies currently in clinical use for these patients. However, in a mouse model, combination therapy with a PI3K inhibitor and a MEK inhibitor is active against *KRAS*-mutant tumors [40], and agents in these families are in the early stages of clinical testing.

3.2 EGFR

EGFR is frequently involved in carcinogenesis and is an important regulator of growth in human cells. Activating mutations of *EGFR* have been described in patients with adenocarcinoma, and are present in 10% of adenocarcinomas in the U.S. and in higher numbers of Asian patients [44]. These mutations, specifically those in exons 19 and 21 activating the kinase domain of the enzyme, are associated with responsiveness to EGFR inhibition [44, 45]. These activating mutations are very rare in squamous cell carcinoma, but EGFR amplification is commonly seen [41]. About 5% of squamous cell carcinomas have mutations in the extracellular domain of EGFR, also known as the variant III *EGFR* mutation; these mutations do not confer sensitivity to EGFR inhibition and may confer resistance [46].

3.3 EML4-ALK

The *EML4-ALK* translocation was the first chromosomal translocation described in lung cancer. A rearrangement on chromosome 2 creates a constitutively activated anaplastic lymphoma kinase (ALK) that drives growth [47]. This translocation is found in about 3–7% of adenocarcinomas and does not occur in patients with either *KRAS* or *EGFR* mutations [47–49]. These patients are sensitive to treatment with crizotinib, an ALK/c-Met inhibitor [50]. This drug was recently FDA approved, along with a companion diagnostic test.

3.4 Other Molecular Changes

KRAS and *EGFR* mutations and *ALK* rearrangements are the most clinically important genetic abnormalities seen in NSCLC, but other changes are also seen. *BRAF* mutations are found in a small percentage of lung adenocarcinomas, 2% in one study [51]. BRAF inhibitors have been successful in the treatment of metastatic melanoma [52]; drugs in this class are also being investigated for NSCLC. The phosphatidylinostil 3′-kinase (PI3K) pathway includes Akt and mTOR, and tumor cells have increased activation of this pathway relative to normal cells [53]. The gene *PIK3CA*, which encodes the catalytic unit of PI3K, is mutated in approximately 5% of NSCLC [54]. There are a number of inhibitors of this pathway in clinical development, including mTor inhibitors, Akt inhibitors, and PI3K inhibitors.

Mutations in *DDR2* have been identified as driver mutations in about 4% of squamous cell carcinomas [55]. This gene encodes for a kinase with roles in cell adhesion and proliferation [56]. Patients with this mutation may be more sensitive to treatment with dasatinib [55]. Also, amplifications of *FGFR1*, encoding for a fibroblast growth factor receptor, have been described in over 20% of squamous cell carcinomas [57].

Aberrant angiogenesis is one of the hallmarks of cancer [58], and vascular endothelial growth factor (VEGF) is a regulator of angiogenesis in both normal tissue and in malignancy [59]. VEGF and VEGF receptor are aberrantly expressed in lung cancer, and this expression may be associated with poor prognosis [60, 61]. VEGF-A is expressed more commonly in adenocarcinoma than squamous cell carcinoma [62]. Bevacizumab, a monoclonal antibody against VEGF receptor, has been shown to be effective in the first-line treatment of adenocarcinoma in combination with chemotherapy [59]. It is not used in squamous carcinomas due to an increased risk of serious bleeding [63].

Most lung cancers have alterations in pathways responsible for DNA repair. p53, a critical regulator of the cell cycle, apoptosis, and DNA repair, is an important tumor suppressor, and it is thought that more than half of all human cancers have mutations in *TP53* [64, 65]. *TP53* is mutated in the majority of NSCLCs, both adenocarcinoma and squamous cell carcinoma, and mutations are more common in smokers [41]. Though multiple attempts have been made to target p53 pharmacologically, there are no proven therapies targeting this important protein.

Our understanding of the molecular biology of lung cancer continues to evolve. The Tumour Sequencing Project examined 623 genes for mutations in 188 adenocarcinomas and identified a number of mutated genes not previously known to be associated with lung cancer [66]. Tumors with higher grade had more mutations than lower grade tumors, and smokers had more mutations than nonsmokers. Another study determined copy number alterations in lung adenocarcinoma. Many of the sites indentified as consistently gained or lost in lung cancer have not been linked to a specific gene, suggesting that we have yet to discover many of the genes involved in lung carcinogenesis [67].

4 Personalizing Treatment of Lung Cancer

The standard treatment for advanced lung cancer is platinum-based combination therapy, but response rates are low and long term survival is rare [68, 69]. It seems unlikely that we will be able to improve outcomes substantially with a one-size-fits-all approach; we must learn to personalize therapy.

For patients with *EGFR* mutations and *ALK* rearrangements, targeted therapies are the standard front-line treatment [45, 50]. These treatment regimens are well tolerated and are associated with high response rates and extended time to progression, though they are not curative. Unfortunately, less than 15% of patients with adenocarcinoma have one of these genetic alterations; for the remainder of our patients, we do not yet have personalized therapy. A number of recent studies have used our improving understanding of the molecular biology of lung cancer to try to personalize therapy.

At M.D. Anderson we are working towards personalized lung cancer therapy with our biomarker-integrated approaches of targeted therapy for lung cancer elimination (BATTLE) program. In our first BATTLE trial [70], patients with advanced lung cancer had a CT-guided core biopsy, and this tissue was analyzed to create a biomarker profile. This profile helped to determine which of four targeted therapies a patient would receive. The hypothesis is that individual tumors are driven by a dominant signaling pathway, and by identification and targeting of that pathway we may be able to improve outcomes. Bayesian adaptive randomization was used, increasing the chances that an individual patient would receive a therapy from which he is predicted to derive benefit.

This trial demonstrated the feasibility of a biopsy-mandated approach in advanced lung cancer. Preliminary findings, such as a relatively high disease control rate with sorafenib in patients with *KRAS* mutations, have provided hypotheses for further studies.

The BATTLE program is expanding, and accrual is ongoing for two more studies. Both follow similar designs, with biopsies mandated on enrollment. The BATTLE-2 study enrolls patients with previously treated lung cancer to receive one of four targeted therapies or combinations; the BATTLE-Front Line trial enrolls patients with previously untreated lung cancer to receive combinations of chemotherapy and biologic therapy.

5 Personalizing Prevention of Lung Cancer

In treating lung cancer, it is not likely that any single agent will be effective in every patient. The same is true in lung cancer prevention. We propose reverse migration as a method to personalize chemoprevention. Reverse migration is the application to the prevention setting of concepts and ideas that have been developed in advanced cancer. Concepts like risk assessment, biomarker analyses, targeted

Table 3 The reverse migration of tamoxifen [13]

	Treatment setting	Results
Tamoxifen vs DES, 1981 [71]	Metastatic disease[a]	Response rate 33% with tamoxifen
EBCTCG meta-analysis, 1998 [72]	Adjuvant treatment/tertiary prevention	47% decrease in breast cancer recurrence
B-24, 1999 [73]	DCIS/secondary prevention	43% decrease in invasive breast cancer
BCPT, 1998 [25]	Healthy women/primary prevention	49% decrease in invasive breast cancer

BCPT Breast cancer prevention trial, *DES* diethylstilbestrol, *DCIS* ductal carcinoma in situ, *EBCTCG* Early Breast Cancer Trialists' Collaborative Group
[a]Hormone receptor status was not measured prior to enrollment on trial

therapeutics, surrogate endpoints, and predictive markers have been more thoroughly explored in the treatment of cancer, but all are potentially important for cancer prevention as well.

An example of reverse migration is the development of tamoxifen, as described in Table 3. Breast cancer has long been known to be hormonally driven in some patients. Tamoxifen is a selective estrogen receptor modulator, and has been used in the treatment of metastatic breast cancer for over 30 years [71]. It has also been shown to be effective in reducing the risk of recurrent cancer following surgical resection of a breast tumor [72], that is, in the tertiary prevention setting. As secondary prevention, tamoxifen decreases the risk of invasive breast cancer in patients with ductal carcinoma in situ, a premalignant lesion [73]. Tamoxifen is also effective as primary prevention, decreasing the risk of breast cancer in healthy postmenopausal women [25].

Though tamoxifen is the most thoroughly studied example of reverse migration, other examples of this strategy are emerging. In multiple myeloma, lenalidomide, an immunomodulating agent, is an effective treatment [74] and is also being studied for use in smoldering myeloma, the precursor stage to this malignancy. For patients with metastatic basal cell carcinoma, a hedgehog inhibitor, GDC-0449, can result in impressive responses [75]; for patients with Gorlin's syndrome, who are genetically predisposed to basal cell carcinoma, this same hedgehog inhibitor can suppress development of cancer [76]. As in breast cancer, antihormonal agents are effective for prostate cancer in the settings of advanced malignancy [77], localized disease [78], and chemoprevention [27]. In breast cancer, PARP inhibitors, which interfere with DNA repair and induce synthetic lethality, are associated with tumor response in patients with metastatic breast cancer patients and germline BRCA mutations [79]. Women with BRCA1 or BRCA2 mutations are at very high lifetime risk of breast cancer, up to 85% depending on the population studied, and PARP inhibitors may be useful in the chemoprevention setting for these women.

It is now time to begin using a reverse migration approach for the prevention of lung cancer. We are learning more about the biology of lung cancer every day, and genetic analysis of individual tumors is becoming less expensive, more accurate, and quicker [80]. Our targeted treatments for metastatic cancer are more tolerable than traditional cytotoxic chemotherapy; patients can be treated with targeted therapeutics like erlotinib for extended periods of time. Also, we now have experience with the type of biopsy-mandated, biomarker-driven clinical trials that will be necessary to make personalized chemoprevention a reality.

5.1 Personalizing Tertiary Chemoprevention

Tertiary chemoprevention is an obvious setting to develop chemopreventive agents using a reverse migration strategy. Patients with resected early stage lung cancer are at a high risk of recurrence and second primary tumors [35, 36]. In this group, an important concept is the identification of molecular targets, not only in the tumor but also in the surrounding tissue.

At MD Anderson we have an extensive program to identify risk factors for tumor recurrence following resection of early stage lung cancers, with a long term goal of developing new therapeutic approaches to adjuvant treatment and prevention. In one arm of these project, 49 patients with resected early stage NSCLC were enrolled to a prospective clinical trial in which they underwent serial bronchoscopies with biopsies yearly for 3 years [81]. Primary endpoints were recurrence and second primary tumors. Analysis is ongoing to determine which markers in the bronchial epithelium correlate with recurrence, but preliminary results show that activation of the PI3 kinase pathway puts patients at a higher risk of recurrence. This pathway may be a therapeutic target, and inhibitors of this pathway are currently in clinical development.

Another part of this project is a retrospective analysis of 370 resected early stage lung tumors. The expression of 23 prespecified biomarkers, selected from preclinical work as being important in carcinogenesis, were measured and correlated with outcomes, including recurrence free survival and overall survival [82]. Using these markers, a risk model was created, where patients could be classified as low, intermediate, or high risk based on expression of these markers.

We propose an idea for a BATTLE-type study in the adjuvant/tertiary prevention setting [13]; see Fig. 1 for a schema. In this study, patients undergoing surgical resection for early stage lung cancer would have biomarker analyses performed on both the tumors and the adjacent epithelium. Based on the molecular abnormalities found, patients would be assigned to a targeted treatment. For example, in patients with resected tumors bearing EGFR mutations, adjacent, histologically normal bronchial epithelium frequently harbors mutations as well [83, 84]. Therefore, these patients would receive EGFR inhibitors, which are effective in the setting of advanced disease [45]. The current data for EGFR inhibitors in the adjuvant setting is mixed [85, 86]; a phase III trial (RADIANT) is ongoing which should address this issue. Patients with *ALK* rearrangements would receive crizotinib. Patients with overexpression of cyclin D1 could receive a combination of erlotinib and bexarotene; overexpression predicts response to this combination in the metastatic setting [87]. Patients with high levels of inflammatory markers, such as COX-2,

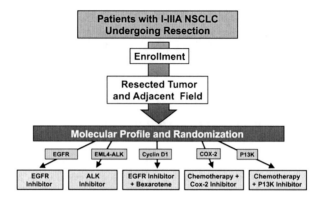

Fig. 1 Schema for a personalized trial of tertiary chemoprevention/adjuvant treatment. The tumor and adjacent field will undergo biomarker analysis, and patients will be grouped based on marker status. Treatment will be assigned based on biomarkers [13]

could receive anti-inflammatory medications, and patients with alterations of the PI3K pathway could receive PI3K/Akt inhibitors. Primary endpoints would be recurrences and second primary tumors. Secondary endpoints would be tolerability, biomarker modulation, and correlation of biomarker modulation with outcome.

5.2 Personalizing Primary and Secondary Chemoprevention

To personalize chemoprevention successfully for patients with no history of cancer, we must be able to identify those at highest risk, perform screening studies appropriately, and treat those at high risk for cancer with targeted chemoprevention agents.

Risk assessment models have long incorporated demographic data, such as age, smoking history, and family history of cancer. Now, we are learning to incorporate biomarkers into these models. Biomarker assessment can be noninvasive, such as blood draws or buccal swabs, or invasive, like bronchoscopic biopsies. Recent epidemiologic studies have identified single nucleotide polymorphisms (SNPs) that can be used to create a risk prediction model that is more accurate than one incorporating only clinical characteristics [88]. Other studies have incorporated bronchoscopic biopsies into risk assessment models. A study by Gustafson et al. found that activation of the PI3 kinase pathway correlated with dysplasia [89]. Treating with an inhibitor of PI3 kinase, *myo*-inositol, effectively down-regulates this pathway and reverses dysplasia in some patients. This pathway, as discussed earlier, was also identified in an MD Anderson study as predicting increased risk of tumor recurrence following resection [81].

Screening for lung cancer is another opportunity to reduce mortality in patients at risk for lung cancer. Until recently there were no proven screening tests for lung cancer. Recently published results from the National Lung Screening Trial, however, show that yearly low-dose CT scans can decrease lung cancer mortality and all-cause mortality in a group of patients at high risk for lung cancer [90]. Enrolled participants were between 55 and 74 years of age and had at least a 30 pack-year history of cigarette smoking. Lung cancer mortality was decreased by 20% with CT scans compared to chest X-rays. Though these results are impressive, CT screening for lung cancer has not yet been universally accepted, possibly due to concerns regarding costs. Improved risk assessment could improve the cost–benefit ratio by avoiding scans in patients at lower risk for cancer.

Recent chemoprevention studies have used targeted agents and incorporated biomarker based endpoints. In two recently reported studies, patients were treated with celecoxib, a COX-2 inhibitor [91, 92]. COX-2 is a modular of inflammation, and its overexpression is seen in premalignant lung lesions [93] and predicts a worse outcome in surgically resected early stage NSCLC [94, 95]. Both studies revealed a decrease in Ki-67, a marker of proliferation, in bronchial epithelium with celecoxib treatment [91, 92]. In addition, one study identified a biomarker predicting benefit from celecoxib [92]. A study reported by Keith et al. found that oral iloprost, a prostacyclin analog, reduced bronchial dysplasia in former smokers, though current smokers did not benefit [96]. A recent meta-analysis of randomized trials studying aspirin for cardiovascular disease prevention provided further evidence of the benefit of anti-inflammatory treatment – aspirin treatment was associated with a decrease in mortality from a number of cancers, including lung adenocarcinoma [97].

These studies are notably different from some of the older studies in lung cancer prevention. Studies like CARET and ATBC [4, 5] took agents without clear targets identified in epidemiologic studies directly into large clinical trials, enrolling thousands of patients, with endpoints related to cancer incidence. These studies were expensive and unsuccessful. Studies like those described above have utilized targeted agents and have incorporated correlative analyses. The number of patients enrolled is relatively small, and surrogate endpoints are used. With clinical trials like these, we can learn more about how these drugs work, and for whom they work, before bringing them to large, definitive trials.

In addition to any pharmacologic prevention treatments, smoking cessation remains critically important. Several clinical trials have shown that if patients continue smoking, chemoprevention can be ineffective and even harmful. In the CARET study, current smokers were harmed by treatment with beta-carotene and retinol, while former smokers had a trend towards lower lung cancer incidence with treatment [5]. In the Intergroup Lung Trial, current smokers had an increased risk of death with isotretinoin treatment, while former and never smokers had a trend towards benefit [7]. More recently, Keith et al. noted an improvement in endobronchial dysplasia in former smokers treated with iloprost, while current smokers had no histologic improvement [96]. The reasons for the lack of efficacy of chemoprevention in patients who continue tobacco use is likely due to interactions between tobacco carcinogens and chemopreventive agents. Beta-carotene may lead to induction of certain cytochrome P450 enzymes, causing increased bioactivation of tobacco-associated procarcinogens [98]; retinoids might have similar effects. Increased oxidative stress from supplementation is another hypothesized mechanism [99]. Future chemoprevention studies should focus on those patients who have already quit smoking – current smokers should be referred to intensive tobacco cessation programs.

6 Conclusions

There are significant barriers to the reverse migration approach, but also rewards for overcoming them. Though we frequently describe signaling pathways as simple, step-wise progressions, in cells they are often complex and there is crosstalk between pathways. Therefore, inhibiting a particular pathway may require multiple pharmacologic agents and may have unintended consequences. In addition, it has frequently been difficult to match biomarkers and targeted agents – we often cannot predict which patients will respond to therapy. Risk prediction is still difficult; models incorporating SNPs are only marginally more accurate than those incorporating clinical risk factors alone [88].

There are also practical barriers. Clinical trials are expensive, and the costs of clinical trials incorporating biopsies and biomarkers are even higher. There are regulatory issues involved when using biomarkers in clinical trials. Also, there is the issue of infrastructure. These types of trials require cooperation between many groups, including pathologists, statisticians, radiologists, pulmonologists, and oncologists, among others. There are issues regarding patient enrollment. Many healthy patients are not willing to enroll on clinical trials requiring biopsies, though we have been able to complete enrollment successfully on these types of trials in the past.

If we are able to overcome these challenges, there are significant rewards. Effective chemoprevention of lung cancer could potentially save many lives, and reverse migration may represent a more efficient and effective pathway to that goal. Our goal is to bring personalized therapy to every patient, from those at risk for cancer to those with metastatic disease (Fig. 2). For patients at lower risk for cancer, counseling on lifestyle changes could be offered; for those at higher risk, screening programs and chemoprevention studies should be considered.

Lung cancer is a molecularly heterogeneous disease, and should no longer be treated with a "one-size-fits-all" approach. In our BATTLE program for cancer treatment, we attempt to identify the molecular drivers of a patient's tumor so that we can hijack these drivers with molecularly matched agents. This approach is applicable not only to the treatment of advanced malignancy, but also to the prevention of cancer in patients at risk. Using what we have learned regarding the biology and treatment of lung cancer, we are now ready to take the BATTLE to the chemoprevention setting.

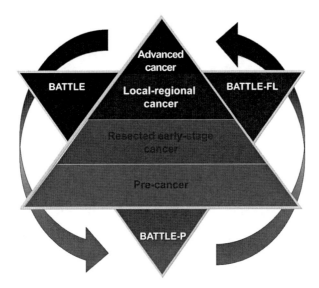

Fig. 2 Schema for a comprehensive BATTLE strategy against all stages of carcinogenesis, from pre-malignancy to advanced cancer. BATTLE (*upper left*) represents our trial program in advanced, previously treated lung cancer. BATTLE–frontline (BATTLE–FL, *upper right*) is our currently accruing trial in previously untreated advanced stage NSCLC. BATTLE–prevention (BATTLE–P, *bottom*) is our developing program in adjuvant treatment and prevention [13]

Grant Support This work was supported by National Cancer Institute grant P01-CA091844 (to W.K. Hong), Department of Defense grant W81XWH-06-1-030302 (to W.K. Hong), and National Foundation for Cancer Research (NFCR) grant LF01-065-1 (to W.K. Hong).

References

1. Siegel R, Ward E, Brawley O, Jemal A (2011) Cancer statistics, 2011: the impact of eliminating socioeconomic and racial disparities on premature cancer deaths. CA Cancer J Clin 61:212
2. Sporn MB, Dunlop NM, Newton DL, Smith JM (1976) Prevention of chemical carcinogenesis by vitamin A and its synthetic analogs (retinoids). Fed Proc 35:1332
3. Hong WK, Endicott J, Itri LM et al (1986) 13-cis-Retinoic acid in the treatment of oral leukoplakia. N Engl J Med 315:1501
4. The Alpha-Tocopherol Beta Carotene Cancer Prevention Study Group (1994) The effect of vitamin E and beta carotene on the incidence of lung cancer and other cancers in male smokers. N Engl J Med 330:1029
5. Omenn GS, Goodman GE, Thornquist MD et al (1996) Effects of a combination of beta carotene and vitamin A on lung cancer and cardiovascular disease. N Engl J Med 334:1150
6. Karp DD, Lee SJ, Shaw Wright GL et al (2010) A phase III, intergroup, randomized, double-blind, chemoprevention trial of selenium supplementation in resected stage I non-small cell lung cancer. J Clin Oncol, 2010 ASCO Annual Meeting Proceedings 28: CRA7004

7. Lippman SM, Lee JJ, Karp DD et al (2001) Randomized phase III Intergroup trial of isotretinoin to prevent second primary tumors in stage I non-small-cell lung cancer. J Natl Cancer Inst 93:605
8. van Zandwijk N, Dalesio O, Pastorino U, de Vries N, van Tinteren H (2000) EUROSCAN, a randomized trial of vitamin A and N-acetylcystein in patients with head and neck cancer or lung cancer. J Natl Cancer Inst 92:977
9. Lippman SM, Klein EA, Goodman PJ et al (2009) Effect of selenium and vitamin E on risk of prostate cancer and other cancers: the selenium and vitamin E cancer prevention trial (SELECT). JAMA 301:39
10. Steinbach G, Lynch PM, Phillips RKS et al (2000) The effect of celecoxib, a cyclooxygenase-2 inhibitor, in familial adenomatous polyposis. N Engl J Med 342:1946
11. Arber N, Eagle CJ, Spicak J et al (2006) Celecoxib for the prevention of colorectal adenomatous polyps. N Engl J Med 355:885
12. Bertagnolli MM, Eagle CJ, Zauber AG et al (2006) Celecoxib for the prevention of sporadic colorectal adenomas. N Engl J Med 355:873
13. Gold KA, Kim ES, Lee JJ et al (2011) The BATTLE to personalize lung cancer prevention through reverse migration. Cancer Prev Res (Phila) 4:962
14. Slaughter DP, Southwick HW, Smejkal W (1953) Field cancerization in oral stratified squamous epithelium: clinical implications of multicentric origin. Cancer 6:963
15. Belinsky SA, Liechty KC, Gentry FD et al (2006) Promoter hypermethylation of multiple genes in sputum precedes lung cancer incidence in a high-risk cohort. Cancer Res 66:3338
16. Spira A, Beane JE, Shah V et al (2007) Airway epithelial gene expression in the diagnostic evaluation of smokers with suspect lung cancer. Nat Med 13:361
17. Mao L, Lee JS, Kurie JM et al (1997) Clonal genetic alterations in the lungs of current and former smokers. J Natl Cancer Inst 89:857
18. Kidd JG, Rous P (1938) The carcinogenic effect of a papilloma virus on the tarred skin of rabbits. J Exp Med 68:529
19. Auerbach O, Forman JB, Gere JB et al (1957) Changes in the bronchial epithelium in relation to smoking and cancer of the lung; a report of progress. N Engl J Med 256:97
20. Auerbach O, Stout A, Hammond EC, Garfinkel L (1961) Changes in bronchial epithelium in relation to cigarette smoking and in relation to lung cancer. N Engl J Med 265:253
21. Soria J-C, Kim ES, Fayette J et al (2003) Chemoprevention of lung cancer. Lancet Oncol 4:659
22. Scott KL, Nogueira C, Heffernan TP et al (2011) Proinvasion metastasis drivers in early-stage melanoma are oncogenes. Cancer Cell 20:92
23. Perou CM, Sorlie T, Eisen MB et al (2000) Molecular portraits of human breast tumours. Nature 406:747
24. Lippman SM, Hawk ET (2009) Cancer prevention: from 1727 to milestones of the past 100 years. Cancer Res 69:5269
25. Fisher B, Costantino JP, Wickerham DL et al (1998) Tamoxifen for prevention of breast cancer: report of the National Surgical Adjuvant Breast and Bowel Project P-1 Study. J Natl Cancer Inst 90:1371
26. Barrett-Connor E, Mosca L, Collins P et al (2006) Effects of raloxifene on cardiovascular events and breast cancer in postmenopausal women. N Engl J Med 355:125
27. Thompson IM, Goodman PJ, Tangen CM et al (2003) The influence of finasteride on the development of prostate cancer. N Engl J Med 349:215
28. The FUTURE II Study Group (2007) Quadrivalent vaccine against human papillomavirus to prevent high-grade cervical lesions. N Engl J Med 356:1915
29. Chang MH, Chen CJ, Lai MS et al (1997) Universal hepatitis B vaccination in Taiwan and the incidence of hepatocellular carcinoma in children. Taiwan Childhood Hepatoma Study Group. N Engl J Med 336:1855
30. Sandler RS, Halabi S, Baron JA et al (2003) A randomized trial of aspirin to prevent colorectal adenomas in patients with previous colorectal cancer. N Engl J Med 348:883

31. Giardiello FM, Hamilton SR, Krush AJ et al (1993) Treatment of colonic and rectal adenomas with sulindac in familial adenomatous polyposis. N Engl J Med 328:1313
32. Baron JA, Cole BF, Sandler RS et al (2003) A randomized trial of aspirin to prevent colorectal adenomas. N Engl J Med 348:891
33. Vogel VG, Costantino JP, Wickerham DL et al (2010) Update of the National Surgical Adjuvant Breast and Bowel Project Study of tamoxifen and raloxifene (STAR) P-2 trial: preventing breast cancer. Cancer Prev Res (Phila) 3:696
34. Peto R, Doll R, Buckley JD, Sporn MB (1981) Can dietary beta-carotene materially reduce human cancer rates? Nature 290:201
35. Pairolero PC, Williams DE, Bergstralh EJ et al (1984) Postsurgical stage I bronchogenic carcinoma: morbid implications of recurrent disease. Ann Thorac Surg 38:331
36. Thomas P, Rubinstein L (1990) Cancer recurrence after resection: T1 N0 non-small cell lung cancer. Lung Cancer Study Group. Ann Thorac Surg 49:242
37. Morgensztern D, Waqar S, Subramanian J, Gao F, Govindan R (2009) Improving survival for stage IV non-small cell lung cancer: a surveillance, epidemiology, and end results survey from 1990 to 2005. J Thorac Oncol 4:1524
38. Wistuba II (2007) Genetics of preneoplasia: lessons from lung cancer. Curr Mol Med 7:3
39. Auerbach O, Gere JB, Pawlowski JM et al (1957) Carcinoma-in-situ and early invasive carcinoma occurring in the tracheobronchial trees in cases of bronchial carcinoma. J Thorac Surg 34:298
40. Engelman JA, Chen L, Tan X et al (2008) Effective use of PI3K and MEK inhibitors to treat mutant Kras G12D and PIK3CA H1047R murine lung cancers. Nat Med 14:1351
41. Herbst RS, Heymach JV, Lippman SM (2008) Molecular origins of cancer: lung cancer. N Engl J Med 359:1367
42. Le Calvez F, Mukeria A, Hunt JD et al (2005) TP53 and KRAS mutation load and types in lung cancer in relation to tobacco smoke: distinct patterns in never, former, and current smokers. Cancer Res 65:5076
43. Linardou H, Dahabreh IJ, Kanaloupiti D et al (2008) Assessment of somatic k-RAS mutations as a mechanism associated with resistance to EGFR-targeted agents: a systemic review and meta-analysis of studies in advanced non-small-cell lung cancer and metastatic colorectal cancer. Lancet Oncol 9:962
44. Lynch TJ, Bell DW, Sordella R et al (2004) Activating mutations in the epidermal growth factor receptor underlying responsiveness of non-small-cell lung cancer to gefitinib. N Engl J Med 350:2129
45. Mok TS, Wu Y-L, Thongprasert S et al (2009) Gefitinib or carboplatin-paclitaxel in pulmonary adenocarcinoma. N Engl J Med 361:947
46. Ji H, Zhao X, Yuza Y et al (2006) Epidermal growth factor receptor variant III mutations in lung tumorigenesis and sensitivity to tyrosine kinase inhibitors. Proc Natl Acad Sci USA 103:7817
47. Soda M, Choi YL, Enomoto M et al (2007) Identification of the transforming EML4-ALK fusion gene in non-small-cell lung cancer. Nature 448:561
48. Wong DW-S, Leung EL, Kam-Ting K et al (2009) The EML4-ALK fusion gene is involved in various histologic types of lung cancer from non-smokers with wild-type EGFR and KRAS. Cancer 115:1723
49. Shaw AT, Yeap BY, Mino-Kenudson M et al (2009) Clinical features and outcome of patients with non-small-cell lung cancer who harbor EML4-ALK. J Clin Oncol 27:4247
50. Kwak EL, Bang YJ, Camidge DR et al (2010) Anaplastic lymphoma kinase inhibition in non-small-cell lung cancer. N Engl J Med 363:1693
51. Schmid K, Oehl N, Wrba F et al (2009) EGFR/KRAS/BRAF mutations in primary lung adenocarcinomas and corresponding locoregional lymph node metastases. Clin Cancer Res 15:4554
52. Chapman PB, Hauschild A, Robert C et al (2011) Improved survival with vemurafenib in melanoma with BRAF V600E mutation. N Engl J Med 364:2507

53. West KA, Linnoila IR, Belinsky SA, Harris CC, Dennis PA (2004) Tobacco carcinogen-induced cellular transformation increased activation of the phosphatidylinositol 3′-kinase/Akt pathway in vitro and in vivo. Cancer Res 64:446
54. Yamamoto H, Shigematsu H, Nomura M et al (2008) PIK3A mutations and copy number gains in human lung cancer. Cancer Res 68:6913
55. Hammerman PS, Sos ML, Ramos AH et al (2011) Mutations in the DDR2 kinase gene identify a novel therapeutic target in squamous cell lung cancer. Cancer Discov 1:78
56. Labrador JP, Azcoitia V, Tuckermann J et al (2001) The collagen receptor DDR2 regulates proliferation and its elimination leads to dwarfism. EMBO Rep 2:446
57. Weiss J, Sos ML, Seidel D et al (2010) Frequent and focal FGFR1 amplification associates with therapeutically tractable FGFR1 dependency in squamous cell lung cancer. Sci Transl Med 2:62ra93
58. Hanahan D, Weinberg RA (2000) The hallmarks of cancer. Cell 100:57
59. Sandler A, Gray R, Perry MC et al (2006) Paclitaxel-carboplatin alone or with bevacizumab for non-small-cell lung cancer. N Engl J Med 355:2542
60. Mattern J, Koomagi R, Volm M (1996) Association of vascular endothelial growth factor expression with intratumoral microvessel density and tumour cell proliferation in human epidermoid lung carcinoma. Br J Cancer 73:931
61. Seto T, Higashiyama M, Funai H et al (2006) Prognostic value of expression of vascular endothelial growth factor and its flt-1 and KDR receptors in stage I non-small cell lung cancer. Lung Cancer 53:91
62. Bonneson B, Pappot H, Holmstav J, Skov B (2009) Vascular endothelial growth factor A and vascular endothelial growth factor receptor 2 expression in non-small cell lung cancer patients: relation to prognosis. Lung Cancer 66:314
63. Johnson DH, Fehrenbacher L, Novotny WF et al (2004) Randomized phase II trial comparing bevacizumab plus carboplatin and paclitaxel with carboplatin and paclitaxel alone in previously untreated locally advanced or metastatic non-small-cell lung cancer. J Clin Oncol 22:2184
64. Levine AJ (1997) p53, The cellular gatekeeper for growth and division. Cell 88:323
65. Foulkes WD (2007) p53 – master and commander. N Engl J Med 357:2539
66. Ding L, Getz G, Wheeler DA et al (2008) Somatic mutations affect key pathways in lung adenocarcinoma. Nature 455:1069
67. Weir BA, Woo MS, Getz G et al (2007) Characterizing the cancer genome in lung adenocarcinoma. Nature 450:893
68. Schiller JH, Harrington D, Belani CP et al (2002) Comparison of four chemotherapy regimens for advanced non-small-cell lung cancer. N Engl J Med 346:92
69. Scagliotti GV, Parikh P, Von Pawel J et al (2008) Phase III study comparing cisplatin plus gemcitabine with cisplatin plus pemetrexed in chemotherapy-naive patients with advanced-stage non-small-cell lung cancer. J Clin Oncol 26:3543
70. Kim ES, Herbst RS, Wistuba II et al (2011) The BATTLE trial: personalizing therapy for lung cancer. Cancer Discov 1:44
71. Ingle JN, Ahmann DL, Green SJ et al (1981) Randomized clinical trial of diethylstilbestrol versus tamoxifen in postmenopausal women with advanced breast cancer. N Engl J Med 304:16
72. Early Breast Cancer Trialists' Collaborative Group (1998) Tamoxifen for early breast cancer: an overview of the randomised trials. Early Breast Cancer Trialists' Collaborative Group. Lancet 351:1451
73. Fisher B, Dignam J, Wolmark N et al (1999) Tamoxifen in treatment of intraductal breast cancer: National Surgical Adjuvant Breast and Bowel Project B-24 randomised controlled trial. Lancet 353:1993
74. Weber DM, Chen C, Niesvizky R et al (2007) Lenalidomide plus dexamethasone for relapsed multiple myeloma in North America. N Engl J Med 357:2133

75. VonHoff DD, LoRusso PM, Rudin CM et al (2009) Inhibition of the hedgehog pathway in advanced basal-cell carcinoma. N Engl J Med 361:1164
76. Tang JY, Mackay-Wiggan JM, Aszterbaum M et al (2011) An investigator-initiated, phase II randomized, double-blind, placebo-controlled trial of GDC-0449 for prevention of BCCs in basal cell nevus syndrome. In: Proceedings of the AACR 102nd Annual Meeting
77. Loblaw DA, Virgo KS, Nam R et al (2007) Initial hormonal management of androgen-sensitive metastatic, recurrent, or progressive prostate cancer: 2006 update of an American Society of Clinical Oncology practice guideline. J Clin Oncol 25:1596
78. Horwitz EM, Bae K, Hanks GE et al (2008) Ten-year follow-up of radiation therapy oncology group protocol 92–02: a phase III trial of the duration of elective androgen deprivation in locally advanced prostate cancer. J Clin Oncol 26:2497
79. Fong PC, Boss DS, Yap TA et al (2009) Inhibition of poly(ADP-ribose) polymerase in tumors from BRCA mutation carriers. N Engl J Med 361:123
80. Chin L, Andersen JN, Futreal PA (2011) Cancer genomics: from discovery science to personalized medicine. Nat Med 17:297
81. Kadara H, Saintigny P, Fan YH et al (2011) Gene expression analysis of field of cancerization in early stage NSCLC patients towards development of biomarkers for personalized prevention. American Association of Cancer Research 2011 Annual Meeting Abstract: 3674
82. Gold KA, Lee JJ, Ping Y, et al (2011) Biologic risk model for recurrence in resected early-stage non-small cell lung cancer. J Clin Oncol 2011 ASCO Annual Meeting Proceedings 29:7053
83. Tang X-M, Varella-Garcia M, Xavier AC et al (2008) Epidermal growth factor receptor abnormalities in the pathogenesis and progression of lung adenocarcinomas. Cancer Prev Res 1:192
84. Tang X-M, Shigematsu H, Bekele BN et al (2005) EGFR tyrosine kinase domain mutations are detected in histologically normal respiratory epithelium in lung cancer patients. Cancer Res 65:7568
85. Goss GD, Lorimer I, Tsao MS et al (2010) A phase III, randomized, double-blind, placebo-controlled trial of the epidermal growth factor receptor inhibitor gefitinib in completely resected stage IB-IIIA non-small cell lung cancer. J Clin Oncol ASCO Annual Meeting Proceedings 28:LBA7005
86. Janjigian YY, Park BJ, Zakowski MF et al (2011) Impact on disease-free survival of adjuvant erlotinib or gefitinib in patients with resected lung adenocarcinomas that harbor EGFR mutations. J Thorac Oncol 6:569
87. Dragnev KH, Ma T, Cyrus J et al (2011) Bexarotene plus erlotinib suppress lung carcinogenesis independent of KRAS mutations in two clinical trials and transgenic models. Cancer Prev Res (Phila) 4:818
88. Spitz MR, Amos CI, D'Amelio A Jr, Dong Q, Etzel C (2009) Re: Discriminatory accuracy from single-nucleotide polymorphisms in models to predict breast cancer risk. J Natl Cancer Inst 101:1731
89. Gustafson AM, Soldi R, Anderlind C et al (2010) Airway PI3K pathway activation is an early and reversible event in lung cancer development. Sci Transl Med 2:26ra25
90. Aberle DR, Adams AM, Berg CD et al (2011) Reduced lung-cancer mortality with low-dose computed tomographic screening. N Engl J Med 365:395
91. Kim ES, Hong WK, Lee JJ et al (2010) Biological activity of celecoxib in the bronchial epithelium of current and former smokers. Cancer Prev Res 3:148
92. Mao JT, Roth MD, Fishbein MC et al (2011) Lung cancer chemoprevention with celecoxib in former-smokers. Cancer Prev Res (Phila) 4:984
93. Koki AT, Khan NK, Woerner BM et al (2002) Characterization of cyclooxygenase-2 (COX-2) during tumorigenesis in human epithelial cancers: evidence for potential clinical utility of COX-2 inhibitors in epithelial cancers. Prostaglandins Leukot Essent Fatty Acids 66:13
94. Lu C, Soria JC, Tang X et al (2004) Prognostic factors in resected stage I non-small-cell lung cancer: a multivariate analysis of six molecular markers. J Clin Oncol 22:4575

95. Khuri FR, Wu H, Lee JJ et al (2001) Cyclooxygenase-2 overexpression is a marker of poor prognosis in stage I non-small cell lung cancer. Clin Cancer Res 7:861
96. Keith RL, Blatchford PJ, Kittelson JK et al (2011) Oral iloprost improves endobronchial dysplasia in former smokers. Cancer Prev Res 4:793
97. Rothwell PM, Fowkes FG, Belch JF et al (2011) Effect of daily aspirin on long-term risk of death due to cancer: analysis of individual patient data from randomised trials. Lancet 377:31
98. Paolini M, Cantelli-Forti G, Perocco P et al (1999) Co-carcinogenic effect of beta-carotene. Nature 398:760
99. Goralczyk R (2009) Beta-carotene and lung cancer in smokers: review of hypotheses and status of research. Nutr Cancer 61:767

Natural-Agent Mechanisms and Early-Phase Clinical Development

Janet L. Wang, Kathryn A. Gold, and Scott M. Lippman

Abstract The evolution of chemoprevention research continues in exciting new directions. Large chemoprevention trials in unselected patients have often been negative, but this trend promises to be reversed by more-focused and novel trial designs emphasizing the identification of molecular targets and predictive biomarkers. Phase 0 designs, blood and tissue-based biomarkers, and surrogate endpoints are examples of important features of new prevention-trial design. Breakthroughs in the identification of novel mechanisms of carcinogenesis have contributed to a better understanding of key signaling pathways in cancer development. There has been substantial progress in elucidating molecular targets of promising synthetic and natural agents such as epigallocatechin gallate, indole-3-carbinol, myo-inositol, and deguelin, raising great optimism that biomarkers predicting efficacy, such as those associated with metformin effects, will be identified. This review will highlight several promising natural agents and how early clinical development may elucidate their role in personalized cancer chemoprevention.

Keywords Biomarkers · Chemoprevention · Clinical trial design · Targeted therapies

---

J.L. Wang
Baylor College of Medicine, Houston, TX, USA

K.A. Gold
Department of Thoracic/Head and Neck Medical Oncology, The University of Texas MD Anderson Cancer Center, Unit 0432, 1515 Holcombe Blvd, Houston, TX 77030, USA

S.M. Lippman (✉)
Director, Moores Cancer Center, University of California, San Diego, 3855 Health Sciences Drive, La Jolla, CA 92093, USA
e-mail: slippman@ucsd.edu

Contents

1 Introduction 242
2 Natural Agents with Potential for Personalized, Molecular-Targeted Prevention 243
   2.1 Green Tea: Extract and EGCG 243
   2.2 Indole-3-Carbinol 245
   2.3 Myo-Inositol 245
   2.4 Metformin 246
   2.5 Deguelin 248
3 Summary and Conclusions 249
References 250

1 Introduction

Chemoprevention has witnessed exciting new breakthroughs and developments that promise to help realize the clinical potential of the field. Historically, potential new chemopreventive agents were identified from epidemiologic data [1]. Relying too heavily on this approach, however, can lead to many years of development before an agent is fully evaluated. Reverse migration, that is, employing agents, targets, and study designs already used in the advanced cancer setting, has been proposed as a means of expediting the development of chemopreventive agents [2]. Tamoxifen use in breast cancer is the most prominent example of reverse migration to date, from its use in metastatic disease to adjuvant treatment to the prevention of breast cancer [2–6].

Cancer-preventive agents are also being developed de novo. It will be key to identify mechanisms of action, and to guide carefully the early development, of many newly identified natural agents in order to avoid replicating the negative results of many previous large-scale chemoprevention trials [7–9].

During carcinogenesis, mutations accumulate and normal cell-cycle functions are disrupted, transforming normal cells into premalignant lesions, such as intraepithelial neoplasia (IEN), and eventually into invasive carcinoma. These changes can evolve slowly over many years, presenting a prime opportunity for intervention. Cancer chemoprevention aims to treat and reverse, or at least delay, these processes leading to malignancy [10]. Personalized, molecular-targeted research, both in cancer therapy and prevention, has begun to include natural agents along with synthetic agents. Identifying new molecular targets or predictive biomarkers of clinical benefit is a critical aspect of chemopreventive drug development, as is the elucidation of novel molecular mechanisms and their involvement in various signaling pathways that promote tumorigenesis.

The key natural agents reviewed here – green tea (extract and the derivative epigallocatechin gallate [EGCG]), indole-3-carbinol (I3C), myo-inositol, metformin, and deguelin – have a body of evidence on their molecular mechanisms and/or predictive markers that promotes their potential for personalized cancer prevention. The growing elucidation of agent mechanisms, predictive markers, and molecular targets is making the development of personalized chemoprevention, with either natural or other agents, feasible.

2 Natural Agents with Potential for Personalized, Molecular-Targeted Prevention

2.1 Green Tea: Extract and EGCG

Promising preclinical data (such as elegant new mechanistic findings on the peptidyl prolyl *cis/trans* isomerase Pin1) and clinical data (e.g., promising results from randomized trials in head and neck and prostate premalignancy) on green tea and its polyphenolic flavonoid constituent EGCG are a model of molecular-targeted research in this field. Green tea has long been of interest for chemoprevention. It is extracted from the leaves of the *Camellia sinensis* plant and contains multiple active phytochemicals such as EGCG, which is the most abundant and bioactive flavonoid [11]. Green tea polyphenols have produced chemopreventive effects in a number of very promising preclinical and clinical studies.

Several epidemiologic and clinical studies have demonstrated an association between green tea and a reduced risk of prostate cancer. The incidence of prostate cancer is much lower in Asians (higher green tea intake) vs Western populations (lower green tea intake) [12]. In 2008, Kurahashi et al. [13] reported on the green tea consumption of 49,920 men from 1990 to 2004 in the Japan Public Health Center-based Prospective Study. New cases of prostate cancer were diagnosed in 404 men and consumption of green tea was associated with a dose-dependent decrease in the risk of advanced prostate cancer. A double-blinded, placebo-controlled trial reported by Bettuzzi et al. [14] also showed a reduced incidence of prostate cancer associated with a green tea supplement (200 mg orally, three times per day) for 1 year (vs placebo) in men with pre-existing high-grade prostatic IEN. After 1 year of follow-up, 3% of the green tea group compared with 30% of the placebo group were diagnosed with prostate cancer. No significant side effects or adverse events were noted. This preventive effect persisted at a 2-year follow-up.

A number of subsequent laboratory studies showed that EGCG exerted multiple widespread antitumor effects [15]. In vitro, EGCG affects many signal transduction pathways leading to growth inhibition, apoptosis, antiangiogenesis, and inhibition of metastases through the inhibition of a number of receptor tyrosine kinases, most notably epidermal growth factor receptor (EGFR), insulin-like growth factor 1 receptor (IGFR), and vascular endothelial growth factor receptor (VEGFR) [16–18]. EGCG also modulated transcription factors such as nuclear factor kappa B (NF-κB) and AP-1 in cell culture as well as in animal studies [19]. Activation of NF-κB and AP-1 led to activation of multiple genes and the inhibition of apoptosis and cell-cycle regulation. Elevated activity of both NF-κB and AP-1 are seen in a number of malignancies. NF-κB has also been shown to antagonize the tumor suppressor gene *p53*. NF-κB and AP-1 promote cell-cycle transition by upregulating cyclin D1 and c-jun, which leads to further expression of antiapoptotic genes.

EGCG's antineoplastic effects have been further translated into animal studies. Wang et al. showed that intraperitoneal or oral green tea polyphenol fractions significantly inhibited tumor growth in mice with UV-induced skin papillomas [20]. Lu et al. [21] demonstrated that topically administered EGCG induced apoptosis and inhibited malignant proliferation in mice with UVB-induced skin tumors.

In clinical trials, EGCG has been studied substantially in head and neck cancer patients. Pisters et al. [22] reported a phase I trial to determine the maximum tolerated dose of green tea extract (GTE) in patients with advanced malignancy. The majority of the 49 patients in this trial had non-small-cell lung cancer (NSCLC) or head and neck cancer. Patients received varying doses of GTE either once or three times daily for 4 weeks up to a maximum of 6 months. Dose-limiting toxicities were mostly neurologic (tremors, restlessness, insomnia) and gastrointestinal (abdominal bloating, constipation, nausea) and were primarily thought to be secondary to the 7% caffeine component of the GTE. No major clinical responses were observed, however.

Tsao et al. [23] conducted a translational phase II randomized placebo-controlled trial (RCT) of GTE at varying doses (500 mg/m^2, 750 mg/m^2, and 1,000 mg/m^2 3 times daily for 12 weeks) in 36 evaluable patients with high-risk oral premalignant lesions. Clinical response was greater in the combined arms of higher (750 mg/m^2 and 1,000 mg/m^2) doses (58.5%) compared with the lowest-dose (500 mg/m^2; 36.4%) or placebo (18.2%) arms ($P = 0.03$). These results suggested a dose-response clinical effect. The histologic response rate was 21.4% (in all combined GTE arms) compared with 9.1% (placebo), although the difference was not statistically significant ($P = 0.65$). Treatment was well-tolerated, with higher dosing causing increased insomnia, diarrhea, and oral/neck pain but no grade-4 toxicity. Baseline stromal VEGF levels correlated with clinical ($P = 0.04$) but not histologic response, and downregulation of VEGF and cyclin D1 correlated with clinical response. Other baseline biomarkers (epithelial VEGF, p53, Ki-67, cyclin D1, and p16 promoter methylation) were not associated with response or outcome.

A recent breakthrough study of EGCG illuminates a new molecular target for this agent and may help inform future clinical trials. In their elegant new mechanistic study, Urusova et al. [24] discovered that EGCG inhibits the peptidyl prolyl *cis/trans* isomerase Pin1, which is required for EGCG effects on cell growth, c-Jun activation, and NF-κB- and activator protein 1 (AP-1)-mediated transcription regulation. By mediating isomerization of phosphorylated target proteins, it is thought that Pin1 is an important regulator of pathways involved in cellular transcription and differentiation. Pin1 was also found to have crucial interactions with AP-1, NF-κB, NF of activated T cells 1 (NFAT1), and β-catenin. Pin1 is highly active in breast, colorectal, prostate, and thyroid tumors, and inhibition of Pin1 triggers apoptosis and suppresses transformation. EGCG was shown to inhibit AP-1 and NF-κB activity by downregulating Pin1 phosphorylation of key substrates in these pathways. The identification of Pin1 as a key substrate of EGCG provides an attractive tool to use in future chemoprevention research and provides a promising molecular target for the development of EGCG analogs, e.g., those targeting Pin1 more specifically. These analogs could be developed and then tested in phase 0 trials, such as that recently conducted with an I3C compound derived from cruciferous vegetables and discussed in more detail below.

2.2 Indole-3-Carbinol

I3C is a naturally occurring phytochemical shown to protect against chemically induced carcinogenesis in animal studies. It is a product of the breakdown of glucosinolates, which are found at high levels in cruciferous vegetables. Acid-condensation products of I3C are ligands of the aryl hydrocarbon receptor. I3C alters expression of cytochrome P450 (CYP) enzymes that regulate estrogen metabolism. Increasing CYP1A1 causes it to function as a potent inducer of 2-hydroxylation of estradiol, thus increasing the antiproliferative metabolite 2-hydroxyestrone and decreasing 16-alphahydroxyestrone. 16-Alphahydroxyl estrone is carcinogenic, increasing cell synthesis and the hyperproliferation of epithelial cells [25].

In addition to being antiestrogenic, I3C induces tumor growth arrest and apoptosis through a variety of other mechanisms that target multiple signaling pathways including phosphoinositide 3-kinase (PI3K) AKT, NF-κB, MAPK kinases, the cyclin-dependent kinase (CDK) inhibitors p21 and p27, and cyclin D1 [26].

As a naturally occurring AKT inhibitor, I3C is of great interest for chemoprevention. AKT/protein kinase B is a major target downstream of PI3K. Elevated AKT signaling is thought to play a critical role in tumor activities including cellular proliferation and suppression of apoptosis. I3C is normally unstable at physiological conditions but has more stable analogs, including SR13668, which has been tested in a phase 0 trial.

Phase 0 studies are designed mainly to evaluate pharmacokinetic and pharmacodynamic properties of an agent at very small doses rather than identifying a maximum tolerated dose, as in a phase I study, or determining efficacy, as in phase II and III studies. A phase 0 study of the I3C analog SR13668 was developed to determine the optimal oral bioavailability of this agent in 18 healthy adult participants [27]. The participants received a single dose of five different formulations. In a short period of time, the lead formulation of SR13668 was identified for further clinical testing.

This phase 0 trial was the first such trial ever conducted in chemoprevention. The phase 0 design allows shorter study durations, smaller, nontherapeutic dosing, and fewer patients vs phase I, II, or III designs, thus promising to conserve resources and expedite identification of more promising candidates within a shorter timeframe. Phase 0 trials are an exciting prospect that may accelerate the development of new chemopreventive drugs with more-specific targeting of key pathways [10].

2.3 Myo-Inositol

Found in various foods such as whole grains, seeds, and fruits, the promising natural chemopreventive agent myo-inositol is an isomer of glucose and a precursor of several second messengers in the phosphatidylinositol cycle such as diacylglycerol, inositol-1,4,5-triphosphate, and phosphatidylinositol-3,4,5-biphosphate. Myo-inositol is a component of the compound inositol hexaphosphate within its source foods, and inositol hexaphosphate is hydrolyzed by the enzyme phytase in the gastrointestinal tract to free myo-inositol. Myo-inositol has been shown to inhibit phospho-Akt [28] and to regulate PI3K expression signatures [29] in the lungs of smokers, suggesting the potential of this agent for cancer prevention.

Initial studies in humans were targeted towards psychiatric disorders such as depression, obsessive–compulsive disorder, and panic attacks. Both preclinical and clinical studies, however, have suggested its benefit as a chemopreventive agent [30].

Myo-inositol has shown great benefit in mouse studies. Wattenberg et al. [31] reported that myo-inositol given at 1% of the diet to female A/J mice prior to receiving various carcinogens inhibited pulmonary adenoma formation by 64%. Myo-inositol combined with dexamethasone further reduced tumor formation (by 86%). Wattenberg et al. [32] subsequently showed that even low doses of myo-inositol (0.3% of the diet) could reduce pulmonary adenoma formation by 53%. The combination of myo-inositol with inhaled budesonide was significantly more effective than was either agent alone, reducing adenoma formation by 79%. Myo-inositol was also effective when administered in the post-initiation period.

The parent compound of myo-inositol, inositol hexaphosphate, inhibits tumor formation in human breast, colon, liver, and prostate cell lines. Proposed mechanisms for these actions include inhibition of the PI3K pathway and cellular proliferation, as well as induction of cellular differentiation and apoptosis. Huang et al. [33] found that inositol hexaphosphate significantly inhibited tumor promoter-induced cell transformation, AP-1 activity, and extracellular signal-regulated kinase (ERK) activation in a dose-dependent manner. It blocked PI3K activity in JB6 mouse cells and impaired PI3 activity in vitro.

Based on these promising preclinical data, Lam et al. conducted a phase I, open-label, dose-escalation trial of myo-inositol (at four doses) to assess the safety, tolerability, and maximum tolerated dose of myo-inositol in 26 high-risk current or former smokers with bronchial dysplasia [30]. Bronchoscopy-guided biopsies were taken before and after treatment. There was a statistically significant increase in the regression rate of dysplastic lesions in patients taking 18 g/daily compared with placebo subjects from a proximally completed clinical trial with the same inclusion/exclusion criteria – 91% vs 48% ($P = 0.014$). Reductions in blood pressure were also observed. Aside from some minor flatulence and diarrhea, no serious adverse effects were noted.

2.4 Metformin

The biguanide metformin (1,1-dimethylbiguanide hydrochloride) is derived from the French lilac plant (*Galega officinalis*) and is commonly used as an anti-hyperglycemic agent for type 2 diabetes mellitus. Very strong preclinical and clinical data supporting metformin's potential role for chemoprevention [34] include findings of effects on AMP-activated protein kinase (AMPK) and IGF-1R signaling and of germline markers predicting metformin benefit [35, 36]. Epidemiologic studies have shown that diabetics have an increased risk of cancer and cancer mortality, compared with normoglycemic individuals, but there is also increasing evidence that in diabetics using metformin, cancer mortality is reduced compared with nondiabetic patients [37]. Metformin's primary actions in diabetic treatment are the inhibition of hepatic glucose production and reduction of insulin resistance in peripheral tissues, leading to decreased circulating glucose levels.

The first evidence for the potential of metformin for cancer chemoprevention came from retrospective studies of patients with diabetes. Libby et al. [38] reported that cancer was diagnosed in 7.3% of patients taking metformin versus in 11.6% of diabetics never exposed to metformin (unadjusted hazard ratio of 0.46) in a large retrospective study of over 8,000 diabetic patients. Even after adjustment for sex, age, body mass index, hemoglobin A1C, smoking, and use of other drugs, there continued to be a significantly reduced risk of cancer associated with metformin (hazard ratio = 0.63). Retrospective data from the Zwolle Outpatient Diabetes project Integrating Available Care (ZODIAC) study, which assessed primary outcomes of diabetic complications, also showed that cancer-related mortality was reduced in patients taking metformin [39]. The study enrolled 1,353 patients with type 2 diabetes mellitus from 1998 to 1999. At a median follow-up of 9.6 years, 122 of 570 total deaths were the result of malignancies. The adjusted hazard ratio for cancer mortality of patients taking metformin versus those on other antidiabetic agents was 0.43.

Metformin helps to regulate insulin and IGF levels. These molecules can promote proliferation and increased survival. The IGF-1 signaling pathway has two receptors (IGF-1R and IGF-2R) and their respective ligands. IGF-1 is produced mostly in the liver in response to stimulation by growth hormone. IGF-2 is a major fetal growth factor produced in a variety of tissues. Both receptors are capable of clustering and autophosphorylation, leading to downstream signaling. IGF-1 has been shown to inhibit chemotherapy-induced apoptosis by activating the PI3K/AKT/mTOR pathway and may play a role in chemoresistance [40]. Metformin can reduce levels of IGF-1 by increasing peripheral insulin sensitivity, which decreases hyperinsulinemia, resulting in the negative feedback of insulin on IGF binding proteins. mTOR activation via metformin inhibition of IGF-1 may be clinically predictive of response to metformin. It has been suggested that basal levels of mTOR pathway activation may predict benefit of metformin in patients [41].

Laboratory studies demonstrate that metformin also inhibits cancer cells via other pathways [42, 43]. In human prostate, colon, and breast cancer cell lines, metformin activates the AMPK pathway, which inhibits protein synthesis and gluconeogenesis during cellular stress. AMPK is typically activated when ATP concentrations drop and $5'$AMP levels increase in response to nutrient deprivation, hypoxia, or metformin administration. AMP binds directly to AMPK, causing a conformational change that leads to exposure of a threonine in the activation loop of the alpha subunit. Metformin is thought to exert antitumor effects primarily by activating liver kinase B1 (LKB1) and thus its downstream target AMPK [44]. LKB1 is a serine–threonine kinase encoded by the tumor suppressor gene *STK11*, which is commonly mutated in Peutz–Jeghers syndrome, which predisposes affected patients to multiple cancers. Phosphorylation of LKB1 is required for its translocation from the nucleus to cytoplasm, where subsequently it is activated by metformin leading to activation of AMPK. Glucose-lowering effects of metformin are also thought to be mediated by activation of LKB1 in the liver. Activated AMPK has been shown to phosphorylate signaling molecules binding to mTOR, thus down-regulating cell survival and proliferation [45].

LKB1 may potentially serve as a predictive marker of metformin inhibition of tumorigenesis. Loss or decreased expression of LKB1 can occur via epigenetic silencing. In breast cancer, LKB1 expression is absent or decreased in a significant percentage of cell lines and primary tumors, and the absence or loss correlates with cell line growth. These findings suggest that tumors with lost or decreased LKB1 expression are more likely to be resistant to metformin, thus suggesting the potential benefit of patient stratification by LKB1 status for future metformin cancer prevention trials [46].

Metformin's anticancer effects have been assessed in recent clinical studies. Jiralerspong et al. [47] conducted a retrospective analysis of metformin in a group of diabetic and nondiabetic patients who had received neoadjuvant chemotherapy for early-stage breast cancer between 1990 and 2007. Sixty-eight diabetic patients took metformin, 87 did not, and the complete pathological tumor response rate was significantly higher in the metformin group (24%) than in the non-metformin group (8%; $P = 0.007$). The response rate in the nondiabetic group was 16%. A recent early-phase clinical trial suggests activity in colorectal neoplasia as well [48].

Bodmer et al. conducted a case-control analysis of metformin in 22,621 females with type 2 diabetes taking oral hypoglycemic agents; 305 of these women were diagnosed with breast cancer. Women taking metformin for >5 years had an adjusted breast-cancer odds ratio of 0.44 compared with those not taking metformin, thus showing that long-term metformin use was associated with decreased breast-cancer risk.

A large ongoing phase III trial is randomizing nondiabetic patients with resected breast cancer to either metformin or placebo (NCT01101438). The primary endpoint is disease-free survival; secondary endpoints include overall survival. This and other trials should help to define the role of metformin in cancer prevention.

2.5 Deguelin

Promising cancer preventive agent deguelin is a rotenoid isolated from the African plant *Mundulea sericea* (Leguminosae). It is a heat shock protein-90 (Hsp90) inhibitor and has been found to have potent angiogenic and antiapoptotic properties. Deguelin binds directly to the ATP pocket of Hsp90α, leading to decreased expression in multiple client proteins via ubiquitin-mediated degradation. Deguelin affects several proteins promoting tumorigenesis such as p53, cyclin-dependent kinase 4, MAPK1/2, hypoxia-inducible factor (HIF)-1α, and AKT [49, 50]. This agent was initially identified as a mitochondria complex I, nicotinamide adenine dinucleotide (NADH) dehydrogenase inhibitor implicated in the pathophysiology of Parkinson's disease [51]. Due to concern that high doses may lead to a Parkinson's like syndrome, a liposomal delivery system was designed in order to increase bioavailability and reduce dosing to acceptable toxicity levels. The liposomal formulation has cytotoxic activity against premalignant bronchial epithelial cells and NSCLC cell lines in vitro. In vivo testing demonstrated the antitumor activity of intranasal or intratracheal administration in two different mouse models (lung tumors either induced by a tobacco carcinogen or by oncogenic K-ras [52]). Deguelin has also been shown to have activity against breast, gastric, and prostate cancer cell lines [53–55].

3 Summary and Conclusions

Natural agents are emerging as candidates with strong potential for targeted cancer chemoprevention. Initial enthusiasm based on epidemiologic data for antioxidant natural agents has unfortunately not translated into significant chemopreventive benefits. The Alpha-Tocopherol, Beta-Carotene (ATBC) Cancer Prevention Study and Beta-Carotene and Retinol Efficacy Trial (CARET) demonstrated that beta-carotene caused excess lung cancer incidence and mortality with no chemopreventive benefit [8, 9]. The Selenium and Vitamin E [prostate] Cancer Prevention Trial (SELECT) showed that selenium, either alone or in combination with vitamin E, did not decrease the rates of prostate cancer [7]. With longer follow-up, vitamin E supplementation was found to increase the incidence of prostate cancer [56]. These three trials enrolled large numbers of unselected patients to receive agents with undefined mechanisms of action; all were unsuccessful.

In the future, smaller studies should be conducted with a focus on targeted agents and biomarker-related endpoints to overcome some of the obstacles seen in these large trials. Identifying molecular targets and developing predictive markers may prove to be breakthroughs in developing more personalized, target-based approaches. EGCG represents a promising new model for further molecular-targeted research of natural agents including I3C, myo-inositol, and deguelin. Continuing clinical studies of metformin are uncovering mechanistic pathways of this drug. Constant progress in developing other natural agents includes a recent phase II trial of strawberry extract showing this agent's ability to reduce dysplasia in esophageal premalignant lesions. Natural agent development is benefitting from novel experimental drug design, such as phase 0 trials, and further elucidation of molecular markers, such as Pin1; continuing work in both avenues will expedite the development of personalized cancer prevention.

References

1. Sporn MB, Dunlop NM, Newton DL, Smith JM (1976) Prevention of chemical carcinogenesis by vitamin A and its synthetic analogs (retinoids). Fed Proc 35:1332
2. Gold KA, Kim ES, Lee JJ et al (2011) The BATTLE to personalize lung cancer prevention through reverse migration. Cancer Prev Res (Phila) 4:962
3. Fisher B, Costantino JP, Wickerham DL et al (1998) Tamoxifen for prevention of breast cancer: report of the National Surgical Adjuvant Breast and Bowel Project P-1 Study. J Natl Cancer Inst 90:1371
4. Fisher B, Dignam J, Wolmark N et al (1999) Tamoxifen in treatment of intraductal breast cancer: National Surgical Adjuvant Breast and Bowel Project B-24 randomised controlled trial. Lancet 353:1993
5. Early Breast Cancer Trialists' Collaborative Group (1998) Tamoxifen for early breast cancer: an overview of the randomised trials. Early Breast Cancer Trialists' Collaborative Group. Lancet 351:1451
6. Ingle JN, Ahmann DL, Green SJ et al (1981) Randomized clinical trial of diethylstilbestrol versus tamoxifen in postmenopausal women with advanced breast cancer. N Engl J Med 304:16
7. Lippman SM, Klein EA, Goodman PJ et al (2009) Effect of selenium and vitamin E on risk of prostate cancer and other cancers: the selenium and vitamin E cancer prevention trial (SELECT). JAMA 301:39
8. The Alpha-Tocopherol Beta Carotene Cancer Prevention Study Group (1994) The effect of vitamin E and beta carotene on the incidence of lung cancer and other cancers in male smokers. N Engl J Med 330:1029
9. Omenn GS, Goodman GE, Thornquist MD et al (1996) Effects of a combination of beta carotene and vitamin A on lung cancer and cardiovascular disease. N Engl J Med 334:1150
10. Kummar S, Doroshow JH (2011) Phase 0 trials: expediting the development of chemoprevention agents. Cancer Prev Res (Phila) 4:288
11. Kim JW, Amin AR, Shin DM (2010) Chemoprevention of head and neck cancer with green tea polyphenols. Cancer Prev Res (Phila) 3:900
12. Jemal A, Bray F, Center MM et al (2011) Global cancer statistics. CA Cancer J Clin 61:69
13. Kurahashi N, Sasazuki S, Iwasaki M, Inoue M, Tsugane S (2008) Green tea consumption and prostate cancer risk in Japanese men: a prospective study. Am J Epidemiol 167:71
14. Bettuzzi S, Brausi M, Rizzi F et al (2006) Chemoprevention of human prostate cancer by oral administration of green tea catechins in volunteers with high-grade prostate intraepithelial neoplasia: a preliminary report from a one-year proof-of-principle study. Cancer Res 66:1234
15. Shin DM (2009) Oral cancer prevention advances with a translational trial of green tea. Cancer Prev Res (Phila) 2:919
16. Li M, He Z, Ermakova S et al (2007) Direct inhibition of insulin-like growth factor-I receptor kinase activity by (−)-epigallocatechin-3-gallate regulates cell transformation. Cancer Epidemiol Biomarkers Prev 16:598
17. Masuda M, Suzui M, Weinstein IB (2001) Effects of epigallocatechin-3-gallate on growth, epidermal growth factor receptor signaling pathways, gene expression, and chemosensitivity in human head and neck squamous cell carcinoma cell lines. Clin Cancer Res 7:4220
18. McLarty J, Bigelow RL, Smith M et al (2009) Tea polyphenols decrease serum levels of prostate-specific antigen, hepatocyte growth factor, and vascular endothelial growth factor in prostate cancer patients and inhibit production of hepatocyte growth factor and vascular endothelial growth factor in vitro. Cancer Prev Res (Phila) 2:673
19. Surh YJ (2006) NF-κB and AP-1 as molecular targets for chemoprevention with EGCG, a review. Environ Chem Lett 4:137
20. Wang ZY, Huang MT, Ho CT et al (1992) Inhibitory effect of green tea on the growth of established skin papillomas in mice. Cancer Res 52:6657

21. Lu YP, Lou YR, Xie JG et al (2002) Topical applications of caffeine or (−)-epigallocatechin gallate (EGCG) inhibit carcinogenesis and selectively increase apoptosis in UVB-induced skin tumors in mice. Proc Natl Acad Sci USA 99:12455
22. Pisters KM, Newman RA, Coldman B et al (2001) Phase I trial of oral green tea extract in adult patients with solid tumors. J Clin Oncol 19:1830
23. Tsao AS, Liu D, Martin J et al (2009) Phase II randomized, placebo-controlled trial of green tea extract in patients with high-risk oral premalignant lesions. Cancer Prev Res (Phila) 2:931
24. Urusova DV, Shim JH, Kim DJ et al (2011) Epigallocatechin-gallate suppresses tumorigenesis by directly targeting Pin1. Cancer Prev Res (Phila) 4:1366
25. Kim YS, Milner JA (2005) Targets for indole-3-carbinol in cancer prevention. J Nutr Biochem 16:65
26. Aggarwal BB, Ichikawa H (2005) Molecular targets and anticancer potential of indole-3-carbinol and its derivatives. Cell Cycle 4:1201
27. Reid JM, Walden C, Qin R et al (2011) Phase 0 clinical chemoprevention trial of the Akt inhibitor SR13668. Cancer Prev Res (Phila) 4:347
28. Han W, Gills JJ, Memmott RM, Lam S, Dennis PA (2009) The chemopreventive agent myoinositol inhibits Akt and extracellular signal-regulated kinase in bronchial lesions from heavy smokers. Cancer Prev Res (Phila) 2:370
29. Gustafson AM, Soldi R, Anderlind C et al (2010) Airway PI3K pathway activation is an early and reversible event in lung cancer development. Sci Transl Med 2:26ra25
30. Lam S, McWilliams A, LeRiche J et al (2006) A phase I study of myo-inositol for lung cancer chemoprevention. Cancer Epidemiol Biomarkers Prev 15:1526
31. Wattenberg LW, Estensen RD (1996) Chemopreventive effects of myo-inositol and dexamethasone on benzo[a]pyrene and 4-(methylnitrosoamino)-1-(3-pyridyl)-1-butanone-induced pulmonary carcinogenesis in female A/J mice. Cancer Res 56:5132
32. Wattenberg LW, Wiedmann TS, Estensen RD et al (2000) Chemoprevention of pulmonary carcinogenesis by brief exposures to aerosolized budesonide or beclomethasone dipropionate and by the combination of aerosolized budesonide and dietary myo-inositol. Carcinogenesis 21:179
33. Huang C, Ma WY, Hecht SS, Dong Z (1997) Inositol hexaphosphate inhibits cell transformation and activator protein 1 activation by targeting phosphatidylinositol-3′ kinase. Cancer Res 57:2873
34. Pollak M (2010) Metformin and other biguanides in oncology: advancing the research agenda. Cancer Prev Res (Phila) 3:1060
35. Rocha GZ, Dias MM, Ropelle ER et al (2011) Metformin amplifies chemotherapy-induced AMPK activation and antitumoral growth. Clin Cancer Res 17:3993
36. Zhou K, Bellenguez C, Spencer CC et al (2011) Common variants near ATM are associated with glycemic response to metformin in type 2 diabetes. Nat Genet 43:117
37. Decensi A, Puntoni M, Goodwin P et al (2010) Metformin and cancer risk in diabetic patients: a systematic review and meta-analysis. Cancer Prev Res (Phila) 3:1451
38. Libby G, Donnelly LA, Donnan PT et al (2009) New users of metformin are at low risk of incident cancer: a cohort study among people with type 2 diabetes. Diabetes Care 32:1620
39. Landman GW, Kleefstra N, van Hateren KJ et al (2010) Metformin associated with lower cancer mortality in type 2 diabetes: ZODIAC-16. Diabetes Care 33:322
40. Zha J, Lackner MR (2010) Targeting the insulin-like growth factor receptor-1R pathway for cancer therapy. Clin Cancer Res 16:2512
41. Dowling RJ, Zakikhani M, Fantus IG, Pollak M, Sonenberg N (2007) Metformin inhibits mammalian target of rapamycin-dependent translation initiation in breast cancer cells. Cancer Res 67:10804
42. Algire C, Moiseeva O, Deschenes-Simard X et al (2012) Metformin reduces endogenous reactive oxygen species and associated DNA damage. Cancer Prev Res (Phila) 5:536–543

43. Bao B, Wang Z, Ali S et al (2012) Metformin inhibits cell proliferation, migration and invasion by attenuating CSC function mediated by deregulating miRNAs in pancreatic cancer cells. Cancer Prev Res (Phila) 5:355–364
44. Zakikhani M, Dowling R, Fantus IG, Sonenberg N, Pollak M (2006) Metformin is an AMP kinase-dependent growth inhibitor for breast cancer cells. Cancer Res 66:10269
45. Shaw RJ, Kosmatka M, Bardeesy N et al (2004) The tumor suppressor LKB1 kinase directly activates AMP-activated kinase and regulates apoptosis in response to energy stress. Proc Natl Acad Sci USA 101:3329
46. Memmott RM, Dennis PA (2009) LKB1 and mammalian target of rapamycin as predictive factors for the anticancer efficacy of metformin. J Clin Oncol 27:e226, author reply e227
47. Jiralerspong S, Palla SL, Giordano SH et al (2009) Metformin and pathologic complete responses to neoadjuvant chemotherapy in diabetic patients with breast cancer. J Clin Oncol 27:3297
48. Hosono K, Endo H, Takahashi H et al (2010) Metformin suppresses colorectal aberrant crypt foci in a short-term clinical trial. Cancer Prev Res (Phila) 3:1077
49. Oh SH, Woo JK, Yazici YD et al (2007) Structural basis for depletion of heat shock protein 90 client proteins by deguelin. J Natl Cancer Inst 99:949
50. Chun KH, Kosmeder JW 2nd, Sun S et al (2003) Effects of deguelin on the phosphatidylinositol 3-kinase/Akt pathway and apoptosis in premalignant human bronchial epithelial cells. J Natl Cancer Inst 95:291
51. Kim WY, Chang DJ, Hennessy B et al (2008) A novel derivative of the natural agent deguelin for cancer chemoprevention and therapy. Cancer Prev Res (Phila) 1:577
52. Woo JK, Choi DS, Tran HT et al (2009) Liposomal encapsulation of deguelin: evidence for enhanced antitumor activity in tobacco carcinogen-induced and oncogenic K-ras-induced lung tumorigenesis. Cancer Prev Res (Phila) 2:361
53. Thamilselvan V, Menon M, Thamilselvan S (2011) Anticancer efficacy of deguelin in human prostate cancer cells targeting glycogen synthase kinase-3 beta/beta-catenin pathway. Int J Cancer 129:2916
54. Murillo G, Peng X, Torres KE, Mehta RG (2009) Deguelin inhibits growth of breast cancer cells by modulating the expression of key members of the Wnt signaling pathway. Cancer Prev Res (Phila) 2:942
55. Lee H, Lee JH, Jung KH, Hong SS (2010) Deguelin promotes apoptosis and inhibits angiogenesis of gastric cancer. Oncol Rep 24:957
56. Klein EA, Thompson IM Jr, Tangen CM et al (2011) Vitamin E and the risk of prostate cancer: the selenium and vitamin E cancer prevention trial (SELECT). JAMA 306:1549

Index

A
Açai, 12
Adenocarcinomas, 2, 27, 40, 81, 154, 185, 209
    cigarette smoking, 226
Adipokines, 61
Aldoketo reductases (AKRs), 170
Allium spp, 73, 86
17-Allylamino-17-demethoxygeldanamycin
    (17-AAG), 192
Allyl isothiocyanate, 183
Allyl mercaptan (AM), 73, 86
Alpha-tocopherol and beta-carotene (ATBC)
    trial, 225, 249
Alzheimer's disease (AD), 75, 135
4-Aminobiphenyl (ABP), 141, 185
2-Amino-3-methylimidazo[4,5-f]
    quinoline, 184
AMP-activated protein kinase (AMPK), 246
Amphiregulin protein kinase, 11
Amyotrophic lateral sclerosis (ALS), 135
Anacardic acid, 73, 93
Androgen receptor (AR) signaling, 94
Androgens, prostate, 207
Anthocyanins, 9, 94, 216
Antioxidant genes, regulation, 138
Antioxidant response element (ARE), 133,
    138, 168
Antioxidants, 38, 133, 136, 186
Antioxidative stress defense systems, 135
Apicidin, 86
Apoptosis, 10, 89, 190, 208, 243
    caffeine, 67
Apples, 73, 102
    polyphenols, 84, 92, 102
Ascorbic acid (AA), 137

ATR/Chk1 pathway, 61
5-Azacytidine, 148
5-Aza-2-deoxycytidine (5-Aza)
    (decitabine), 92, 148

B
Balsamodendron mukul, 36
B-cell lymphoma 2 (Bcl-2), 7, 42, 49, 85, 88,
    135, 208
Bcl-2-associated X protein (Bax), 7, 42
Berberine, 141
Berries, chemoprevention, 1
Betanin, 82
Bevacizumab, 228
Biomarker-integrated approaches of targeted
    therapy for lung cancer elimination
    (BATTLE) program, 229
Biomarkers, 221, 241
Blackberries (BB), 8
Black raspberries (BRB), 9, 10, 73, 96
Black tea, 61
Blueberries, 12
Brazil nuts, 8
Breast cancer, 24, 81, 102, 205, 213, 230, 242
    genistein, 154
    guggulsterone, 43
    metformin, 246
    γ-TmT, 28
    tocopherols, 21
Broccoli sprout beverages, 170
Burseraceae, 36
Butylated hydroxyanisole (BHA), 140
Butyrate, 73, 99, 102
Butyric acid, 84

C

Caffeic acid, 99, 102
Caffeine, 61, 244
  UVB-induced carcinogenesis, 62
Cambinol, 73, 89
Cancer chemoprevention, 1ff
  expectations, 215
Canola, 180
Carcinogenesis, antioxidant defense
    system, 136
  multistep, 223
Carcinogens, 1
Carotene and retinol efficacy trial
    (CARET), 225, 249
Carotenoids, 12, 137, 222
Caspases, 28, 39, 136, 151
Catalase, 137, 139
Catechins, 139, 153
Catechol-O-methyltransferase (COMT), 213
β-Catenin, 96
Cell cycle regulation, 83
Cell differentiation, 98
Cell signaling, 94
Chaetocin, 73, 87
Chemoprevention, expectations, 215
Chemoprophylaxis, 203, 205
Chlorogenic acid, 99
Chromanol ring, 21, 22
c-Jun, 1, 6
  N-terminal kinase (JNK), 81, 136
Coffee, 63, 68, 73, 102, 144
Colon cancer, guggulsterone, 42
  inflammation, γ-TmT, 27
  tocopherols, 21
Commiphora spp., 36
CpG islands, 76
Crucifers, 73, 88, 103, 152, 165, 179, 244
Curcumin, 73, 83, 93, 139, 150, 205
  DNA hypomethylation, 151
  histone modification, 151
Cyclin D1, 145, 231, 245
Cyclin kinase inhibitor, 208
Cyclohexylmethyl isothiocyanate, 189
Cyclooxygenase-2 (COX-2), 1, 7, 38
  inhibitor, 212, 223
Cysteinyl-glycinease (GCase), 166
Cytochrome P450, 139, 245
Cytokines, 12, 52, 142, 191

D

Daidzein, 82
DDR2, 228
Death-associated protein kinase (DAPK), 91
Deguelin, 241, 248
Dehydroascorbic acid (DHA), 137
Delphinidin, 94
Detour carcinogenesis, 223
Detoxification, 81
Diallyl disulfide (DADS), 73, 86
Diallyl trisulfide (DATS), 147
Dietary phytochemicals, 133
Dihydrocoumarin (DHC), 73, 89
Dihydrotestosterone (DHT), 94
5-α-Dihydroxytestosterone, 207
Diindolylmethane (DIM), 73
Dimethylbiguanide hydrochloride
    (metformin), 246
Disulfiram, 100
DNA adducts, 163
DNA damage, 36, 73, 136
DNA methylation, 73, 75, 133, 148
DNA methyltransferases (DNMT), 73, 75,
    102, 139, 148, 151
DNA repair, 91

E

E-cadherin, 96
EGFR, regulator of growth, 227
Ellagic acid (EA), 8, 82
Ellagitannins, 12
EML4-ALK, 227
Epicatechin, 103, 153, 207
Epicatechin-3-gallate, 153, 207
Epidermal growth factor receptor (EGFR), 243
Epigallocatechin gallate (EGCG), 82, 87, 103,
    111, 115, 139, 143, 153, 205, 207, 241
  DNA methylation, 153
Epigenetics, 73, 147
Epigenomics, 133
Epithionitriles, 181
Esophagus, squamous cell carcinoma, 1
Ethoxyquin, 140
Exo-2-acetyl-exo-6-
    isothiocyanatonorbornane, 184
Extracellular signal-regulated kinases
    (ERKs), 138

F

Farnesoid X receptor (FXR), 38
Field cancerization, 223
Finasteride, 213
Flavonoids, 104, 137, 139, 243
Folate, 73, 79
Fruits, 8

Index

G
Galega officinalis, 246
Gallic acid, 94, 112
Garcinol, 73, 93
Garlic, 147
Gastrointestinal cancer, guggulsterone, 45
Gelsolin, 87
Genistein, 73, 82, 83, 105, 154
Glucoraphanin, 166, 170
Glucosinolates, 152, 165, 179
Glutamate cysteine ligase (Gcl), 137
γ-Glutamyl cysteine synthase (γGCS), 137
Glutaredoxins, 137
Glutathione, 137
Glutathione peroxidases (GPxs), 137
Glutathione *S*-transferases (GSTs), 133, 214
Goji, 12
Grapes, 8
Green tea, 61, 74, 88, 100, 203, 205
    EGCG, 243
    in vivo studies, 208
    polyphenols (GTP), 82, 154
    UVB-induced carcinogenesis, 62
Growth factors, guggulsterone, 52
GSTP1, 81
Guggul plants, 35
Guggulsterone, 35
14-3-3, guggulsterone, 51

H
Head and neck squamous carcinoma (HNSC), guggulsterone, 44
Heat shock proteins, 94, 137, 169, 179, 192, 248
Heat shock transcription factor 1 (HSF1), 191
Helicobacter pylori, 92
Heme oxygenase-1 (HO-1), 38, 133
Hepatitis B/C, 92
Hepatocellular carcinoma (HCC), 46
High fat diet, 70
Histone acetyltransferases (HATs), 77, 133
Histone deacetylases (HDACs), 77, 133, 148, 191
Histone lysine demethylases (HDMs), 78
Histone lysine methylation, 73
Histone lysine methyltransferases (HMTs), 78
Histones, acetylation, 77
    methylation, 78
    modifications, 73, 77, 133, 148
HSP90, 73, 94
Human mutL homolog 1 (hMLH1), 153
Hypermethylation, 147

I
Indole-3-carbinol (I3C), 73, 241, 245
Inducible nitric oxide synthase (iNOS), 1, 7
    guggulsterone, 52
Inflammation, 133
    Nrf2-mediated antioxidant, 139
Insulin-like growth factor 1 receptor (IGFR), 243
Interleukin-8 (IL-8), 12
Isoflavones, 73, 87, 88, 99, 106, 154
Isothiocyanates, 1, 86, 152, 179
Isotretinoin, 225

J
JNK, 138
Jumonji domain-containing (JmjC) histone demethylases, 78

K
Keap1, 133, 163, 179
Keap1–Nrf2 signaling, 167
Keto-γ-methylselenobutyrate (KMSB), 85
KRAS, GTPase, 227

L
Leukotriene B4 (LTB4), 26
Lipoic acid, 137
Liver cancer, guggulsterone, 46
Lung cancer, chemoprevention, 222, 225
    guggulsterone, 46
    molecular pathways, 226
    personalized treatment/prevention, 229
    reverse migration, 221
    tocopherols, 21
    vitamin E, 23, 26
Lycopene, 73, 82, 99, 106, 141, 145, 210
Lysine specific demethylase 1 (LSD1), 78

M
Mahanine, 88
Melilotus officinalis, 89
Mercapturic acids, 163
Metformin, 241, 246
Methyl-CpG binding domain proteins (MBDs), 148
5-Methylcytosine (5mC), 75
O(6)-Methylguanine methyltransferase (MGMT), 153
α-Methylselenopyruvate (MSP), 85

MicroRNAs (miRNAs), 78
Mitogen activated protein kinases (MAPK), 39, 136
   guggulsterone, 46
Multidrug resistance (MDR), 51, 143
Mundulea sericea, 248
Myo-inositol, 241, 245
Myrosinases, 166, 179

N
N-Acetyltransferase (NAT), 166
NAD(P)H:quinone oxidoreductase 1 (NQO1), 133, 179, 187
NF-κB, 7, 38, 73, 92, 179, 243
   guggulsterone, 50
Nitriles, 181
N-Nitrosomethylbenzylamine (NMBA), 4
NNK, 183
Noni, 12
Nordihydroguaiaretic acid (NDGA), 73, 84, 141
NQO1, 133, 179, 187
Nrf2, 82, 133, 138, 163, 179
   anti-inflammation, 141
   epigenomic reactivation, 149
   guggulsterone, 47
Nutri-epigenetics, 73
Nuts, 8

O
Ornithine decarboxylase (ODC), 208
Oxazolidine-2-thiones, 181
Oxidative stress, 133, 135

P
p21, 245
p27, 245
p38, 138
p53, 73
Parametrial fat pads, 69
Parkinson's disease (PD), 135, 249
Parthenolide, 87
Pecans, 8
Periredoxins, 137
Peroxiredoxins (Prdx), 139
Peroxisome proliferator-activated receptor γ (PPARγ), 21
P-glycoprotein, guggulsterone, 51
Phenylethyl isothiocyanate (PEITC), 73, 82, 152, 179
Phenylhexyl isothiocyanate (PHI), 73, 84

Phloretin, 82
Phosphatidylinositol-3-kinase (PI3K), 138, 245
Phytosterols, 9, 35
PI3K, 138
PI3K/Akt, 90
   guggulsterone, 47
Polyamine analogs, 73
Polyphenols, 8, 73, 82, 99, 139, 150, 207, 212, 243
Polyubiquitination, 164
Preneoplastic esophagus (PE), 12
Proliferating cell nuclear antigen (PCNA), 7
Prostaglandin E$_2$ (PGE$_2$), 7, 26
Prostate cancer, green tea, 203, 206
   guggulsterone, 43
Prostate carcinogenesis, γ-TmT, 28
Prostate, chemoprevention, 203
   tocopherols, 21
Prostate-specific antigen (PSA), 207, 213
Protein 38 (p38), 138
Protocatechuic acid, 82

R
RassF1A, 88
Rat esophagus tumor model, 4
Reactive nitrogen species (RNS), 133
Reactive oxygen and nitrogen species (RONS), 22
Reactive oxygen species (ROS), 37, 133, 134
Red raspberries, 8
Responders *vs.* nonresponders, 215
Resveratrol, 73, 90
Retinoic acid, 73
Retinoic acid receptors (RAR), 98, 153
Retinol/*N*-acetylcysteine, 225
Reverse migration, 223
Rofecoxib, 212
Rosmarinic acid, 82
Rotenoids, 248

S
S-Adenosyl-L-homocysteine (SAH), 75
S-Adenosyl-L-methionine (SAM), 75, 151
S-Allylmercaptocysteine (SAMC), 86
Se (selenium), 137, 222
   compounds, 73
Secreted frizzled-related proteins (sFRPs), 96
Se-methionine (SM), 85
Se-methyl-Se-cysteine (SMC), 85
Sinapic acid, 82

Index

SIRT1, 89, 111, 155
Skin cancer, sunlight-induced, 61
Smoking, 75, 165, 206, 222
Sodium selenite, 82
Soy, 73, 82, 104, 112, 143, 154, 214
Squamous cell carcinoma, 1
STATs, guggulsterone, 50
Stilbenes, 12
Strawberries, 8
Stress-activated protein kinase (SAPK), 136
Sulfiredoxin (Srx), 137
Sulforaphane (SFN), 73, 85, 152, 163
   gene expression changes, 169
   Keap1, 168
Sunlight-induced skin cancer, 61
Superoxide dismutases (SODs), 137
Syringic acid, 82

T

Tamoxifen, 213, 224, 230, 242
Targeted therapies, 221, 241
Targeted Therapy for Lung Cancer Elimination (BATTLE) trial, 221
Tea, 61, 74, 203
   black, 61
   green, 61, 74, 88, 100, 203, 205
   polyphenols, 153
   skin carcinogenesis, 65
Terminal deoxynucleotidyl transferase dUTP nick end labeling (TUNEL), 7
Testosterone, 94, 207, 224
12-O-Tetradecanoylphorbol-13-acetate (TPA), 185
Tetrahydroxychalcone, SIRT inhibitor, 90
Thiocyanites, 181
Thioglucosidases, 170, 179
Thioredoxin (Txn), 137
   reductase (Txnd), 137, 187
γ-TmT, 21

Tobacco, 4, 16, 44, 92, 142, 167
   cessation, 222
   lung cancer, 221
Tocopherols, 9, 21, 25, 137, 207, 225
TP53, 228
Transgenic rat for adenocarcinoma of prostate (TRAP) model, 28
Trinitrobenzenesulphonic acid (TNBS)-induced colitis, 39
Tumor necrosis factor alpha (TNF-α), 38, 93, 111, 143, 191
Turmeric (*Curcuma longa*), 73, 83, 93, 103, 139, 143, 150, 205

U

UDP-glucuronyl transferases (UGTs), 133, 137
UVB-induced apoptosis, 61

V

Vascular endothelial growth factor (VEGF), 1, 10, 42, 44, 50, 190, 228
Vascular endothelial growth factor receptor (VEGFR), 228, 243
Vitamin A, 3, 9
Vitamin B complex, 79, 102
Vitamin C, 25, 137
Vitamin D_3, 90
Vitamin E, 9, 21, 25, 137, 207, 225

W

Walnuts, 8
Wnt signaling, 95, 100

X

Xenobiotics, 139
Xenograft tumors, 21, 89, 186, 205, 209